TOPICS IN STEREOCHEMISTRY

VOLUME 10

ADVISORY BOARD

TOPICS IN
STEREOCHEMISTRY

EDITORS

ERNEST L. ELIEL

Professor of Chemistry
University of North Carolina
Chapel Hill, North Carolina

NORMAN L. ALLINGER

Professor of Chemistry
University of Georgia
Athens, Georgia

VOLUME 10

AN INTERSCIENCE® PUBLICATION
JOHN WILEY & SONS
New York • Chichester • Brisbane • Toronto

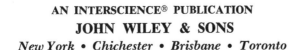

An Interscience® Publication

Copyright © 1978 by John Wiley & Sons, Inc.

Library of Congress Catalog Card Number: 67–13943

ISBN 0–471–04344–3

Printed in the United States of America

10 9 8 7 6 5 4 3 2 1

To the memory of
William Klyne

INTRODUCTION TO THE SERIES

During the last fifteen years several texts in the areas of stereochemistry and conformational analysis have been published, including *Stereochemistry of Carbon Compounds* (Eliel, McGraw-Hill, 1962) and *Conformational Analysis* (Eliel, Allinger, Angyal, and Morrison, Interscience, 1965). While the writing of these books was stimulated by the high level of research activity in the area of stereochemistry, it has, in turn, spurred further activity. As a result, many of the details found in these texts are already inadequate or out of date, although the student of stereochemistry and conformational analysis may still learn the basic concepts of the subject from them.

For both human and economic reasons, standard textbooks can be revised only at infrequent intervals. Yet the spate of periodical publications in the field of stereochemistry is such that it is an almost hopeless task for anyone to update himself by reading all the original literature. The present series is designed to bridge the resulting gap.

If that were its only purpose, this series would have been called "Advances (or "Recent Advances") in Stereochemistry." It must be remembered, however, that the above-mentioned texts were themselves not treatises and did not aim at an exhaustive treatment of the field. Thus the present series has a second purpose, namely to deal in greater detail with some of the topics summarized in the standard texts. It is for this reason that we have selected the title *Topics in Stereochemistry*.

The series is intended for the advanced student, the teacher, and the active researcher. A background for the basic knowledge in the field of stereochemistry is assumed. Each chapter is written by an expert in the field and, hopefully, covers its subject in depth. We have tried to choose topics of fundamental import aimed primarily at an audience of organic chemists but involved frequently with fundamental principles of physical chemistry and molecular physics, and dealing also with certain stereochemical aspects of inorganic chemistry and biochemistry.

It is our intention to bring out future volumes at intervals of one to two years. The Editors will welcome suggestions as to suitable topics.

We are fortunate in having been able to secure the help of an international board of Editorial Advisors who have been of great assistance by suggesting topics and authors for several articles and by helping us avoid duplication of topics appearing in other, related monograph series. We are grateful to the Editorial Advisors for this assistance, but the Editors and Authors alone must assume the responsibility for any shortcomings of *Topics in Stereochemistry*.

E. L. Eliel
N. L. Allinger

June 1978

PREFACE

The first of the four chapers in Volume 10, by Benzion Fuchs, is about the stereochemistry, and, in particular, the conformation of five-membered rings, both carbocyclic and heterocyclic. In contrast to the cyclo-hexane ring and its heterocyclic analogs, which lie in deep energy wells represented by the chair conformations, five-membered rings are known to be highly mobile, conformationally speaking. It was recognized early that cyclopentanes are puckered and that there are two nonplanar confor-mations more symmetrical than the others: the half-chair or C_2 and the envelope or C_s conformation; it was also recognized that cyclopentane rapidly "pseudorotates" between these conformations and an infinity of other, less symmetrical ones. Nevertheless there arose a tendency to over-interpret the behavior of substituted cyclopentanes or heterocyclopentanes in terms of the two symmetrical conformations of the parent compound. In his chapter, Fuchs has critically examined the conformational behavior of a number of such compounds both on a theoretical and on an experimen-tal basis, with the emphasis of the experimental work being on spectro-scopic and diffractional behavior. While the chapter does not purport to be exhaustive, it hopefully deals with enough systems to allow workers in the field to generalize the findings to whatever saturated five-member ring they may have encountered.

Topics in Stereochemistry has sometimes been criticized for its pau-city of coverage of the extensive field of inorganic stereochemistry. This has largely not been by choice of the editors, but has resulted from the difficulty we have encountered in interesting inorganic chemists to con-tribute to a stereochemistry series in which the emphasis has been on the organic side. We are constantly endeavoring to correct this situation and are pleased that, in the present volume, there is a purely inorganic-stereochemical chapter by Y. Saito concerned with the absolute stereo-chemistry of chelate complexes. Determining the absolute configuration of a representative series of chiral compounds is basic to the understand-ing of any area of stereochemistry; the emphasis in Saito's chapter is on the determination of configuration of inorganic complexes, in most cases by the Bijvoet method (X-ray fluorescence). Saito proceeds to discuss the correlation of absolute configuration and optical rotatory dispersion (or circular dichroism), the hope being that the much simpler and quicker technique of ORD-CD may take the place of the relatively tedious crystallographic determinations of configuration, at least when complexes of similar structure are compared. This correlation is so much the more important, since chemical interconversion—so popular in the determination of configuration of organic compounds—is generally not an available option for inorganic species which may differ in the nature of the metal as well as that of the ligand.

The third chapter in this volume, by H. B. Kagan and J. C. Fiaud, deals with asymmetric synthesis. The topic of asymmetric synthesis was covered exhaustively in *Asymmetric Organic Reactions* by J. D. Morrison and H. S. Mosher in 1971. (The hardback edition of this text is now out of print, but fortunately, the book has been reprinted as a paperback by the American Chemical Society.) However, much has happened since 1971— indeed, the subject of asymmetric synthesis has been a very popular one in recent years, with one or more articles or communications appearing in nearly every issue of the major organic journals today. The practical importance of synthesizing chiral compounds directly (avoiding the need for resolution) is self-evident and several efficient methods, both chemical and catalytic, have been developed in recent years (though the literature also abounds with cases which are clearly of no practical utility). Kagan and Fiaud have reviewed and organized the various types of syntheses developed or refined since the appearance of the Morrison-Mosher book in a chapter containing over 300 references.

The fourth chapter, by G. Sullivan, deals with the determination of optical purity by chiral shift reagents. In Volume 2 of *Topics in Stereochemistry* we published a chapter by Raban and Mislow dealing with the general topic of determination of optical purity; this chapter has now become the classic in the field. The one important method which it does not include—because the method had not yet been discovered when the chapter appeared in 1967—is that involving chiral lanthanide complexes. Sullivan has discussed this method with special emphasis on the practical aspects, with the thought that the method will be much used in the future even by investigators who have little prior experience with the use of lanthanide shift reagents and that these investigators will appreciate any help they can get in planning their experiments with maximum opportunity for success.

We are saddened to have to report the death, on November 13, 1977, of William Klyne, one of our advisors and one of the world's experts in the area of stereochemistry. We shall miss his sage advice and his loyal friendship. This volume is dedicated to his memory.

ERNEST L. ELIEL
NORMAN L. ALLINGER

Chapel Hill, North Carolina
Athens, Georgia
June 1978

CONTENTS

CONFORMATIONS OF FIVE-MEMBERED RINGS
  by Benzion Fuchs, Department of Chemistry, Tel-Aviv University, Tel-Aviv, Israel 1

ABSOLUTE STEREOCHEMISTRY OF CHELATE COMPLEXES
  by Yoshihiko Saito, The Institute for Solid State Physics, University of Tokyo, Tokyo, Japan 95

NEW APPROACHES IN ASYMMETRIC SYNTHESIS
  by H. B. Kagan and J. C. Fiaud, Université Paris Sud, Laboratoire de Synthèse Asymétrique, Orsay, France 175

CHIRAL LANTHANIDE SHIFT REAGENTS
  by Glenn R. Sullivan, Department of Chemistry, California Institute of Technology, Pasadena, California 287

Subject Index 331

Cumulative Index, Volumes 1–10 345

TOPICS IN STEREOCHEMISTRY

VOLUME 10

Conformations of Five-Membered Rings

BENZION FUCHS

Department of Chemistry,
Tel-Aviv University,
Tel-Aviv, Israel

I.	Prologue	2
II.	Introduction	2
III.	Theoretical Conformational Analysis	4
	A. Cyclopentane	4
	B. Substituted Cyclopentanes	14
	C. Fused Cyclopentanes	19
	D. Bridged Cyclopentanes	26
	E. Cyclopentene	27
	F. Cyclopentanone	27
	G. Heterocyclic Five-Membered Rings	27
IV.	Experimental Conformational Analysis	29
	A. Cyclopentane	29
	B. Substituted Cyclopentanes	33
	C. Cyclopentanone and Derivatives	46
	D. Cyclopentene and Derivatives	49
	E. Fused Systems	49
	F. Bridged Systems	56
	G. Heterocyclic Systems	59
	1. Oxygen Heterocycles	59
	2. Nitrogen Heterocycles	66
	3. Sulfur Heterocycles	68
	4. Selenium Heterocycles	70
	5. Phosphorus Heterocycles	71
	6. Miscellaneous Heterocyclic Systems	72
	H. Sugars	73
	I. Prostaglandins	75
V.	Concluding Remarks	76
VI.	Addendum	77
	Acknowledgments	78
	References	78

1

I. PROLOGUE

Next to the six-membered ring, the five-membered one is
probably the most widespread and important building block of
Nature's molecular edifices. Thus one can find the five-membered carbocyclic or heterocyclic ring--substituted or fused,
mobile or rigid--in a variety of natural products and their
synthetic derivatives: steroids (D ring), amino acids (prolines),
carbohydrates (furanosides, dioxolane acetals), mono- and
sesquiterpenoids, cyclitols (five-membered), antibiotics
(sarkomycin), alkaloids (pyrrolidines, pyrrolizidines, tropanes),
prostaglandins, and so forth.

Attempts to tackle the problem of cyclopentane conformational analysis appeared as early as the early forties and
steadily developed in the direction of theoretical, physical,
and stereochemical studies. This development has been, however, overshadowed by the extensive treatment given to the
six-membered ring, which led to a tremendous and sophisticated
knowledge of its framework. While much of this information
greatly contributed to the study of other systems, it also
created preconceived notions that became firmly, but not always
justifiably, implanted in extraneous grounds. Indeed, the conformational analysis of the five-membered ring turned out to be
much more complicated than had been anticipated, for reasons
elaborated in this chapter.

Some early, general reviews are available (1-3) on the
subject. At present the information on the conformational
analysis of five-membered rings has accumulated to a level that
requires a unified and critical treatment. This is attempted in
the present chapter.

II. INTRODUCTION

The idea of a nonplanar five-membered ring did not strike
immediate roots in chemical thinking. Even after much and sound
evidence for nonplanarity became available-*vide infra*-one still
could find in the literature rationalizations invoking planar
cyclopentane. Sometimes this assumption had no immediate bearing on the conclusions (4,10), but other times it led to quite
erroneous results, as in the case of the quantitative relationship between the IR frequency and the H···OH distance in
intramolecular hydrogen bonds (5).

Aston and co-workers (6) were the first to rationalize a
puckered cyclopentane; in a study involving calorimetric
measurement coupled with spectroscopic assignments the planar
form was found to be incompatible with experimental entropy
data. Thus (Table 1), when the entropy was calculated on the

TABLE 1

Entropy of Cyclopentane (ideal gas, 1 atm, 230.00°K) (6)

Method	Symmetry	Entropy (cal/deg mol)
Calculated	C_S (σ = 1)	65.65
(from spectroscopic data)	C_2 (σ = 2; d,l)	65.65
	D_{2h} (σ = 10)	61.08
Experimental		
(calorimetry)		65.27 ± 0.15

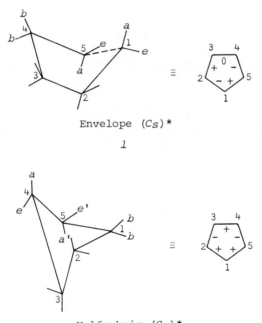

Envelope (C_S)*

1

Half-chair (C_2)*

2

  *The trivial terms *envelope* (C_S) and *half-chair* (C_2) originate with Brutcher and co-workers (10). It had also become customary (2,3,10) to characterize the exocyclic bonds in terms of axial (*a*), equatorial (*e*), pseudoaxial (*a'*), pseudoequatorial (*e'*), and bisectional (*b*) or isoclinal (11). The forms on the right carry the notation introduced by Bucourt (11) to describe the ring torsion angles.

basis of symmetry number (σ) 1, 2, and 10, only the first two
values agreed with the calorimetric result (6,7). This excludes
the planar form (D_{5h}, σ = 10) and is in accord with the puckered forms having one carbon out of plane, 1 (C_S, σ = 1) or two,
one above and the other below the plane, 2 (C_2, σ = 2) but two
enantiomers giving an entropy contribution of $-R \ln 2 + R \ln 2$
= 0) (6,8). Any attempt to rationalize the conformation of
cyclopentane (and its derivatives) needed to account for its
strain energy of 6.2 kcal/mol (12), that is, the difference in
heat of combustion per CH_2 as compared to cyclohexane or open-
chain saturated hydrocarbons. Evidently this could not be
related to the small deviation of the bond angles in cyclopentane from the tetrahedral value, and torsional strain, i.e.,
repulsive forces between vicinal bonds, was invoked (6-9) as
being responsible for the strain energy and for the puckering
of the ring.

III. THEORETICAL CONFORMATIONAL ANALYSIS

A. Cyclopentane

A systematic approach to the problem of cyclopentane conformational analysis (geometry and energy) was made by Pitzer
and co-workers in a series of pioneering and, to this day,
significant papers (7-9). To account for all experimental results, namely heat of formation and particularly the different
entropies derived from thermodynamic and spectroscopic measurements, a puckering of the ring had to be assumed (6,8) as well
as the existence of an additional degree of freedom (7-9), in
the form of a rotation of the puckering around the ring. This
internal rotation, characterized by two coordinates, an amplitude of puckering q and a phase angle f, has no angular
momentum associated with it and therefore was defined as a
pseudorotation. Objections against this concept were rejected
later on both theoretical and experimental grounds (*vide
infra*). A recent chapter (13) on pseudorotation in five-
membered rings provides an excellent account of this phenomenon.

To describe the pseudorotation of puckering, the following expression for the perpendicular displacement z_j of the jth
carbon from the plane of the unpuckered ring was proposed (8,9):

$$z_j = \left(\frac{2}{5} \right)^{1/2} q \cos 2\left(\frac{2\pi}{5}j + f \right) \tag{1}$$

where q is the amplitude and f the phase angle of puckering,
i.e., the normal coordinate of pseudorotation with $f = l\pi/10$
for C_S and $f = (2l + 1)\pi/20$ for C_2 (where l = 0, 1, 2. . .).
The coefficient $(2/5)^{1/2}$ is a normalization factor that gives

$\sum_{j=1}^{5} z_j^2 = q^2$. As f varies from 0 to π the structure goes through all possible C_s and C_2 forms.

The energy of cyclopentane as a function of the degree of puckering was calculated by Pitzer and Donath (9) with respect to angle bending strain, torsional strain, and nonbonded interactions (of the attractive kind only). The calculations were carried out only on the symmetric C_2 and C_s forms which, at a puckering amplitude of 0.48 Å, were found to be of lowest energy, with the envelope form (C_s) preferred by ca. 0.5 kcal/mol, a value that is well within the error of the treatment. As postulated (6,7), the torsional strain appears to be the main factor in this energy scheme. The torsional angles thus calculated are presented, together with later results (*vide infra*), in Table 2.

Starting with a potential function of the type $V = 1/2\ V_{ij}$ (1 + cos ϕ_{ij}) (where the position indices i and $j = i + 1$ can take the values 1 to 5), and using the calculated torsional angles ϕ_{ij} (Table 2), the following expression was developed for the torsional barrier to pseudorotation $C_s \rightarrow C_2$ (9) in monosubstituted five-membered rings:

$$\Delta V_C = 1.45\ V_{12} - 0.55\ V_{23} - 0.91\ V_{34} \qquad [2]$$

i.e., cyclopentane itself is a practically free pseudorotator, since $V_{12} = V_{23} = V_{34} = V_{ethane}$. (Note that ΔV_C is not the energy difference between separate potential minima but the height of the energy barrier restricting pseudorotation.) Thus, if position one in cyclopentane is substituted, the barriers are 2.8 kcal/mol except for $V_{12} = V_{51}$, yielding for $C_s \rightarrow C_2$

$$\Delta V_C = 1.45\ (V_{12} - V_{ethane})\ \text{kcal/mol} \qquad [3]$$

Hence, when a position *in* or *on* the ring is substituted, the potential barriers for rotation around the bonds adjacent to the altered center change, with corresponding inhibition of pseudorotation. As these potential barriers (V_{12}) increase or decrease, the C_s or C_2 form (each one representing either a maximum or a minimum in the pseudorotation potential) should be favored, respectively. Known values of potential barriers were used in an attempt to evaluate the preferred conformation of various cyclopentane derivatives; e.g., for methylcyclopentane $V_{12} = 3.40$ (the torsional barrier of propane) and $\Delta V_C = 0.9$. Hence a preference for the C_s form was predicted. On the other hand, for cyclopentanone $V_{12} = 1.15$ (the torsional barrier in acetaldehyde) and $\Delta V_C = -2.4$; thus a C_2 form was predicted.

The following points emerge from this treatment: (1) The extent of puckering (i.e., its amplitude) oscillates about a

TABLE 2

Torsional Angles in Cyclopentane as Calculated by Various Methods[a]

Torsional angles, deg[b]	Pitzer & Donath (9)		Hendrickson (14)		Lifson & Warshel (16)		Brutcher & Lugar (23)	
	C_S	C_2	C_S	C_2	C_S	C_2	C_S	C_2
$\phi_{12} = \phi_{51}$	46.1	15.2	41.7	13.7	40	13	45.3	15.8
$\phi_{23} = \phi_{45}$	28.6	39.4	25.9	35.5	25	34	29.3	41.0
ϕ_{34}	0.0	48.1	0.0	44.0	0	42	0.0	48.0

[a]Taken from the cited references. When only incomplete data were available (14,16), other values were calculated by using eq. [4] (cf. also ref. 17a,b).
[b]The numbering follows that found in 1 and 2.

stable equilibrium value (0.48 Å) and the puckering rotates
around the ring in a pseudorotation. (2) The energy appears to
depend largely on the degree of puckering but is rather insen-
sitive to change in the phase angle f; i.e., the potential
barriers in the pseudorotation circuit are very small. The
molecule is, consequently, able to invert into its mirror
image without passing through the high-energy [4-5 kcal/mol
(9,20)] planar form. (3) Ten C_S (envelope) and ten C_2 (half-
chair) forms are available in the whole circuit (Fig. 1).
(4) In the absence of substantial potential-energy barriers,
that is, in cyclopentane, intermediate nonsymmetrical forms
in the pseudorotational itinerary have energies similar to
those of C_S and C_2. (5) In monosubstituted cyclopentanes,
conformations obtained by placing the substituent in an unsymmetrical position (in C_S or C_2) have energy values intermediate between those of the symmetric forms.

In his pioneering molecular mechanics computer calculations, Hendrickson (14) minimized the total energy of cyclopentane with respect to bond angle strain and nonbonded
interactions with both attractive and repulsive terms, by
defining one dihedral angle and three internal ones, then
varying the latter independently through a range near the
tetrahedral value, until a conformation of minimum energy was
reached for each value of the dihedral angle. Thus no symmetry
restrictions were imposed, and for all values of ϕ_1 below 45°
cyclopentane conformations of similar energy are obtained, the
C_S and C_2 forms occurring at the two extremes of this conformational spectrum (Table 3).

Following this work, more efforts have been devoted to
improve and refine molecular mechanical techniques (15),
the criteria being the duplication and prediction of experimental results, namely, enthalpies of formation as determined
thermochemically and structural properties as found from
X-ray- or electron-diffraction studies.

In their "consistent force field," Lifson and Warshel
(16) use a set of energy functions to calculate equilibrium
conformations, vibrational frequencies, and excess enthalpies,
optimizing the parameters by comparison with a large number
of available experimental data. This turns out to be gratify-
ing for cyclopentane: a reasonable strain energy value is
obtained and a pseudorotational, equipotential (to less than
5 cal/mol) path is calculated by keeping one torsional angle
ϕ_3 constant and minimizing the energy with respect to the
other variables. This was repeated at small intervals of ϕ_3,
whereby elliptic curves were obtained (Fig. 2) and the interdependency of torsional angles was found to follow an expression similar to that (*vide infra*) obtained earlier by Altona
and co-workers (17) in the framework of extensive investigative efforts, both experimental and theoretical, on five-

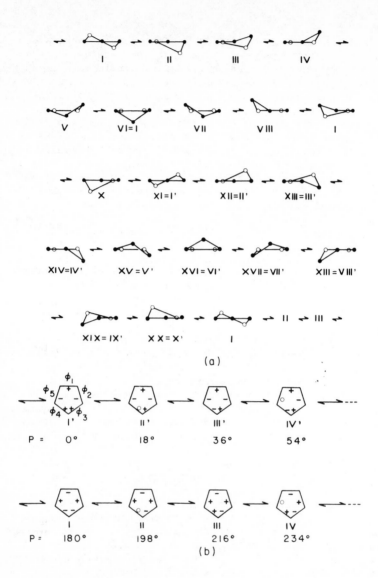

Fig. 1. Pseudorotation circuit of cyclopentane: 10 C_2 forms (odd numbered) and 10 C_S forms (even numbered). Each form consists of 10 symmetric forms (i.e., $P = 2f = 180°$; cf. eqs. [i] and [iv]) away from its mirror image. (a) The full circuit in projections; (b) a portion of the circuit in Bucourt's notation (11) and with corresponding phase angles P. The symbol $\rightleftharpoons$ is chosen to denote pseudorotation.

TABLE 3

Torsional Angles (ϕ), Internal Bond Angles (θ), Torsional Energy (E_t), Bond Angle Strain (E_θ), Nonbonded Interactions $(E_R - E_A)_{HH}$, and Total Energy (ΣE_{HH}) of Cyclopentane as Calculated by Hendrickson (14)[a]

ϕ_{34}	ϕ_{45}	ϕ_{51}	ϕ_{12}	ϕ_{23}	θ_1	θ_2	θ_3	θ_4	θ_5	E_t	E_θ	$(E_R - E_A)_{HH}$	ΣE_{HH}
*0.0	-25.9	41.7	-41.7	25.9	106.0	106.0	103.6	101.7	103.6	7.43	2.89	-0.74	9.58
5.0	-29.8	43.0	-40.0	21.8	106.2	105.6	103.2	101.6	104.2	7.36	2.96	-0.73	9.59
10.0	-33.3	43.8	-37.7	17.2	106.3	105.3	102.6	101.9	104.7	7.36	2.97	-0.72	9.60
15.0	-36.4	43.9	-34.8	12.3	106.4	104.8	102.2	102.4	105.1	7.37	2.95	-0.71	9.61
20.0	-39.2	43.4	-31.2	7.0	106.3	104.2	102.0	102.7	105.6	7.38	2.95	-0.70	9.62
25.0	-41.6	42.3	-27.1	1.3	106.0	103.6	101.8	103.2	106.0	7.33	3.00	-0.70	9.63
30.0	-43.4	40.4	-22.0	-5.0	105.8	102.8	101.8	104.0	106.1	7.30	3.04	-0.69	9.64
35.0	-44.0	36.5	-14.9	-12.5	105.2	102.2	102.2	104.8	106.3	7.34	2.99	-0.67	9.66
40.0	-43.5	30.6	-5.7	-21.3	104.2	101.6	103.0	105.6	106.2	7.30	3.03	-0.64	9.69
*45.0	-39.4	18.8	9.3	-33.7	102.4	101.4	104.4	106.1	105.2	7.09	3.27	-0.61	9.75
50.0	-40.9	16.3	14.7	-40.0	100.7	100.4	103.7	106.2	103.8	6.25	4.53	-0.64	10.14
55.0	-43.8	16.2	17.7	-44.8	99.0	99.0	102.6	106.1	102.3	5.50	6.31	-0.66	11.16
60.0	-47.2	17.0	19.7	-48.6	97.0	97.8	100.8	106.4	100.8	4.94	8.66	-0.61	13.00
65.0	-49.9	17.0	22.3	-52.4	94.8	96.8	98.6	107.1	98.8	4.58	11.76	-0.45	15.88

[a]The asterisks mark the location of the C_s form (upper) and C_2 form (lower) in the table. Note that the pseudorotation in one and the same circuit extends only between these two marks, and that in this range the energy is both lowest and nearly constant.

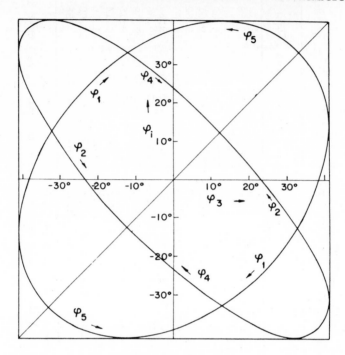

Fig. 2. Plot of ϕ_i (i = 1-5) vs. ϕ_3 in the course of cyclopentane pseudorotation (taken from ref. 16a by permission).

membered rings, carried out in the Leiden laboratories
(17,19).

These authors (17a,b) plotted the torsional angles ϕ_{j+i}
(i = 1-4) against ϕ_j from Pitzer's (9) and Hendrickson's data
(14) as well as their own standardized experimental data for
steroidal D rings (17a), obtaining similar graphs (Fig. 3),
that is, nearly perfect ellipses. Analyzing these data, they
found the torsional angles to be interrelated by a function
of type

$$\phi_{j+1} = \phi_0 \cos\left(\frac{4}{5}\pi j + P\right); \quad j = 1, 2, \ldots, 5 \qquad [4]$$

where ϕ_0 is the torsional angle for maximum puckering and the
phase angle is $0 \leq P \leq 2\pi$. Thus alternating C_S and C_2 forms
exist for every $P = l \times \pi/10$ (l = 0, 1, 2,...) in the course of
the pseudorotation circuit, or in total there are 10 C_S and 10
C_2 forms, each form having its mirror image at a distance of
$P = \pi$ (cf. Fig. 1).

A useful expression for the phase angle of pseudorotation
P for any ring in a defined pseudorotation circuit, as shown

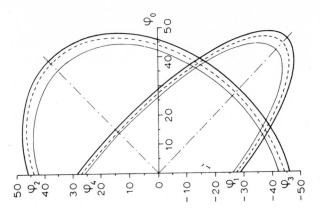

Fig. 3.   Another representation of the interdependency
of torsional angles during pseudorotation. Thin line, theoretical model of ref. 9; heavy line, theoretical model of ref.
14; dotted line, standard steroid ring D as evaluated in ref.
17a from various X-ray studies (see Sect. IV-C) (taken from
ref. 17a by permission).

in Figure 1b, was also derived (17a,c):

$$\tan P = \frac{\phi_3 + \phi_5 - \phi_2 - \phi_4}{3.0777\ \phi_1} \qquad [5]$$

To deepen the understanding of these phenomena it should
be emphasized that, as a consequence of the puckering, the
bond angles (θ) are smaller than the theoretical value (108°).
Their interdependence (θ_{j+1} vs. θ_j) is indeed described by a
graph similar to that shown in Figure 2, hence by similar
cosine functions. Moreover, since the θ's change with a
periodicity twice that of the torsional angles, a plot of θ_j
vs. ϕ_j (16,17) yields a graph of Lissajous curves (Figure 4)
from which one can see that the periodical angular change
associated with the pseudorotation lies between ca. 102.5 and
106.5°, the largest internal angle being associated with the
"tip" of the half-chair and the smallest angle with the "flap"
of the envelope. The value of ϕ_0 calculated by Lifson and
Warshel (16) is 42°, as compared to 44° (Hendrickson's value
(14) and 48° [Pitzer and Donath (9)]. The 20 forms (5 enantiomeric C_2 pairs and 10 C_S forms) of cyclopentane are presented
in Figure 1a, and a portion of the pseudorotational circuit
emphasizing the change in the *sign* of the torsional angles
when gradually changing the phase angle, that is, moving from
one form (C_2 or C_S) to another (C_S or C_2, respectively), is
depicted in Figure 1b.

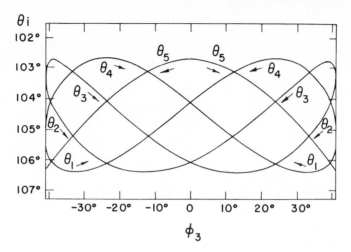

Fig. 4. Plot of θ_i (i = 1-5) vs. ϕ_5 during pseudorotation (taken from ref. 16 by permission).

Allinger and collaborators have used (20a) and later improved (20b) a force field that yielded good heats of formation *inter alia* for cyclopentane and some of its derivatives. The cyclopentane envelope and half-chair emerged as equienergetic with ϕ_0 = 44° (20) and, interestingly, the steric energy in either conformer was found to comprise a ratio of bending, torsional, and van der Waals strain of ca. 1:1:1, in contrast to all other rings. We attribute significance to this result in explaining cyclopentane behavior.

The striking fact that the same results are obtained from different sources, that is, force fields with different potential functions and on the other hand purely structural data, indicates that the outcome largely reflects the peculiar geometrical features of the five-membered ring, as contended by Lifson and Warshel (16) and implied by Altona's group (17,18), rather than energy arguments. Dunitz (21) has elaborated on this aspect, providing mathematical proof of the empirical expression (eq. [1]) put forward by Pitzer and co-workers (7-9), showing that, for infinitesimal displacements of a regular pentagon from planarity, a direct linear relationship between torsion angles and displacements is obtained, thus allowing the amplitudes and phase angles to be rigorously derived.

Other interesting points were also stressed (21), such as the fact that, independently of the phase angle, the algebraic sum of the torsional angles in the five-membered ring vanishes; that is, $\Sigma \ \phi_{jk}$ = 0. Furthermore, an approximate method of evaluating the puckering amplitude q from a knowledge of bond or torsion angles was derived (21):

$$\Sigma \; \phi_{jk}^2 \, (\deg^2) \; = \; 6 \times 10^4 \; q^2$$

and

$$\Sigma \; \theta_j \, (\deg) \; = \; 240 \; q^2 \quad (\text{for } 0.1 < q < 0.3)$$

Cremer and Pople have most recently (22a) generalized expression [1] for any ring, in a mathematical formulation whereby a unique mean plane ($z = 0$) is defined so that displacements perpendicular to this plane satisfy the following expression for cyclopentane:

$$z_j = \left(\frac{2}{5}\right)^{1/2} q \cos\left[\frac{F + 4\pi(j - 1)}{5}\right] \qquad [6]$$

where F in eq. [6] corresponds to $2f$ in eq. [1] and there are 10 envelope (C_S) forms, for $F = 0, 36, 72°, \ldots,$ and 10 half-chair (C_2) forms, for $F = 18, 54, 90°. \ldots$ One important feature of this procedure is that it can be readily carried over to the general five-membered ring with different bond lengths and angles.

In a recent investigation by Brutcher and Lugar (23), force-field calculations of cyclopentane using bond-bending, torsional, and (H, H) nonbonded interactions with apparently judicious choice of parameters yielded results largely similar to those of previous methods (Table 2) except for unusually small bond angles at the "flap" of the envelope (96.3°) and at the "tip" of the half-chair (102.8°) (cf. Fig. 4 and preceding discussion). Whether and when this is true is hard to say.

We conclude this section by considering the quantum mechanical effort that has been invested in cyclopentane study. It appears that extensive and in-depth investigations along these lines are still prohibitive and, therefore, only limited information confirming experimental or molecular mechanics results is available. Usually, mainly because of the computational limitations, only the symmetric forms have been considered. Thus Hoffmann (24) started out by performing extended Hückel calculations on, strangely, Brutcher's maximally puckered forms. Nevertheless, a reasonable stability order (planar < half-chair < envelope) was found; reasonable results were also obtained in a modified approach of this type (25).

In another, Hückel MO approach (26) resonance integrals between nonneighboring atoms (24) were deleted. Acceptable relative results were obtained for cyclopentane as well as for its mono- and dimethyl derivatives.

Other semiempirical methods such as MINDO (27) were also used to provide similarly limited information, in particular on relative energies.

An energy gradient method using extended Hückel theory was applied to cyclopentane (28). Good agreement with experimental and molecular mechanical results was obtained, concerning free pseudorotation, puckering amplitudes, and barrier to planarity. Moreover, the independence of the energy on the phase angle P was nicely confirmed.

An *ab initio* calculation was also reported (29), with the expected relative energies of the symmetric cyclopentane conformations, but a rather high (8.27 kcal/mol) energy difference between the puckered and planar forms.

A group theoretical analysis of cycloalkane conformations has been published (30). A freely pseudorotating cyclopentane emerged from this approach also.

Finally, in a most recent *ab initio* MO study (22b) using a carefully defined geometrical model (22a) (*vide supra*), Cremer and Pople studied the potential surfaces in the pseudorotation of five-membered rings. For cyclopentane, equienergetic C_S and C_2 forms emerge, which are, however, flatter ($q_0 = 0.37$ Å) than experimental and molecular-mechanics results indicate. (For barriers to pseudorotation see Fig. 5).

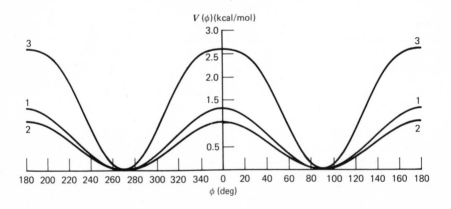

Fig. 5. Pseudorotation potential calculated (4-31G basis MO study) for oxolane (1), 1,3-dioxolane (2), and cyclopentanone (3) (taken from ref. 22b by permission).

B. Substituted Cyclopentanes

The method that Pitzer and Donath (9) used to predict the conformations of monosubstituted cyclopentanes was outlined in Sect. III-A). It should, however, be noted that this rational-

ization, for all its originality, took into account only the
symmetrical C_S and C_2 forms of the ring and, although the
authors pointed out that intermediate forms exist, some
subsequent workers were biased in discussing conformations of
five-membered rings in terms of symmetrical envelopes and
half-chairs only, without paying due attention to the possible
(and probable) occurrence of intermediate, nonsymmetrical
forms.

Monosubstituted cyclopentanes were initially taken to
exist in the symmetric C_S (envelope) form with the substituent
equatorial on the "flap" of the envelope 3 (9). This was also
the outcome of Allinger's (20) force-field calculation for
methylcyclopentane. Lugar, however, calculated (23) conformation
4 to be most stable, along with envelopes carrying the substituent in "bisectional" conformation. As the size of the
substituent increased to isopropyl, the equatorial form 3
became favored (23).

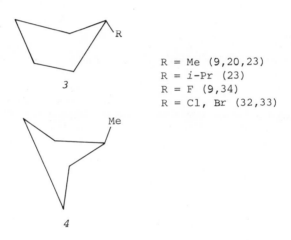

$$R = Me \quad (9,20,23)$$
$$R = i\text{-}Pr \quad (23)$$
$$R = F \quad (9,34)$$
$$R = Cl, \ Br \quad (32,33)$$

A similar controversy exists in the 1,3-dimethylcyclo-
pentanes. After configurational reassignment (see Sect. IV-B)
the cis isomer had been concluded to be of lower enthalpy
(12c), and this greater stability was rationalized (9,114)
by invoking a 2,5-diequatorial envelope form 5. Force-field
calculations (20,23), however, indicate that the cis-2,4-
diequatorial half-chair 6 is the lowest form (by ca. 0.3
kcal/mol) over the best choices of the trans form, namely 1,3-
bisectional-equatorial 7 or 2,5-diequatorial 8. The situation
is, therefore, far from being straightforward, and simple-
minded analogies drawn from the six-membered ring behavior
are certainly not desirable. For one reason, as has been
predicted(9), the axial-equatorial energy difference should
be smaller in cyclopentane than in cyclohexane [A value (2)].

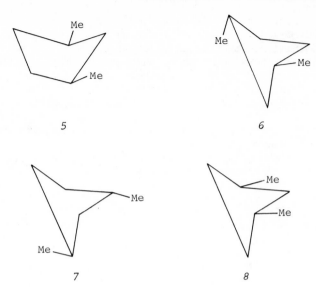

This was indeed substantiated in subsequent experimental
investigations (see Sect. IV).

Some simple, Pitzer-type (9,32,33) calculations of halogenocyclopentanes have shown low positive potential barriers to
pseudorotation [ΔV_C = 0.7 for F (10) and ΔV_C = 1.1 for Cl and
Br (32,33)], indicating that the corresponding envelope (C_S)
forms 3 are somewhat preferred. It is, however, agreed that the
energy wells are rather flat and that consequently, a multitude
of conformations may coexist. In a more sophisticated and
recent CNDO/2 study (34) of fluorocyclopentane, a single
minimum pseudorotational potential was again predicted, with
an equatorial fluorine. Altona and co-workers (33) have performed further calculations (based only on torsional and nonbonded strain terms) on *trans*-1,2-dichloro- and *trans*-1,2-

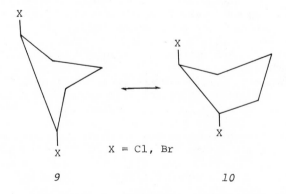

dibromocyclopentanes. The notion of "pseudolibration" was
introduced (33) to describe the oscillation of the puckering
amplitude in a limited phase angle range of an energy minimum
in a restricted pseudorotation. They concluded that the *trans*-
1,2-dihalogenocyclopentanes occur as diaxial forms *9* (slightly
favored) and *10* with the halogens occupying the most puckered
part of the ring and the pseudorotation barrier being ca. 1.7
kcal/mol. The Leiden group also performed extensive experimental
work on these and related substrates to substantiate these
findings (cf. Sect. IV-B).

Ouannes and Jacques (39) used a Pitzer-type approach and
a Hendrickson procedure to calculate energies of substituted
cyclopentane and cyclopentanone derivatives. Thus the 20 symmetrical forms in the pseudorotational circuit (Fig. 1) were
considered, with the substituents in the different positions
on the ring. Reasonable results were obtained and confirmed
experimentally (Fig. 6; cf. Sect. IV-B.).

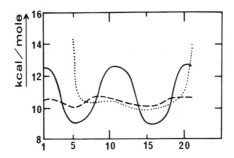

Fig. 6.   Pseudorotation potential of 3-methylcyclopentan-
one (solid line), *trans*-3-methylcyclopentanol (broken line),
and *cis*-3-methylcyclopentanol (dotted line) (taken from ref.
39 by permission).

Fuchs and Wechsler (35), who performed energy calculations
for *cis*- and *trans*-1,3-dichlorocyclopentanes (using torsional,
nonbonding, and dipole-dipole interaction terms), interpreted
the results in terms of a (Boltzmann) population distribution
of the thermal equilibrium mixtures of all possible C_S and C_2
forms in the pseudorotational circuits (Figs. 7 and 8). This
distribution was applied to theoretical dipole moments of all
conformers as well as to coupling constants to yield values
in good agreement with the experimental values for *cis*-1,3-
dichlorocyclopentane but more modestly so for the trans isomer.
The results indicate a rather broad energy well for the cis
isomer with conformation XI preferred, together with VIII, IX,
and X (Fig. 7a) and an even flatter potential for the trans
isomer having *V* at the bottom, accompanied by the I-VIII forms

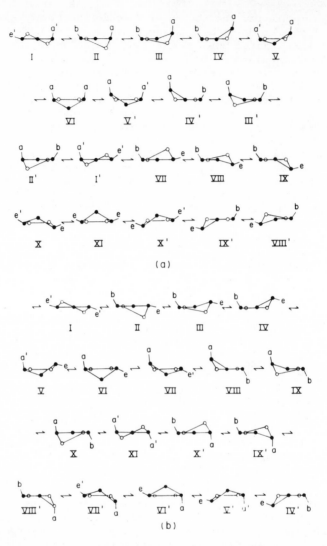

Fig. 7.   Pseudorotation circuits for cis (*a*) and trans (*b*)
1,3-disubstituted cyclopentanes (taken from ref. 35 by permission).

(Fig. 7b). In both cases intermediate unsymmetrical forms
must exist. These results, in conjunction with those of the
previously described studies as well as a wealth of experimental evidence (*vide infra*), seem to indicate a complex situation
in simply substituted cyclopentanes, with no exclusively

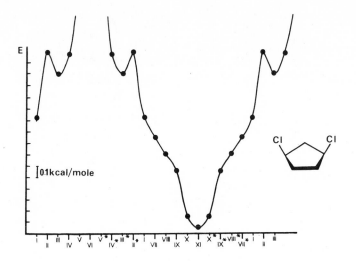

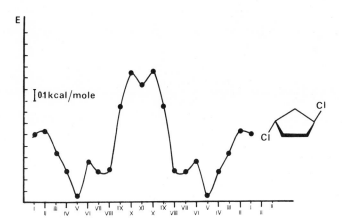

Fig. 8. Pseudorotation potential of *cis-* and *trans-*1,3-
dichlorocyclopentanes (taken from ref. 35 by permission).

populated conformations. Consequently it was suggested (35)
that to discuss such cases in terms of well-defined "envelopes"
or "half-chairs" with "equatorial" or "axial" substituents
probably leads to gross oversimplification and, in the hands of
the uninitiated, to erroneous interpretations.

C. Fused Cyclopentanes

The basic systems that have been theoretically dealt with
in this series are *cis-*bicyclo[3.1.0]hexane (*11*) (37) and *cis-*

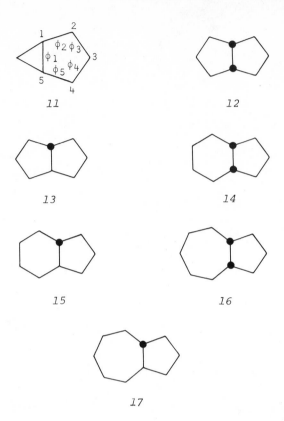

11   12

13   14

15   16

17

and *trans*-bicyclo[3.3.0]octane (pentalanes *12, 13*) (20,37),
cis- and *trans*-bicyclo[4.3.0]nonane (hydrindanes *14, 15*) (17,
20,37,38), and *cis*- and *trans*-bicyclo[5.3.0]decane (perhydro-
azulenes *16, 17*) (37).

Boyd and co-workers (37) have used force-field calculations with energy minimization to investigate, among other
compounds, strained cyclic hydrocarbons, the criterion of
success being, as is usually the case, the agreement of calculated heats of formation (hence strain energies) with experimental values (see Table 4 for structural parameters).

Significantly, cyclopentane, as well as some of its
derivatives, seems to present some difficulties in Boyd's work
(37). Although their strain energies generally compare well
with experimental values, *trans*-pentalane (*13*) is a notable
exception, the calculated strain energy exceeding by over 6
kcal/mol the experimental value. Also, for the cis isomer *12*,
Boyd's valence-force calculations failed to converge, and only
by using a Urey-Bradley force field (37) was reasonable agreement with experiment obtained. It is worth pointing out that
a puckering amplitude of q = 0.38 Å [cf. that of ref. (9),

TABLE 4
Calculated Valency Angles (θ)[a] and Torsion Angles (ϕ)[b] (deg) in Fused Cyclopentanes According to Boyd (37)

Compound	ϕ_1	ϕ_2	ϕ_3	ϕ_4	ϕ_5	θ_3	θ_4	θ_5	θ_1	θ_2
11 (C_s)	0	-15	24	-24	15	106	106	108	108	106
12 (C_{2v})	0	93[d]	-36	-36	93[d]	103	106	106	106	106
13 (C_{2h})[c]	51	-28	13	17	-25	107	102	102	103	102
14	47	-23	10	14	87[d]	107	105	106	103	106
15 (C_2)	63	-24	12	15	-22	107	104	103	103	104
16	36	-29	11	11	-29	106	107	103	104	106
17	87[d]	-40	35	-17	8	105	103	104	105	107

[a]The numbering of internal angles starts at the fusion, clockwise, as in 11.
[b]The torsion angles are also numbered clockwise, starting at the fusion, as in 11.
[c]This structure was treated according to C_s symmetry (Table XII in ref. 37) rather than C_{2h} as a double half-chair 18 requires.
[d]These values are probably misprints (37); in any case they may be regarded as erroneous. Note also that for 13-15 and 17 $\Sigma\phi = 0$ (21).

21

0.48Å] was calculated for cyclopentane. Finally, a number of
calculated geometries (Table 4) are in agreement with other
theoretical as well as experimental findings (cf. Sect. IV-E).

Allinger and co-workers (20) calculated acceptable heats
of formation for *trans*-pentalane (*13*) and *cis*-pentalane (*12*).
The latter was calculated (20b) to occur as C_{2v} (crown *19*) and
C_S (*20*) conformations in a roughly 1:2 ratio, in agreement with
spectroscopic evidence (*vide infra*). For hydrindane, the trans
isomer was calculated (no structural details were given) (20a)
to occur in the C_2 conformation *21* and the cis isomer as *22*
with higher enthalpy by 1.66 kcal/mol, which exceeds the exper-

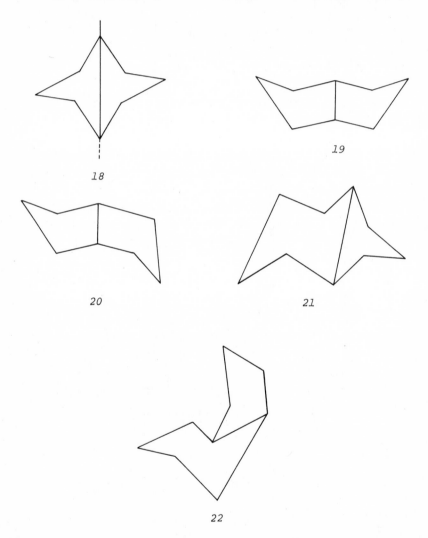

18

19

20

21

22

imental difference by 0.6 kcal/mol (20). Notably, the Schleyer
force field (15) gives good agreement with experimental
enthalpies of formation for *12* and *13*. Bucourt and Hainaut
(38) calculated, *inter alia, cis*- and *trans*-hydrindane geometries by plotting the energies of the six-membered ring in
conjunction with those of the five-membered one with small
torsional gradients. Reasonable structures were obtained although, notably, the torsional angles at the fusion of *trans*-
hydrindane, which were found to be 70° in the cyclohexane
moiety and 50° in the cyclopentane partner, proved to be in
error for reasons outlined below (cf., however refs. 11 and
53).

For the best-known fused cyclopentane system, ring D of
steroids, extensive theoretical and experimental efforts have
yielded a rather detailed understanding of its conformational
features.

Brutcher and Bauer (41a) had adopted essentially a Pitzer-
type approach in evaluating energies, using maximally puckered
cyclopentane models. They defined one of the torsional angles
to be ϕ = 60°, all other geometrical parameters being adjusted
correspondingly by vector analytical techniques. The rationale
offered for this assumption was the need for fitting the
torsional angles of cyclopentane to that of cyclohexane in
trans-fused six-five systems as in the steroid ring D (41b).

Although Brutcher's approach provided interesting insight
into cyclopentane geometry, the assumption of ϕ_{max} = 60°
turned out to be erroneous because (cf. Sect. IV-E) the
deformability of the diequatorial bonds in cyclohexane had
been underestimated. For that reason this approach was
abandoned following a review of X-ray-diffraction data (41c)
and evidence from other laboratories.

The most informative investigations in this field have
again come from the Leiden group (17,18). As a result of their
detailed steroid structure studies, Altona and co-workers
(17a,b) developed their theoretical description of the five-
membered ring (see Sect. III-A). The maximal torsional angle
in all steroid rings D was found to be ca. 47° and, by using
essentially eq. [4], ring D was calculated to exist only in a
few instances as a C(14) envelope *23* or a half-chair *25*, never
as a C(13) envelope *24*. The majority of molecules have ring D
in forms intermediate between envelope and half-chair. As in
trans-hydrindane, pseudorotation is limited to about one tenth
of its phase angle itinerary (360°) because ring C does not
allow the torsional angle at the fusion ϕ_D (17-13-14-15) to
continue closing. The fact that the Brutcher (41) and Bucourt
(38) assumptions are thus refuted is reasonably explained (17a)
in that trigonal symmetry does not hold, the sum of the torsional angles on each side of the trans junction being smaller
than 120° and dependent on the degree of substitution. Thus,
for example, 2β,3α-dichloro-5α-cholestane *26* has ϕ_D = 47.0° and

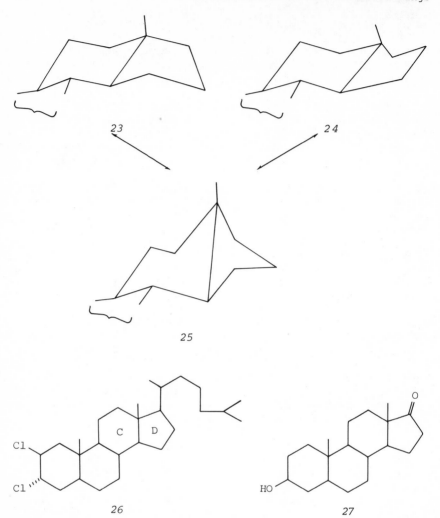

$\phi_C = 60.0°$, that is, $\phi_D + \phi_C = 108.6°$ and the D ring in a conformation intermediate between half-chair and C(13) envelope (see also Sect. IV-E).

In subsequent valence-force calculations using modified Allinger (20) and Boyd (37) force fields, Altona reported (18) on some steroid systems, with calculated parameters in good agreement with X-ray diffraction results, for example, androsterone (27) (Table 5) (18). The Allinger force field itself (40a) yielded satisfactory structures of keto-D-rings and related hydrindanones. Bucourt and Cohen (36) also performed successful calculations on related structures.

TABLE 5

Calculated Internal Bond Angles (θ) and Torsional Angles (ϕ) (deg)
in Ring D of Androsterone (27) (18)[a]

θ_{13}	θ_{14}	θ_{15}	θ_{16}	θ_{17}	$\phi_{13,14}$	$\phi_{14,15}$	$\phi_{15,16}$	$\phi_{16,17}$	$\phi_{17,18}$
99.8	104.5	103.2	105.5	108.0	43.6	-36.9	15.0	12.4	-34.2

[a]Compare with the experimental values given in Table 14.

D. Bridged Cyclopentanes

This type of system is attractive insofar as it apparently ensures rigidity, thereby offering a testing ground for five-membered rings of fixed conformation, albeit with considerable added strain.

The most extensively investigated system of this sort is undoubtedly norbornane (*28*) and its various relatives. The two interlocked cyclopentanes make it particularly interesting. Early calculations at various levels of sophistication have been reported (42-44). Recent force field calculations (15,20) have attained good heats of formation with a bridge bond angle $\theta(C_1C_7C_4)$ of 93° and C_{2v} symmetry.

The latter was confirmed for norbornane or its symmetrically mono- or disubstituted derivatives *29* in valence-force-field calculations by Altona and Sundaralingam (45). The interesting finding, however, was (45) that an asymmetrically substituted molecule, for example, *30*, may depart from C_{2v}

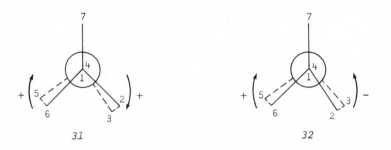

$$28 \qquad\qquad 29 \qquad\qquad 30$$

(a) R = R' = H; R" = Me	R = OH
(b) R' = H; R = R" = Me	(a) R' = Me
(c) R = R" = H; R' = Me	(b) R' = CH_2NMeBz

symmetry as a result of strain induced by certain substituents, in two possible modes of twist as exemplified in the projection formulas *31* and *32* (18,45). Twisting angles of up to 14° were

$$31 \qquad\qquad\qquad 32$$

calculated and good agreement with X-ray-diffraction data was obtained (45; see Sect. IV-F).

E. Cyclopentene (33)

Relatively little theoretical effort has been invested
in this system (46,47,53). It is worth mentioning that it falls
into the category of pseudo-four-membered rings (13) with two
out-of-plane ring puckering modes, one involving twisting around
the double bond and the other an envelope-type ring bending.

33   *34*

Pseudorotation is evidently strongly restricted to the corresponding phase angles, the largest ring torsional angle (at
the flap) being calculated to be 20-25° (46,53).

F. Cyclopentanone (34)

This molecule, with its restricted pseudorotation, has
been shown (9) to prefer the C_2 (half-chair) conformation by
0.9 kcal/mol. Subsequent analyses of the parent molecule (22,31)
and substituted derivatives (10,39) as well as fused ones (18)
have largely confirmed this feature, and experimental structural
work has substantiated it (Sect. IV-6). As expected, a flat
minimum is indicated with a small puckering amplitude, q = 0.22
Å (22).
Allinger and co-workers (40b) have reported interesting
results from their improved force-field calculations on ketones
in general and cyclopentanone in particular. A C_2 conformation,
in agreement with experimental structural data, was arrived at
with a 3.22 kcal/mol barrier to pseudorotation.

G. Heterocyclic Five-Membered Rings

As postulated earlier (9), heteroatoms are bound to cause
restriction of pseudorotation in the five-membered ring as a
result of the adjacent nonethanelike torsional barriers (Sect.
III-A). Various investigators have performed more detailed
studies, most of them in connection with spectroscopic work
(13) (see Sect. IV-G).
Oxolane *35a* had been predicted, through use of eq. [2] (9),
to be a restricted pseudorotator with a preferred half-chair
(C_2) conformation and a barrier to pseudorotation of 2.5
kcal/mol. Later more sophisticated calculations (22,31) largely
confirmed these findings, as did experiments (Sect. IV-G),
albeit with lower barriers to pseudorotation (Fig. 5). Thus,

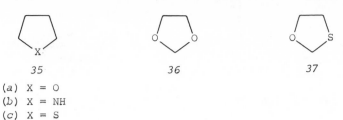

35   *36*   *37*

(a) X = O
(b) X = NH
(c) X = S

using a Hendrickson (14) procedure, Seip (48) calculated that a C_2 form was preferred though it is energetically quite close to a C_S form. The geometrical data were in very bad agreement with experiment (Sect. IV-G), and we attribute this to the rather crude parameterization of the force field (48). An *ab initio* MO study (22) led to reasonable structures (C_2 preferred by less than 1 kcal/mol over C_S with a 1.3 kcal/mol barrier).

Pyrrolidine (*35b*) was predicted (9) to occur as a C_2 form with rather low barrier. Very little further information exists. A theoretical analysis (49) in connection with proline-dimer conformation appears to lead to similar conclusions.

Thiolane (*35c*) was initially calculated (9) to have a quite restricted pseudorotation ($\Delta V_C = -3.0$ kcal/mol) with a preferred C_2 form. A Hendrickson-type calculation (50) similar to that for *35a* (48), yielded reasonable agreement with the previous results as well as with experiment.

Calculations have also been performed on some dihetero- and trihetero-substituted five-membered rings. According to an MO study (22) 1,3-dioxolane (*36*) has a preferred C_2 conformation with low barrier (1 kcal/mol, Fig. 5) and a 0.27 Å degree of puckering. The conformation of 1,3-oxathiolane (*37*) has been calculated (51) by energy minimization but using a rather crude force field ($E_\theta + E_\phi + E_{vdw}$), notably without electrostatic-

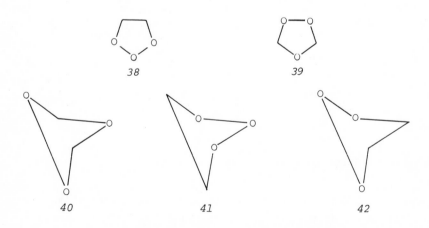

38   *39*

40   *41*   *42*

interaction terms. Shallow minima were obtained with envelope
conformations having the sulfur atom in position β to the "flap"
and a ca. 3 kcal/mol barrier to pseudorotation.

Various 1,2,3- and 1,2,4-trioxolanes, (38,39) have been
analyzed. A molecular-mechanics calculation on 39 of the type
mentioned for oxolane (48) showed it to be in a C_2 form,
although the C_S form was only ca. 1 kcal/mol higher in energy.
A CNDO/2 calculation of the ozonides of type 38 and 39 (52),
performed in connection with olefin ozonization mechanisms, was
reasonably successful (in contrast to EHT). The half-chair 40
was calculated to be the preferred conformation for 39, whereas
for 38 two lowest-energy half-chair forms, 41 and 42, were
predicted.

IV. EXPERIMENTAL CONFORMATIONAL ANALYSIS

This section deals with the conformational behavior of
five-membered rings, as determined by various physical methods,
and with the chemical consequences of this behavior.

A. Cyclopentane

It was the early thermodynamic data for cyclopentane
(6,56,57), which were subsequently confirmed and strengthened
(58,59), that led to the interpretation in terms of puckering
(6) and, with the aid also of spectroscopic data, to the post-
ulation of pseudorotation (7-9,13) (cf. Sect. III-A).

The conformation of cyclopentane can be regarded as a
consequence of its out-of-plane ring vibration. Since any
(near-)planar N-membered cyclic molecule has n-3 out-of-plane
ring vibrations, the two associated with cyclopentane can be
described as a radial and a pseudorotational vibration. The
first is in fact the oscillation of the puckering amplitude
associated with the displacement of a ring atom (parallel to
the fivefold axis) about an equilibrium value q_0. The second
vibration, however, is unique in that it involves the displacement of successive atoms in a wavelike motion around the ring.
The out-of-plane displacement of an atom j is, therefore,
associated with the amplitude q_0 and a phase angle f, as
expressed in eq. [1] (7). The actual out-of-plane displacement
was assigned a vibration of fundamental frequency $\nu_q = 288$ cm^{-1}
(Raman line), whereas the pseudorotation, whose energy levels
are given by

$$E = \frac{n^2 h^2}{8\pi^2 m q^2} = n^2 B \quad (n = 0, \pm 1, \pm 2) \qquad [7]$$

was considered *equivalent* to a real rotation of frequency
around 165 cm^{-1} (7-9).

Objections to pseudorotation (60-62) were soon overcome through heat-capacity studies (58,59) and, although some further Raman studies (63) were inconclusive concerning pseudorotation, a mid-IR spectral study (33-4000 cm^{-1}) (64) provided evidence for pseudorotational structure in a CH_2 deformation band at 1460 cm^{-1}. A constant B of 2.54 cm^{-1} was obtained leading to $q_0 = 0.48$ Å, in excellent agreement with the early value (9). IR and Raman selection rules leading to consistent assignments were subsequently deduced (65), and recalculation (66) of energy levels actually led to an identical value of q_0 (0.48 Å) and to a barrier to planarity of ca. 5.52 kcal/mol, slightly higher than the Pitzer and Donath value (4.80 kcal/mol).

The value for $q_0 = 0.47$ Å was reconfirmed in a Raman spectral study of gaseous cyclopentane (67) using a per-turbation technique (68). The barrier to planarity was also evaluated as 5.21 kcal/mol, again in excellent agreement with calculated values (9,14,16).

Turning to electron diffraction (ED), after a number of early investigations (70-72), a recent study (73) is very informative and instructive concerning the structure and con-formation of cyclopentane. Thus an equilibrium puckering q_e = 0.438 Å was obtained (and defended), significantly lower than the spectroscopic values and some calculated ones (Table 6). A rationale was offered (73) for this discrepancy, but in view of the more accurate recent spectroscopical results (67) the matter is not settled.

TABLE 6
Ring Puckering Amplitude in Cyclopentane (q, Å)

Calculated: 0.427 (16); 0.49 (14); 0.44 (17); 0.42 (23)
Spectroscopic: 0.479 (64), 0.48 (66), 0.47 (67)
Thermodynamic: 0.48 (9)
Electron diffraction: 0.438 (73)

Another interesting feature (73) is the significant lengthening of the C-C bond in cyclopentane (1.546 Å) as com-pared to those in n-alkanes (1.533 Å). This was rationalized (73) as being caused by increased nonbonded repulsion between carbon atoms, the mean nonbonded C-C distance being 2.444 Å in cyclopentane vs. 2.545 Å in n-alkanes, and by bond-bond repulsion encountered during or near eclipsing. The geometrical parameters of cyclopentane as calculated in this ED study (73) are presented in Table 7.

It is now widely accepted that NMR spectroscopy is the most informative spectroscopic method in both static and

TABLE 7

Internal Angles (θ) and Torsional Angles (ϕ) in Cyclopentane Calculated for a Puckering Amplitude q = 0.435 Å (73)

Symmetry	Valency angles[a] (deg)					Torsional angles[a] (deg)				
	θ_1	θ_2	θ_3	θ_4	θ_5	ϕ_{34}	ϕ_{45}	ϕ_{51}	ϕ_{12}	ϕ_{23}
C_s	102.13	103.95	106.13	106.13	103.95	0.0	25.01	40.26	40.26	25.01
C_2	106.43	105.26	102.68	102.68	105.26	42.29	34.34	13.16	13.16	34.34

[a]The numbering is that shown in formulas *1* and *2*; compare also Tables 2 and 3.
[b]Phase angle, as used in eq. [4] (equivalent to 2*f* in eq. [1]).

dynamic conformational analysis. The five-membered ring is no
exception to this assertion although there are some reserva-
tions.

Cyclopentane itself exhibits, as expected, a singlet at
1.51 ppm (CCl$_4$) (74). It stands to reason that no DNMR study
is possible in this case due to the low barrier (near kT) to
pseudorotation (*vide supra*) (69). Another primary objective is,
however, correlation of vicinal coupling constants with dihedral
angles. This has been attempted on cyclopentane or, rather, on
1,1,2,2,3,3-cyclopentane-d_6 (*43*) (as well as some substituted
derivatives) (75), by measuring ^{13}C-satellite spectra with
deuterium decoupling. Thus an *AA'BB'* spectrum of *43* was obtain-

43

ed and analyzed by interative simulation techniques. The
coupling constants thus derived were J_{trans} = 6.30 Hz and J_{cis}
= 7.90 Hz. The problem was now approached using the R-value
method, which is summarized in Figure 9 and which has been
used with considerable success in six-membered rings (76,77)
and some larger rings as well (75). As is evident from Figure
9, the main advantage of using R is that one does not need to
be concerned with the Karplus constant A because it is canceled
out (75). Unfortunately the result for cyclopentane is R = 0.80,
indicating an average dihedral angle of 40° in contrast to the
experimental ED value (73) or the calculated (14) value of 27°
(73). This failure was tentatively but reasonably attributed
to the breakdown of the threefold symmetry of projection
angles in cyclopentane, that is, to appreciable deviation of χ
from 120° (Fig. 9) (cf. Sect. III-C for a similar argument in
fused systems).

The effort (75) was, however, not in vain, since one can
reason back that the change in χ influences J_{trans} but has no
effect on $J_{cis} = A \cos^2\phi$. Hence a value of A = 9.95 is extracted by using this relation and the average ϕ of 27° (73,14).
Use of this A constant in other systems led to torsional angles
that are in fair agreement with known values (*vide infra*).

In a similar, independent study Lipnick (78) used deuterated cyclopentanes for NMR analysis. For *43*, at room temperature, he quotes (78a) J_{trans} = 6.33 Hz and J_{cis} = 8.18 Hz and
substitutes the resulting R_{exp} = 0.77 for an expectation ratio
<R> correlated to the maximal torsional angle (ϕ_m) via an

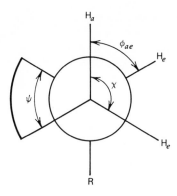

Fig. 9. Newman projection of a CH_2-CHR grouping in a
cyclopentane derivative, with expressions for the torsion
angles, ^3J, and R for the general case, and for X = 120°
(taken from ref. 35 by permission).

$$^3J_{HH} \cong A \cos^2\phi \quad \text{(Karplus)}$$

$$\phi_{aa} = \chi + \Psi$$

$$\phi_{ee} = \chi - \Psi$$

$$R = \frac{J_{trans}}{J_{cis}} = \frac{\frac{1}{2}[\cos^2(\chi + \Psi) + \cos^2(\chi - \Psi)]}{\cos^2\Psi}$$

$$\phi_{ae} = \phi_{ea} = \Psi$$

$$\Psi = \text{arc cos } [\frac{1 - \cos 2\chi}{2(R - \cos 2\chi)}]^{\frac{1}{2}}$$

$$\text{For } \chi = 120° \quad R = \frac{3 - 2\cos^2\Psi}{4\cos^2\Psi} \text{ and } \Psi =$$

$$\text{arc cos } [\frac{3}{2 + 4R}]^{\frac{1}{2}}$$

equation of type [4]. Unfortunately the result ($\phi_m = 60°$) is
again in poor agreement with experiment, probably because here
also a projection angle of 120° was assumed. It might be
expected that, once this assumption is disposed with, reasonable
results would be obtained.

B. Substituted Cyclopentanes

In this class the halogenocyclopentanes in particular
have received early and considerable attention.

As mentioned in Sect. III-B, simple calculations for
monohalogenocyclopentanes (9,32,33) had led to the conclusion
that the barriers to pseudorotation should be small (ca. 1
kcal/mol) and that a multitude of conformations may exist. The
IR (32,79) and Raman(79) spectra have been analyzed. The Brussels
group (32), in a liquid-phase study, emphasized the diffuse

character of the ν(C-X) bands (582 cm^{-1} for X = Cl, 512 cm^{-1} for Br, and 475 cm^{-1} for I, and of the band in the 600-700 cm^{-1} region. These bands sharpen somewhat in the solid. They reasonably attributed this observation to the large number of conformations that must exist as a consequence of the low pseudorotation barriers.

The Leiden group, however, interpreted their findings (Table 8) by suggesting that the C-Cl(Br) stretching frequencies indicate a preponderance of axial halogen, albeit with a considerable amplitude of pseudolibration. Simple (Karplus-type) NMR considerations appeared to support this contention (79).

Ekejiuba and Hallam (81) have performed IR and Raman spectroscopic phase-dependence studies on all halogenocyclopentanes in the 4000-200 cm^{-1} region as well as temperature- and solvent-dependence studies in the IR (800-400 cm^{-1}, 1200-950 cm^{-1}) (Table 8). From low-temperature spectra they assigned the ν(C-X) doublets of the fluoride and chloride to axial and equatorial conformers, with the former predominating, whereas the bromide and iodide allegedly occur as axial conformers only.

The Durig group (82) also dealt with chloro- and bromo-cyclopentane in phase-dependent IR and Raman (33-400 cm^{-1}) spectra. In this range the radial (ring-puckering) modes are seen in the liquids as doublets (268 and 195 cm^{-1} for RCl, 231 and 190 cm^{-1} for RBr) that reduce to singlets in the solid (174 cm^{-1} for R Cl, 152 cm^{-1} for RBr). The C-X bending modes similarly occur as doublets that reduce to singlets in the solid. It is again concluded that two types of conformers exist, equatorial and axial, with the latter preferred (and exclusive in the solid); low barriers to pseudorotation are reaffirmed (82a).

An exclusive axial conformation has been assigned to the chorine in solid chlorocyclopentane (at 77°K), by comparing the experimental (584 cm^{-1}) and calculated (585.4 cm^{-1}) C-Cl stretching frequencies for a C_S (envelope) form with axial Cl at the flap (83).

Fluorocyclopentane has been further investigated both by vibrational (84) and NMR spectroscopy (85). The IR and Raman spectra (1300-1330 cm^{-1}) revealed an unsplit absorption at 972 cm^{-1} (cf. Table 8) and radial and pseudorotational modes which indicate a low (<1 kcal/mol) barrier. The NMR spectrum is interpreted in terms of a single minimum pseudorotational potential, apparently with equatorial fluorine, as theoretically predicted (34) and in contrast to the other halogenocyclopentanes. The argument rests on the fact that in temperature-dependent spectra the signal width of both ^{19}F and the α-proton remain unchanged over a +100 to -112°C range. In chloro-, bromo-, and iodocyclopentane considerable line-width narrowing occurs upon cooling, which is taken to confirm an axial conformation.

As matters stand it might appear that the conformation of

TABLE 8

Carbon-Halogen Stretching Frequencies (cm^{-1}, for liquid)

Cyclopentane	Buys et al. (79)	Ekejiuba & Hallam (81)	Altona (80)[a] axial	equatorial
Fluoro-		968, 975		
Chloro-	590	594, 614	(611), 685[b]	742
Bromo-	514, 709	516	(535), 660[b]	686
Iodo-		478	(486), 548[b]	
1-Chloro-1-methyl-	542		560	650
1-Chloro-1-ethyl-	540 (610)		560, 612	650
1-Bromo-1-methyl-	498		505	640
1-Bromo-1-ethyl	490		505, 588	640

[a]An empirical set of carbon-halogen stretching frequencies correlated with the environment and orientation of C-X (80).
[b]The values in parenthesis were taken as a comparison with the observed values (79); it appears, though, that the higher values are the ones with which to compare (80).

halogenocyclopentanes is fairly well understood. We feel, however, that this state of affairs has been overinterpreted on various occasions, as for example in the assertion that "most investigators agree that the chloro-, bromo-, and iodo-cyclopentane molecules are bent, with the halide in the axial position of the flap" (85).

It appears that these conclusions and assignments should be regarded and used with caution. The terms "axial" and "equatorial," in the context of conformational analysis of flexible five-membered ring compounds, have qualitative meaning at best and, in our opinion, should be taken to represent a range of phase angles in the pseudorotational circuit: the lighter the substitution, the larger this range of "pseudolibration" (33).

An attempt at quantitative analysis of the conformation of monosubstituted and 1,1-disubstituted cyclopentanes by NMR spectroscopy of specifically deuterated derivatives has been published (78). For all the careful and considerable experimental effort, the approach via the R-value method is plagued by the assumption that the projection angles are 120° (*vide supra*). Nevertheless some significant results were obtained. Thus the deuterium decoupled $AA'BB'$ spectra of *44* (R = OH, OAc,

44

Cl, Br, Ph) as well as those of some 1,1-disubstituted derivatives were analyzed, some of them as a function of temperature, and data on chemical shifts and coupling constants were obtained. The expression for R as defined in Figure 9 (76,77) (with $\psi = 120°$!) was used to obtain an average torsional angle ϕ_{av} in the range 32 to 40° for all the above substrates. This was taken to indicate almost free pseudorotation with only slight preference for envelope like conformations, which is probably correct qualitatively even though the above numerical result is unwarranted (75). A detailed study in this series was made of methylcyclopentane (78c), for which the experimental vicinal couplings were compared with those calculated for the whole pseudorotational circuit (still assuming $\psi = 120°$). A restricted pseudorotation circuit was deduced with a number of preferred conformations having equatorially bonded methyl.

Attempts to investigate the conformation of cyclopentanol and derivatives have been made by examining the behavior of

the C-O stretching bands around 1000 cm^{-1} (86-88). The
approaches consisted in inducing shifts of the C-OH absorption
bands by varying solvent polarity (87) or by using gaseous HCl
in an inert solvent to form the oxonium species C-$\overset{+}{O}H_2$ (86);
tentative conformational assignments were then made (86,87).
The results, however, were subsequently shown (88) to be
internally inconsistent, since the shifts were rather erratic.

Another approach was based on the absorptions in the O-H
stretching frequency region, in a study of temperature, solvent,
and concentration dependence. Thus cyclopentanol was reported
(89) to give a rather poorly resolved doublet at 3630 and
3626 cm^{-1}, assigned to the equatorial and axial hydroxyl with
the former predominant. It is, however, doubtful that this
method can be of general diagnostic value.

For disubstituted cyclopentanes, a large amount of
information is available to be put together to give a unified
picture.

It is perhaps adequate at this point to mention early
contributions, mainly of a sterochemical nature. Thus Chiurdoglu
(90) has synthesized and attempted to differentiate according
to chemical behavior a number of stereoisomeric 1,2-dialkyl-
cyclopentanes. Vavon and co-workers (91) have studied stereoisomeric 2-substituted cyclopentanols, and have concluded from
chemical equilibrium studies that the trans isomers are thermodynamically more stable and that they are more slowly oxidized
by chromic acid. This is, of course, in good agreement with
results in six-membered rings and straightforward to interpret.

Hückel and co-workers (86,92-94) were among the first to
investigate the influence of cyclopentane conformation on the
chemical behavior of its derivatives. They were soon driven to
the disappointing statement that "Konstellationsunterschiede
gibt es in der Cyclopentanreihe nicht" (92) and that the configuration of the stereoisomers directs the outcome of reactions
such as the solvolysis of tosylates (93).

In 1,2-dialkylcyclopentanes the trans isomers appear to be
thermodynamically preferred (96). Thus for cis $\overset{\rightarrow}{\leftarrow}$ trans-1,2-
dimethylcyclopentane, $\Delta G°$ lies in the range -1.73 to -1.94
kcal/mol (97,98). The same trend has been found for the 1,2-
dicarbomethoxy- (99) and 1,2-diphenylcyclopentanes (100).
There is little doubt that this preference for the 1,2-trans
geometry stems from vicinal steric interference of substituents
and not from ring-conformational effects.

In this series vicinal dihalogen cyclopentane derivatives
also have received much attention. Thus in Brutcher's group
(101) the dipole moments of cis-1,2-dibromo- and cis-1-chloro-
2-bromocyclopentane were measured (Table 9) and compared with
calculated values for model C_2 and C_S molecules of Pitzer and
Donath-type (9) geometry having the two cis-vicinal halogens in
various positions on the ring. The apparent torsional angles for

TABLE 9

Dipole Moments of Dihalogenocyclopentanes

Cyclopentane	μ (D)	References
cis-1,2-Dibromo-	2.92[a,e]	95
cis-1-Chloro-2-bromo-	3.04[a,e]	95
trans-1,2-Dibromo-	1.51[b,e]; 1.56[a,d]	96, 97
trans-1,2-Dichloro-	1.70[a,e]; 1.64[a,e]; 1.67[a,d]	95-97
trans-1-Chloro-2-bromo-	1.59[a,d]	97
trans-1,2-Dibromo-1-methyl-	1.34[a,d]	97
cis-1,3-Dichloro-	2.76[a,d]; 2.73[c,d]	35
trans-1,3-Dichloro-	1.44[a,c,d]	35
cis-1,3-Dicyano-	4.70[a,d]; 4.51[c,d]	35
trans-1,3-Dicyano-	2.87[a,d]; 2.78[c,d]	35

[a]Benzene.
[b]Dioxane.
[c]Carbon tetrachloride.
[d]25°C.
[e]30°C.

the *cis*-1,2-dibromo- and *cis*-1-chloro-2-bromo-derivatives were
calculated as 59 ± 2° and 56 ± 2°; when strain energy (torsional
and nonbonded) was taken into account to eliminate high-energy
forms, conformations *45* and *46* [(a) μ = 2.90 D; (b) μ = 2.99 D]
were deduced, with a slight preference for *45*.

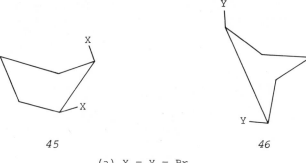

45   *46*

(a) X = Y = Br
(b) X = Cl; Y = Br

Essentially similar approaches were used by many subsequent investigators with added refinements or other supporting methods. The Leiden group dealt extensively with the *trans*-
1,2-dihalogenocyclopentanes. We have already mentioned (Sect.
III-B) their theoretical calculations relating to this system
(33). Experimental dipole moments (Table 9) were compared (103)
with calculated ones for the various *trans*-1,2 positions in C_s,
and the results were taken to indicate a solvent-dependent
equilibrium *aa* $\rightleftarrows$ *ee*, with the former, diaxial conformations
9 and *10* strongly preferred (especially when X = Br). When an
empirically adjusted Karplus relationship was used with NMR
measurements and calculated coupling constants, these assignments seemed to be confirmed. The authors, however, caution
(103) against overlooking the relatively low pseudorotational
barriers of this still flexible system.
      In this framework the same group (104) found a correlation
between dipole moments μ and vicinal coupling constants *J* of
the form

$$\frac{d\mu^2}{dJ} = \frac{(\mu_{ee}^2 - \mu_{aa}^2)}{(J_{ee} - J_{aa})} \qquad [8]$$

(where the subscript *ee* and *aa* refer to vicinal diequatorial
and diaxial substituents, respectively).
      Thus, for a series of conformationally inverting ring
compounds having similar polar substituents and similar
geometry, such as the series just described, this correlation
is linear. The method appears useful in providing a test for

isogeometry of a series of compounds and making possible the
evaluation of Karplus constants and conformational equilibrium
constants for compounds in such series. However, little use
was made of it, probably because a large number of compounds
and parameters are necessary to ensure reliability.

The above-mentioned assignment of predominant diaxial
forms *9* and *10* for the *trans*-1,2-dihalogenocyclopentanes has
apparently also withstood an IR and Raman spectral examination
(105,106) with the only additional provision that the dichloro
derivative seems to undergo enhanced pseudolibration over a
continuum of forms, more so than was previously thought.

Trihalogenocyclopentanes were also investigated, namely
1,1,2-trichloro-1-chlorocyclopentane (107a,108) and *trans*-1,2-
dibromo-1-chlorocyclopentane (107b,108), using IR (107) and NMR
spectroscopy (108) as well as dipole moment correlation for the
latter. The data appear to indicate a conformational equilibrium
with preponderance of polar forms (107). These were assigned
the diequatorial conformations *47* and *48* (108).

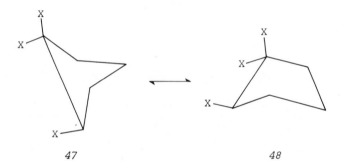

47   *48*

All these assignments are of a qualitative character [cf.
also (118) for NMR spectral data on 2-hydroxycyclopentane-
carboxylic acid derivatives] and should only be used to describe
certain trends in spatial arrangement, in contrast to the
well-defined meaning of such assignments in six-membered rings.
Only quantitative methods of structural study can provide
accurate data for flexible five-membered rings.

One such investigation is the three-dimensional X-ray-
diffraction study of racemic and optically active *trans*-1,2-
cyclopentanedicarboxylic acids (*49*) (109). The results (Fig.
10) are very interesting because they provide not only
accurate geometrical parameters but also excellent agreement
of the latter with theoretical approaches (14). The racemic
pair occurs in a half-chair (C_2) conformation with $\phi_0 = 40°$,
whereas the optically active form is very close to an envelope
with $\phi_0 = 41°$.

An X-ray-diffraction study coupled with an ESR investigation
of cyclopentane-1,1-dicarboxylic acid(*50*) (110) showed that the

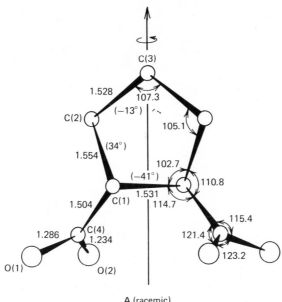

A (racemic)

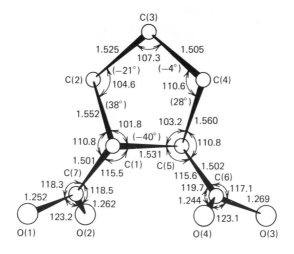

B (optically active)

Fig. 10. Geometry of *trans*-1,2-cyclopentanedicarboxylic acid as obtained from X-ray studies (taken from ref. 109 by permission).

49   50   51

3-carbon atom undergoes rapid thermal movement, an interesting
case of pseudolibration (33) occurring even in the solid. On
the other hand phenylcyclopentane-1-carboxylic acid (51) was
found (111) to exist in the crystal as a rigid five-membered
ring of intermediate geometry between C_2 and C_S with $q_0 = 0.42$
Å. The torsional angles in 51 (111) compare well with those
calculated by Adams and co-workers (73) for a phase angle $P = 10°$ (cf. eq [4] albeit with $q_0 = 0.435$ Å.

We turn now to 1,3-disubstituted cyclopentanes, an interesting class of derivatives from the point of view of stereochemistry and conformational analysis.

The stereoisomeric 1,3-dimethylcyclopentanes had become
notorious for the ambiguities encountered in their behavior;
stereospecific synthesis has led to reversal of their original
configurational assignments (112) and the cis isomer thus
turned out to have a lower heat content (97), in disaccord
with the Auwers-Skita rule (2,113). This apparently abnormal
behavior has been rationalized (114,113b,c) by invoking a cis-
diequatorial conformation (114,9), but much later theoretical
calculations (cf. Sect. III-B) have shown that other forms are
equally possible.

Another case of mistaken assignment is that of the stereoisomeric 1,3-dihydroxycyclopentanes. The original assignments
(115) had to be reversed subsequently (116,117).

Fuchs and Haber (119,88,120) have initiated a study of
1,3-disubstituted cyclopentanes, reasoning that in these molecules spurious steric interference should be at a minimum (in
contrast with 1,2-disubstituted compounds) and ring conformational effects should come into play in accounting for physical
and chemical behavior.

Various cyclopentanols substituted in the 3- and 3,4-
positions (88,119-123) as well as other 1,3-disubstituted
cyclopentanes (35,36,125-128) have been studied using NMR
spectroscopy (88,121,35), dipole moments (35), CrO_3 oxidation
kinetics (120-123), relative stabilities of stereoisomers
(120,122,124), and chemical behavior (120,122,126,127).

The following conclusions were reached: (1) NMR chemical
shifts of methyls in 3-position show practically no(through
space) deshielding effects by cis-hydroxyl as expected (88);
(2) CrO_3 oxidation rates (Table 10) are extremely close for
cis-trans isomers, and the most crowded 3,3,4,4-tetramethyl-

TABLE 10

Relative Oxidation Rates of Substituted Cyclopentanols (25°C)

Cyclopentanol	$k$rel[a]$_{cis}$	$k$rel$_{trans}$	k_{cis}/k_{trans}	Reference
3-Methyl-	1.8	1.4	1.3	88,122
3-t-Butyl-	1.7	1.45	1.2	122
3-Phenyl-	0.8	0.4	2.0	121
trans-3,4-Diphenyl-	0.65			121
3,3-Diphenyl-	1.35			121
cis-3,4-Diphenyl-	1.75	0.8	2.2	121
trans-3,4-Dimethyl-	2.0			88
3,3-Dimethyl-	3.0			123
3,3,4,4-Tetramethyl-	8.1			123

[a]$k$cyclopentanol = 1.0 (from each respective investigation; cf. the discussion in ref. 123).

TABLE 11

Relative Stabilities of Stereoisomeric 1,3-Disubstituted
Cyclopentanols (cis-trans)

1,3-Substituents	$\Delta G°$ (kcal/mol)	$\Delta H°$ (kcal/mol)	$\Delta S°$ (e.u.)	Reference
Me, Me[a]	0.5[b]	0.54	0.0	97,98
Me, OH[c]	-0.2			120
t-Bu, OH[c]	-0.05			122
CO_2H, CO_2H[d]	~0			125
CO_2Me, CO_2Me[e]	-0.23	0.08	1.0	124
CO_2Et, CO_2Et[e]	-0.27	-0.01	0.8	124
CO_2Me, CO_2Et[e]	-0.24	0.07	1.0	124
CN, CN[e]	-0.08	0.00	0.2	124

[a] Gas thermochemical 25-300°C.
[b] Compare, however, ref. 96b for $\Delta G° \approx 0$ by equilibration methods.
[c] Al(i-PrO)$_3$ i-PrOH/84°C.
[d] 20% HCl, H_2O, 100°C.
[e] t-BuOK, t-BuOH, all $\Delta G°$ at 49°C.

cyclopentanol is oxidized 8 times faster than cyclopentanol as compared to a factor of ca. 50 for 3,3,5-trimethylcyclo-hexanol vs. cyclohexanol (123); (3) stereoisomeric pairs of 1,3-disubstituted cyclopentanes (35,88,94,119-126) exhibit very similar physical properties (boiling points, refractive indices, IR and NMR spectra, GLPC retention times), so much so that they are extremely difficult to analyze and to separate; (4) relative cis-trans stabilities are close to 1 (Table 11), the free energies being near zero in slight favor of the trans epimers (except for the 1,3-dimethyl derivative); (5) acetolysis rates are strikingly similar for stereoisomeric tosylates (94,127,128) as well as similar to cyclopentyl tosylate itself (Table 12).

TABLE 12
Acetolysis Rates of 3-Substituted Cyclopentyl Tosylates

Tosylate	Relative rate	Reference
Cyclopentyl	1^{a-c}	127,128
3-Methyl (cis+trans)	0.91^b	127
cis-3-t-Butyl	1.18^b	127
trans-3-t-Butyl	1.11^b	127
3,3-Dimethyl	0.81^c	128

[a]The relative rates were calculated separately for each investigation.
[b]45°C.
[c]65°C.

All these results must certainly be manifestations of the flexibility of the five-membered ring and can be explained (123-35) by its tendency to minimize nonbonded interactions and its ability to do so by pseudorotational movement, thereby smoothly distributing the strain through small torsional and valency angle changes.

It is interesting that even the well-known conformation fixator, the t-butyl group, (2,3) exercises no appreciable physicochemical effect on remote substituents on the ring in appropriately substituted cyclopentane derivatives (Tables 10-12) (see also ref. 18). However, it has been calculated (23) that a t-butyl group actually effectively inhibits pseudo-rotation and distorts the ring badly both by flattening it and by changing its bond angles.

The 1,3-dichlorocyclopentanes serve to confirm some of these points. In a study of experimental (Table 9) vs. theo-retical dipole moments (35) the latter were submitted to a Boltzmann distribution as obtained from energy calculations (cf. Sect. III-B). The same treatment was given the theoretical NMR coupling constants for the grouping $-CHCl-CH_2-CHCl-$, which were calculated using a Karplus relationship (cf. Fig. 9) for all appropriate dihedral angles in the pseudorotational circuit (Fig. 7). Good agreement between such calculated values and observed ones was obtained for the cis isomer, and modest agreement for the trans isomer. The overall picture is, however, one of a multitude of conformations existing in a shallow minimum for the cis and an even flatter double minimum for the trans isomer.

C. Cyclopentanone and Derivatives

Many spectroscopic studies of cyclopentanone have been reported (128-134) but only those of immediate conformational relevancy are discussed here.

It should be recalled that theoretical considerations indicate preference of the half-chair (C_2) form with restricted pseudorotation and $\Delta V_C = -2.4$ kcal/mol (9) or, from molecular mechanics, 3.22 kcal/mol (40). Indeed far-IR and Raman spectroscopic studies indicate pseudorotation barriers in this range, namely, 2.8 ± 0.7 (129), 3.72 (130), and 2.15 kcal/mol (131) (cf. also Fig. 6). Similar conclusions can be drawn from various other recent vibrational spectroscopic investigations (132-134,67,68) although some authors (132) derived selection rules by treating the molecule as a dynamic C_{2v} system that undergoes continuous change, via a quasi-equipotential path, between two limiting C_2 forms. Early microwave studies (136, 137) indicated a puckered and pseudorotating cyclopentanone, and a more recent study (138) confirmed a twisted C_2 conformation 52.

The most informative recent work on cyclopentanone is an ED study (139), from which a value of $q_0 = 0.38$ Å (corresponding to a maximal torsion angle $\phi_0 = 37.2°$) was extracted, flatter than cyclopentane itself as expected (40). Internal bond and torsional angles are depicted in formula 52.

Accurate geometrical parameters for an interesting related compound, chiral spiro(4,4)nonane-1,6-dione (53)

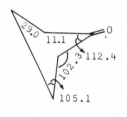

52

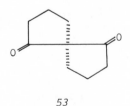

53

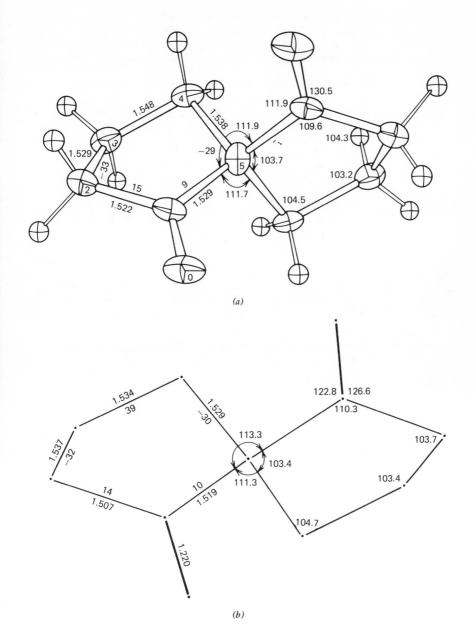

Fig. 11. Experimental (X-ray) (a) and calculated (valence-force) (b) geometries of (S)-(-)-spiro[4.4]nonane-1,6-dione (taken from ref. 140 by permission).

were obtained (140) from an X-ray analysis (and valence-force calculation). The structural parameters are given in Figure 11, from which one sees that both rings adopt a conformation intermediate between envelope and half-chair but closer to the latter.

Early NMR studies on cyclopentanone were largely inconclusive as to its conformation in solution (74,141). Recently attempts to approach the problem using the R value method (75,78) failed quantitatively although a C_2 conformation could be deduced. On the other hand a Karplus approach (75) led to a calculated torsional angle $\phi_0 = 34.5°$, in fair agreement with the accepted experimental value of 37.4° (139).

For substituted cyclopentanones ORD (142) and CD (143) measurements on mono- and polysubstituted cyclopentanones yielded clear-cut, albeit qualitative, confirmation of the existence of an equilibrium mixture of several preferred conformations as previously calculated (39). Even the introduction of a t-butyl group (144) has no influence on the conformational flexibility of the ring as indicated by ORD and CD studies (144).

trans-3,4-Dimethylcyclopentanone has been studied both by vibrational (IR and Raman) (145) and NMR spectroscopy (146). While the former method pointed to a C_2 conformation (145), the latter indicated a net preference of diequatorial methyl groups as in *54* (146-148).

The same trend toward half-chair conformation was found for α-halocyclopentanones in an attempted quantitative treatment (10) of the well-known bathochromic carbonyl shift (149). In conjunction with dipole moments the results were taken to indicate (10) a half-chair form, *55*. A UV study (150) was less convincing.

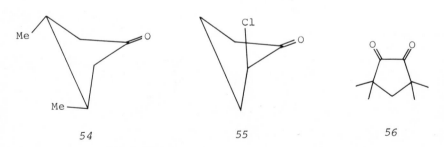

54   55   56

The trend toward flattening in cyclopentanones culminates in 3,3,5,5-tetramethylcyclopentane-1,2-dione (*56*), which was shown, by X-ray diffraction, to be essentially planar (151). On the other hand cyclopenten-3-ones were shown (149) to be nonplanar.

D. Cyclopentene and Derivatives

We do not dwell on the topic of vibrational spectroscopy of this system, since it has been recently and competently reviewed in the class of pseudo-four-membered rings (152). What appears clear is that these compounds, and specifically cyclopentene, occur as severely restricted pseudorotators in an envelope form, with a flap inversion barrier of <1 kcal/mol.

Microwave (MW) spectra qualitatively confirm this situation (153,154) and lead to an estimate of the puckering (out-of-plane) angle of 22 ± 2°.

ED studies (155) indicate a puckering angle of 29 ± 2.5° and the following internal bond angles for *33*: θ_4 = 104.0°, θ_3 = 103.0°, and θ_1 = 111.0°.

Attempts to gain conformational information from vinyl-allylic proton spin coupling were only qualitatively successful (156). Various 3,5-substituted derivatives of *33* were also examined by NMR. Envelope conformations were largely confirmed (158,157), with a value for the puckering angle in *33* of ca. 21°, and in *cis*-3,5-dibromocyclopentene of ca. 19° (158); higher values were proposed for other substituted cyclopentenes (157). Among the benzocyclopentenes 1-mono- and 1,2-disubstituted indanes have been investigated, mainly through NMR spectroscopy (159), with support from dipole moment measurements and IR and Raman spectroscopy (159b). The *trans*-1,2-dihalogen compounds *57* seem to prefer "diaxial" conformations (159a,b). The epimeric 1-indanols have also been studied (159c).

5 7

E. Fused Systems

The simplest system in this series is, of course, bicyclo[3.1.0]hexane (*11*). Its basic conformation is imposed to the extent that the fusion must be cis and the five-membered ring constrained to an envelope, but it can, in principle, choose between a chair or a boat conformation. It turns out that the latter is the actual form of occurrence for virtually all derivatives (*58-65*) that have been investigated structurally.

The most recent information on this system is available from MW studies of *58* (160), *59* (161), *60* (162), and *61* (163),

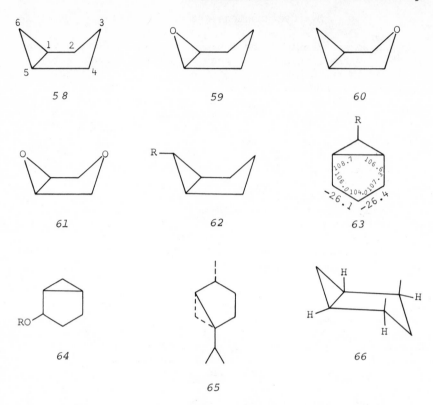

which were also investigated through far-IR studies (164). The latter show single minimum potential functions, implying conformational homogeneity, and the MW spectra give rotational constants in accord with the geometrical data of boat forms 58-61 (Table 13).

TABLE 13

Structural Parameters (deg) of Bicyclo[3.1.0]hexane and Its Analogs, from MW Spectroscopic Studies (160)[a]

Compound	θ_1	θ_2	θ_3	τ
57	108.1	100.8	107.9	35
58	109.6	98.3	108.6	42
59	105.4	101.1	111.3	40
60	107.8	99.7	112.8	41

[a]Internal bond angles (θ) are numbered according to 57 and τ is the puckering angle, that is, $180° - [\angle C_1C_2C_4C_5 - C_2C_3C_4]$.

The most accurate data available, from an X-ray-diffraction analysis of *62* (R = HNSO$_2$C$_6$H$_4$Br), show internal bond and dihedral angles as given in *63* (165) [cf. the good agreement with the calculated structure *11* in Table 4 (37)].

A boat form also emerges from earlier NMR studies on *63* and related compounds (166,167), as well as from dipole moment measurements on *61* (164) and derivatives of *59* (168).

Following earlier NMR (169) and ORD (170) studies on various thujane (*65*) derivatives, recent LIS-NMR (171) and ^{13}C-NMR (172) studies were published. All data indicate boat-like conformations of these derivatives.

The accepted explanation (160,172) for such exclusive preference of the bicyclo[3.1.0]hexane system for the boat conformation is the tendency to avoid the eclipsing strain of the *cis*-1,2 (and 4,5) hydrogens (or substituents) in the chair form *66*. In the boat form *58* all cis-vicinal bonds are gauche to each other.

A representative and interesting system in this class of fused five-membered rings is, obviously, bicyclo[3.3.0]octane (pentalane). The cis (*12*) and trans (*13*) isomers and various derivatives were synthesized and studied by Linstead (173,174) and Granger (175,176). The peculiarity of this system was soon recognized in the fact that the trans isomer *13* is appreciably less stable [$\Delta\Delta H° \simeq$ 6 kcal/mol (12a,37)] than the cis isomer and that the system does not conform to the Conformational Rule (177,178,113). The reason for this behavior (178) seems to lie in the considerable bond angle strain imparted by the trans fusion in *13* (37) as compared to that in the cis isomer *12*. The latter, albeit of smaller molar volume, should easily be able to minimize the nonbonded interactions associated with this situation, whereas the trans isomer *13*, due to its rigidity, cannot distribute and thereby alleviate the imparted strain. This, in our opinion, serves as an additional caveat against interpreting the conformation of *trans*-1,2-disubstituted cyclopentanes in terms of "diaxial" or "maximally puckered" half-chairs or envelopes.

Unfortunately there are no accurate structural data for these systems, as far as we know. IR and Raman spectroscopic studies on the stereoisomeric hydrocarbons and the 3-oxo derivatives have been carried out (179). Qualitative assignments of the double-half-chair conformation *18* to *trans*-pentalane (*13*) (179a) and of the double-envelope (C_s) conformation *20* to the cis isomer (*12*) were made. Similarly the *trans*- and *cis*-bicyclo[3.3.0]hexan-3-ones have been assigned conformations *67* and *68*; the latter assignment does not appear to be definitive, however.

Attempts have also been made to assign preferred conformations to some 2-hydroxy (180) and 2,3-dihydroxy derivatives (181) using NMR coupling constants, but, again, these assign-

67   68

69   70

ments should be regarded with caution due to the problems
inherent in this approach (*vide supra*).

An interesting aspect of this system is that heterosub-
stitution, as in *69* and *70*, improves the stability of the
trans isomer vis-à-vis the cis, as compared with the carbocyclic
pentalane pair (178b,182). The origin of this stabilizing effect
is still obscure.

We turn now to the important steroid D-ring system and the
hydrindane moiety it incorporates. These systems have formed
the subject of a large number of papers, many of them using at
one point or another stereochemical arguments. Only the most
relevant studies can be discussed here because of space limitations. The early developments have been well reviewed (1-3),
and the reader is referred to these texts for literature up to
1963.

It is worth recalling (183) that, while the enthalpy of
trans-hydrindane (*15*) is lower (by 1.04 kcal/mol) than that of
the cis isomer (*14*), the entropy of the latter is higher,
probably because of its enhanced flexibility. Thus the trans
isomer (*15*) is more stable at room temperature, but this
order is reversed at high temperatures (above ca. 200°C).

It has been suggested (184) that the *cis* isomer of 8-
methylhydrindane is slightly more stable (20a).

This brings us to the D-ring in steroids. After the
early theoretical treatment of this system by Brutcher and
Bauer (41a,b), in which the three possible D-ring conformations, *23, 24,* and *25,* were put forward, many investigators
attacked the problem experimentally, using a variety of
methods. NMR coupling constant correlation with the CH-CH
dihedral angle in the D-ring was very popular (see, for
example, refs. 185-188). Altona (18) has pointed out, however,

that such simple correlations (185) are probably in error,
due to the already mentioned departure from trigonal symmetry
in such systems. Conformational arguments have also been used
in chemical studies, such as the stereochemistry of enolization
of 17-ketosteroids (186b) and the equilibration of 2-hydroxy-
A-norcholestanol (189).

Extensive ORD and CD studies of A-nor and D-ring ketones
have been performed (190) with reasoning along lines similar
to those used by Ouannes and Jacques for substituted cyclopentanones (39).

Finally the ultimate method in structural study, X-ray-
diffraction analysis, has been applied to many steroids during
the last three decades with increasing levels of precision
and accuracy (336). The D-ring geometry started receiving
special attention in the early 1960s (41). Brutcher and
Leopold (41c) have examined a number of structural data
available at that time and evaluated torsion angles; they
concluded that their previously calculated "maximally puckered"
model (41a,b) was actually exaggerated.

The Leiden school has invested much investigative effort
in steroid structural analysis (193,17). Special scrutiny of the

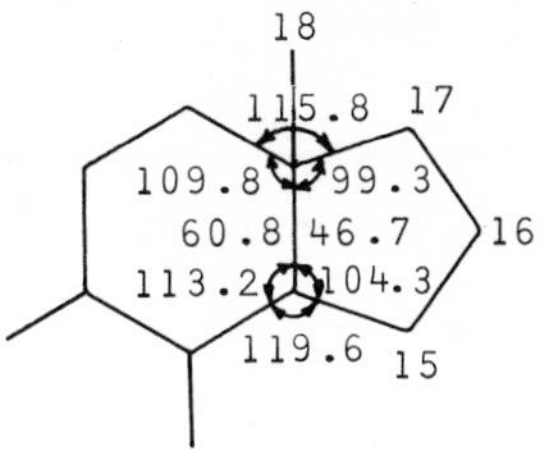

71

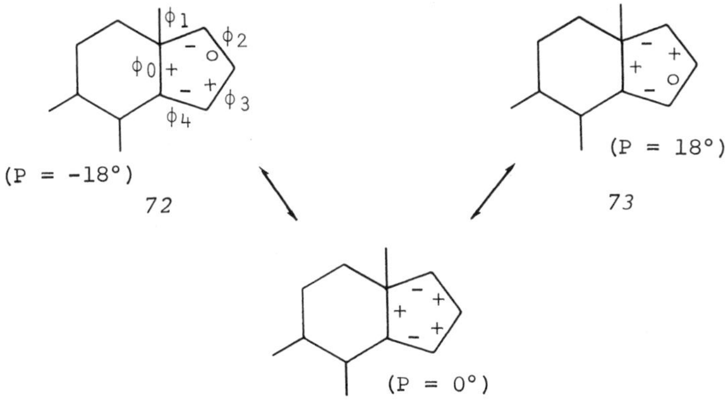

(P = -18°)

72

(P = 18°)

73

(P = 0°)

74

D-ring has also led to significant theoretical results (17,18).
Thus Altona and co-workers (17a) have analyzed a series of 11
literature steroid structures (192-196) having a variety of
structural features in the steroid skeleton but having in
common the C/D trans junction, the angular 18-methyl group, and
various 17-β substituents. The valency angles were taken from
the literature and the torsion angles were calculated. The
weighted averages of torsional and bond angles of interest here
are depicted in formula *71*. Referring to the basic formulas *23*,
24, and *25*, which we have written in equivalent form as *72*, *73*,
and *74*, respectively, and to eq. [4] (ϕ_0 = angle of maximum
puckering), the authors (17a) have evaluated the pertinent
structural features of ring D in the steroids under scrutiny.
Such information for four examples-*26* (193b), *27* (194), *75*
(195), and *76* (196)-along with the standard torsional angles
for the symmetric forms are given in Table 14 and Figure 12.

| 75 | 76 |

The most important conclusions from this research are the
following: (1) no D-ring of exact C_S or C_2 symmetry has yet been
found, although *26* and *27* are very close to C_S and C_2, respectively (this should serve as a caveat in the consideration of
other, nonfused systems as well); (2) the bending of valency
angles about the bridgehead atoms (cf. *71*) accounts for a
large part of the strain and causes breakdown of trigonal
symmetry (the implications of the latter phenomenon have been
discussed here in context of theoretical calculations and NMR
measurements). An interesting discussion of the phenomenon of
conformational transmission in steroids and hydrindane systems
has recently appeared (194). The D-ring conformation has also
been discussed in a recent review in this series (336).

TABLE 14
Internal Bond Angles (θ), Torsion Angles (ϕ), and
Phase Angles (P) (deg) in ring D of Steroids
[from Altona et al (17a)]

	Steroid				Standard	
	26	27	75	76	C_2	C_S
θ_{13}	99.8	99.2	103	102.4		
θ_{14}	104.2	104.3	106	104.1		
θ_{15}	103.6	102.6	102	101.5		
θ_{16}	106.8	106.0	102	110.4		
θ_{17}	103.7	107.8	108	103.9		
$\phi_{13,14}$	47.0	44.9	39.8	42.2	46.7	44.4
$\phi_{14,15}$	-36.1	-38.6	-41.1	-33.5	-37.8	-27.4
$\phi_{15,16}$	10.7	16.6	25.1	10.8	14.4	0
$\phi_{16,17}$	18.1	11.5	- 1.9	14.0	14.4	27.4
$\phi_{17,13}$	-39.1	-34.3	-23.6	-35.3	-37.8	-44.4
P^b	3.95	- 3.95	-19.9	2.2	0	18

[a]Calculated using the weighted average ϕ_0 of 46.7° (cf. 71).
[b]As defined in eq. [4] and 72-74.

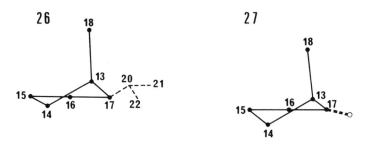

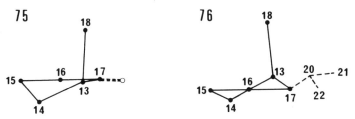

Fig. 12. Ring D geometries in various steroids (cf.
Table 14) (taken from ref. 17a by permission).

F. Bridged Systems

The bicyclo[2.2.1]heptane (norbornane) system is the most extensively investigated one in this class. The parent molecule *28* has been studied by electron diffraction (198–200) (Fig. 13),

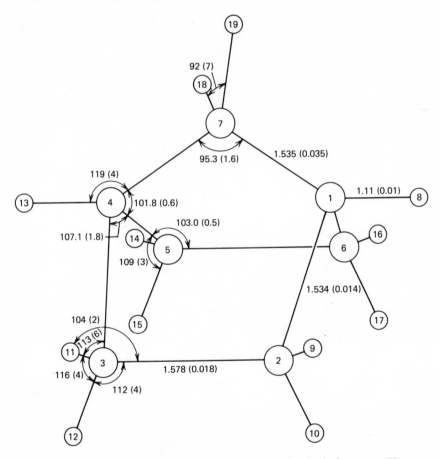

Fig. 13. Norbornane geometry as evaluated from an ED study (taken from ref. 200 by permission).

as has its 1,4-dichloro derivative (199). X-ray-diffraction analysis has been performed on many derivatives (201–204). The torsional angles of norbornane (*28*), 3-*exo*-(*N*-benzyl-*N*-methyl-aminomethyl)-2-*endo*-norbornanol (*30b*) (202), and 1,1'-biapo-camphane (*77*) (203), as determined in structural studies by Altona and Sundaralingam (45), are given in Table 15, along with the values calculated for *28*, *30a*, and *78* by these authors (see

TABLE 15

Experimental and Calculated Torsional Angles (deg) of Norbornane and Two of Its Substituted Derivatives [from Altona and Sundaralingam (45)]

φ[a]	28 exp.[b]	28 calc.	30b exp.[c,d]	30a calc.	77 exp.[c,e]	78 calc.
7-1-2-3	35	36	44	40	33	35
1-2-3-4	0	0	9	5	1	0
2-3-4-7	35	36	30	31	37	36
3-4-7-**1**	55	56	56	54	56	55
4-7-1-2	55	56	61	57	53	54
7-1-6-5	35	36	31	33	33	35
1-6-5-4	0	0	5	4	2	0
6-5-4-7	35	36	39	39	37	36
5-4-7-1	55	56	57	59	56	55
4-7-1-6	55	56	53	56	53	54
6-1-2-3	72	71	64	67	70	72
2-3-4-5	72	71	78	74	72	72
2-1-6-5	72	71	74	73	71	72
6-5-4-3	72	71	68	68	71	72
Twist	0	0	f	f	g	g
p[h]	0, 0		+4, +8[i]			

[a] The numbering is according to formula 28.
[b] ED (200).
[c] X-ray.
[d] From ref. 202.
[e] From ref. 203.
[f] +,+; see formula 31.
[g] -,+; see formula 32.
[h] For phase angles of pseudorotation see eq. [4].
[i] For the left and right five-membered ring in 30b, respectively.

57

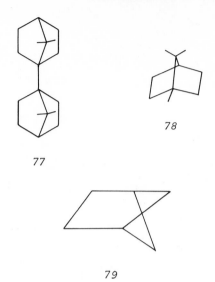

78

77

79

ref. 45 and Sect. III-D). One can see that, indeed, significant
departure from the ideal envelope conformation is induced by
substituents, especially in positions 2, 3, 5, and/or 6. This
ring-twisting effect by substituents is quite additive. Other
derivatives of norbornone, camphane, and norbornene have been
examined, and the findings confirm these trends (45).

These results provide interesting implications regarding
the physical and chemical behavior of norbornane derivatives
as dependent on interactions between substituents on the system.
Thus an NMR study (205) of the influence of substituents (e.g.,
methyl) in norbornanes on the chemical shifts of neighboring
protons indicated a linear dependence of $\Delta v = v_{Me} - v_{H}$ (i.e.,
v with or without a vicinal methyl group) on the measured
distance between H and Me. However, the correlation coefficient
of only 0.89 may indicate that the evaluation of the distances
was probably not accurate as a result of the above mentioned
twisting phenomenon.

The existence of torsional effects, albeit small ones, in
endo-exo equilibration studies has been verified (206). In
another study (207) it has been found that there is no correlation between A (or $\Delta G°$) values of substituents (2) and their
conformational energies in norbornanes. In light of the
preceding discussion one may attribute this, in part at least,
to the different twisting influences of the various substituents
tested (207).

Another bridged five-membered ring that has been looked
at, both experimentally (ED) (208) and theoretically (20b), is
bicyclo[2.1.1]hexane (79), which is shown with some of its
structural data in Figure 14. The two approaches are in quite

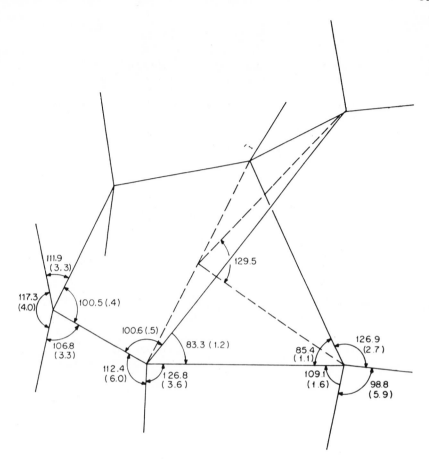

Fig. 14. Bicyclo[2.1.1]hexane geometry as evaluated by ED (taken from ref. 208 by permission).

good agreement [cf., for example, the calculated bridge bond angle of 81.2° (20b)].

G. Heterocyclic Systems

Two reviews are worthy of consultation in this field, one on pseudorotation and vibrational spectra (13), the other on the geometry and conformations of oxygen- and sulfur-containing five-membered rings (19).

1. Oxygen Heterocycles (13,19)

A far-IR spectral study of oxolane (tetrahydrofuran,

35a) provided the first experimental evidence for pseudorotation (209). More detailed far-IR work (210) as well as
MW studies (31b) indicated a low barrier (ca. 0.14 kcal/mol),
that is, essentially free pseudorotation, with the C_2 form
only very slightly preferred (31b); this differs appreciably
from the predicted value of ΔV_C = -2.4 kcal/mol (9) (cf. also
Fig. 5). The puckering amplitude was estimated at 0.44 (209)
and 0.42 Å (66).

Two later studies (212, 213) were inconclusive as to
favored static conformations but confirmed pseudorotation; a
somewhat lower value of q = 0.38 Å was evaluated from ED
data (212).

The conformation of a considerable number of substituted
oxolanes was studied. Thus 3-halogeno- (*80*) and *trans*-3,4-
dihalogeno derivatives (*81*) were investigated using IR and

80   81

Raman spectroscopy coupled with dipole moment data (213). The
authors compared the carbon-halogen stretching frequencies with
the empirical set worked out by Altona (80) (cf. also Table 8)
and arrived at 3-"axial" and 3,4-"diaxial" conformations for *80*
and *81*, respectively, contributing to the extent of 90% for *81*
and almost 100% for *80*. These assignments appear to be supported
by the comparison between experimental and calculated dipole
moments of *80* (Table 16); however, such a comparison is incon-

TABLE 16
Experimental and Calculated Dipole Moments
of *81* and *82* Derivatives (213)

		μ (D)	
		Calc.	
Compound	Exp. (CCl$_4$)	*aa*	*ee*
81 (Cl)	2.10	2.15	1.45
81 (Br)	2.03		
82 (ClCl)	1.05		1.2
82 (ClBr)	1.01	1	
82 (BrBr)	1.07		1.5

clusive when applied to *81*. Simple calculations following an
earlier procedure (33) (without inclusion of dipole-dipole
interactions!) are also in good accord with these assignments,
which are very similar to those of the halogeno- and *trans*-
1,2-dihalogenocyclopentanes (see Sect. IV-B), the main difference being that the energy wells in the pseudorotation potenti-
als of the heterocycles *80* and *81* are steeper than in the
corresponding cyclopentane derivatives.

An interesting point that has not been stressed but is
implicit in this study is that the high preference for an
"axial" conformation of X in *80* can be interpreted as the
tendency of the halogen to be "gauche" to the C-O bond as
exemplified in *82* and in accord with a recently accepted
rationale (19,214). *trans*-2,3-Dichlorotetrahydrofuran was also
assigned a "diaxial" conformation (*83*) based on NMR measurements (215).

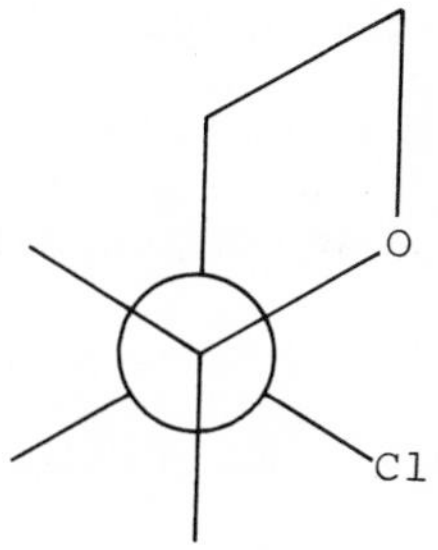

82

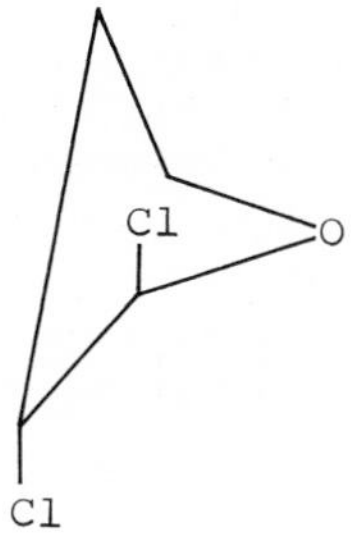

83

An X-ray analysis of the more heavily substituted
oxolane *84* showed it to exist in the crystal in the half-
chair conformation *85* with ϕ_{ext} = 44° (216). Romers and co-
workers (19), using eq [4], have, however, calculated a
maximal torsional angle of ϕ = 40° for *85* (cf. Table XX in

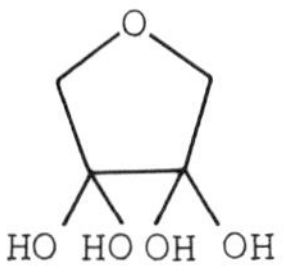

84

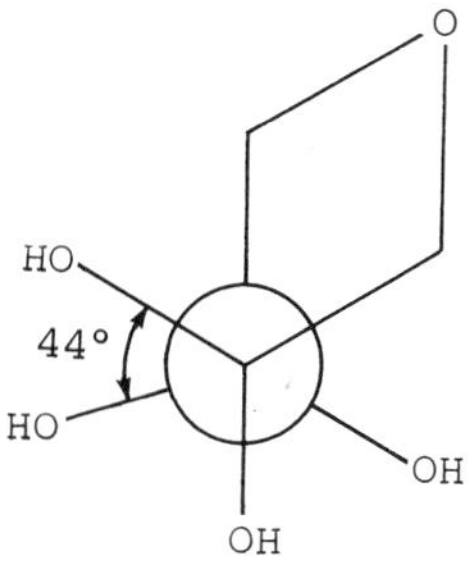

85

ref. 19), but this may again be a manifestation of the lack of trigonal symmetry in five-membered rings.

Other polysubstituted oxolanes, *86*, *87*, and *88*, have been studied by NMR (217,218). In these cases no definite conformations could be assigned but rather flexible states had

86

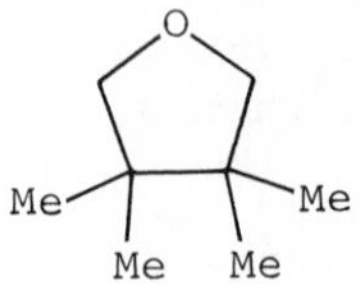

87

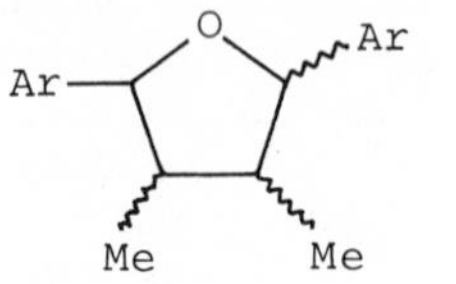

88

to be postulated with a multitude of intermediate forms. Again (218), the indiscriminate application of the Karplus equation leads to results showing the futility of such an approach.

The introduction of one carbonyl group into the oxolane ring not unexpectedly brings about a flattening of the three- or four-atom moiety of which it is part. Thus, D-galactono-γ-lactone (*89*) (219a) was found in the crystal as an envelope with a coplanar $C-O-\overset{\overset{\text{O}}{\|}}{C}-C$ moiety and a puckering amplitude of 0.64 Å. This may be attributed to conjugation of the nonbonded electrons with the carbonyl group as in *90a* (219).

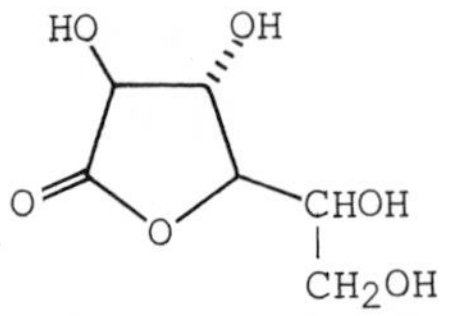

89

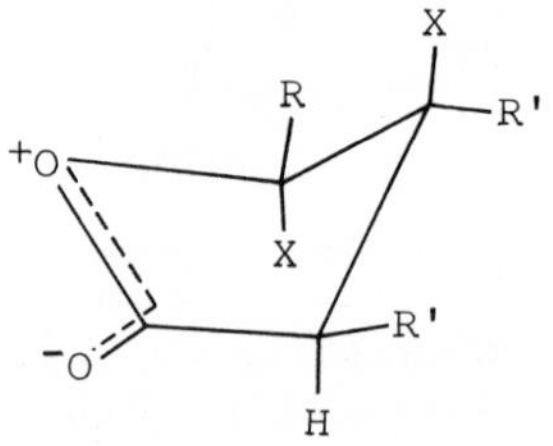

90

(a) R = CHOH–CH$_2$OH
    R' = OH; X = H
(b) R = Me; R' = H; X = Br

The NMR spectra of other substituted γ-butyrolactones have been analyzed accordingly (220). The analysis followed essentially a Karplus approach but, since substituent electronegativity was not taken into account, the correlation with dihedral angles cannot be regarded as quantitative. Nevertheless for *trans*-4,5-dibromo-γ-valerolactone a conformation *90b* with the two bromines nearly "diaxial" is apparently indicated by the coupling constants ($J_{3,4}$ = 5.9, $J_{3',4}$ = -0.2 Hz), similar to other trans-vicinal, dihalogeno five-membered rings (*vide*

supra). From NMR data on oxolan-3-one *91* and its various
derivatives very little conformational information could be
extracted (221) other than the conclusion that the system is
very flexible and amenable to conformational change dependent
on the substitution pattern.

As expected, the flattening effect of the carbonyl
group in this series culminates in succinic anhydride (*92*),
which is very nearly planar in the crystal (222).

    91   *92*

We turn now to one of the more thoroughly studied
five-membered heterocycles, namely, 1,3-dioxolane (*36*) and
its numerous derivatives. The abundance of information in this
system is undoubtedly a result of its easy synthetic accessibility, by acetalization and ketalization of aldehydes or
ketones with appropriately substituted ethylene glycols.

The far-IR spectrum of *36* (13,66,210,223) appears to
indicate once again only slightly (ca. 1.4 kcal/mol) restricted
pseudorotation and an equilibrium puckering amplitude of 0.41 Å
(210) or 0.38 Å (223), apparently smaller than q_e in cyclopentane and oxolane.

Strangely, accurate structural data are rather scarce.
An early X-ray-diffraction study of 2,2'-bi-1,3-dioxolane
shows it to occur in the anti form *93* in the crystal (224)
[whereas in solution the gauche form predominates (19)]. The
two rings are enantiotopic, as indicated by the center of
symmetry found in *93* (224), and the torsion angles, as extracted from crystallographic data (19,225), are $\phi_{1,2} = -28°$,
$\phi_{2,3} = +14°$, $\phi_{3,4} = +5°$, $\phi_{4,5} = -21°$, and $\phi_{5,1} = +30°$; that is,
the conformation is intermediate between C_s (with O_1 on the
flap of the envelope) and C_2 (with O_3 on the "tip" of the
half-chair). The relatively low maximal puckering angle of ca.
30° is qualitatively in accord with vibrational spectroscopic
results, as described above.

The conformations of 1,3-dioxolanes in solution were
investigated mainly by NMR techniques (225-228). Earlier studies
of this sort (229-240) dealt with the analysis of the
O-CHR-CH$_2$-O spectrum primarily in the context of configurational assignments in substituted 1,3-dioxolanes. Thus initial
erroneous assignments for 2,4-disubstituted derivatives *94*
(235) were subsequently reversed (236-240). Stereospecific
synthesis of *cis*-2,4-dimethyl-1,3-dioxolane (*94*) (R = R' = Me)
unequivocally supported these assignments (226).

93   *94*

A study combining NMR spectroscopy ($AA'BB'$ patterns)
with dipole moments of 2-alkoxy-1,3-dioxolanes (*95*) (225)
indicated that the ring geometry depends on the sidechain
orientation, which changes, according to the bulk of R in *95*,
from anti (R-O-C-R') for R = Me, R' = H to gauche for R =
t-Bu, R' = H [cf. also (24)]. Concomitantly a change of J_{cis}
in the $AA'BB'$ spectrum occurs, indicating a change in ring
conformation, with a decrease of the $O-CH_2-CH_2-O$ torsional
angle in the gauche form.

95   *96*

In another, extensive investigation of 2,4-dialkyl-
(*94*) and 2,r-4,cis-5-trialkyl-1,3-dioxolanes (*96*) acid-catalyz-
ed equilibration showed consistently small free energy differences between stereoisomers (226,238,239). Thus (226) for nine
pairs in the *94* series $-\Delta G_{25}^{0}$ for trans $\rightleftarrows$ cis is ca. 0.3
kcal/mol, increasing to ca. 0.5 kcal/mol when R' = t-Bu. Of 14
pairs in the *96* series $-\Delta G_{25}^{0}$ for cis, anti $\rightleftarrows$ cis, syn is in the
range 0.4-0.7 kcal/mol for 10 of them; for the others, all
carrying one or more t-Bu groups, $-\Delta G_{25}^{0}$ decreases and even
changes sign.

In addition to having lower free energy in the *94* series
the cis isomers also exhibit lower refractive index and, with
few exceptions, a shielded H(2) as compared with the trans
isomers, providing a consistent picture similar to the
situation in 1,3-disubstituted cyclopentanes (cf. Sect. IV-B).
Though the vicinal coupling constants are in good agreement
with previously measured values in the series (234,235), they
are largely inconclusive as to conformation. These findings
reemphasize the high flexibility of the five-membered ring and
the negligible magnitude of 1,3-nonbonded interactions (226).
Only in the 2-alkyl-4,5-di-t-butyl derivatives (*96*, R = t-Bu)

can one find evidence for conformational rigidity, probably associated with ring distortion.

NMR studies at 300 MHz of 2-methyl- and 2-trifluoro-methyl-4-halogenomethyl-1,3-dioxolanes (97) have provided (227,228) detailed chemical shifts and coupling constants for ring protons, thus making possible accurate configurational assignments. At the same time attempts were made at conformational analysis based on the original Karplus relationship or on an R-value-type ratio (76). While some of the conformational assignments seem qualitatively reasonable, one should keep in mind the limitations of these methods caused by substituent electronegativity on the one hand and the presumable lack of trigonal projection symmetry on the other (cf. ref. 75 and Sect. IV-A, B). Ethylene carbonate, a dioxolane bearing a carbonyl group at C(2), exists as a puckered half-chair 98

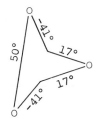

97   98

(a) R = Me
    X = F, Cl, Br, (227)
(b) R = CF$_3$; X = Cl (228)

in the crystal (243). Similar findings were subsequently reported (244) for the gas phase on the basis of MW studies.

Finally 1,2,4-trioxolane (39) was investigated by ED (245) using a C$_2$ model (40), as suggested by a rather crude theoretical calculation (48). A considerably puckered ring was found (99). It is difficult to assert the origin of this enhanced puckering.

99

2. *Nitrogen Heterocycles*

Pyrrolidine (*35b*) appears to exist as a restricted
pseudorotator with a barrier of ca. 0.3 kcal/mol (cf. ref. 9)
derived from thermodynamic studies. In a recent ESR study of
radicals generated from pyrrolidine and some *N*-alkyl derivatives
(*267*) an approximate analysis assuming a $B \cos^2 \phi$ relationship
($B \simeq 5.0$ mT) suggested an equilibrating half-chair geometry *100*.

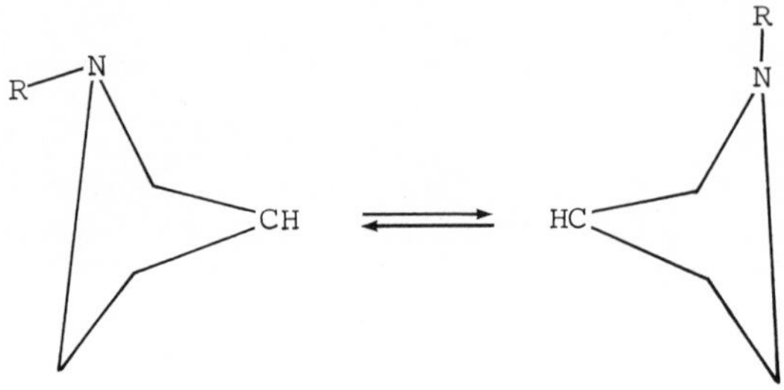

100

Most of the structural work in this area has been
spurred by investigations of amino acids and peptides. Following an early X-ray-diffraction analysis of *trans*-4-hydroxy-
pyroline (*101*) (*247*), numerous such studies were performed.
The torsional angles in three sample structures are given in
Table 17. One can see that all are distorted envelopes with
atoms 4, 2, or 3 on the flap, depending, apparently, on the
substitution pattern in each case. In another recent structure
(*104*) (*250*), a nearly perfect envelope is found with atom 3

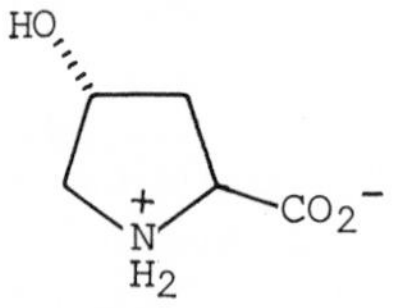

101

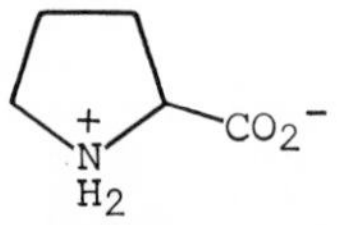

102

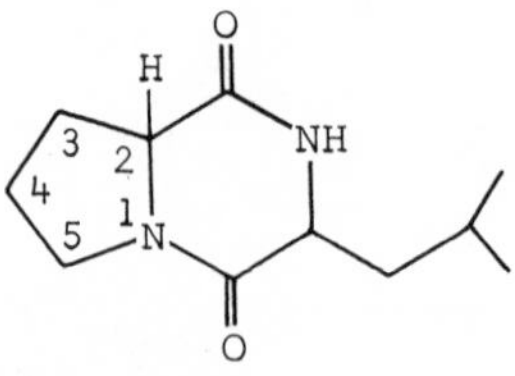

103

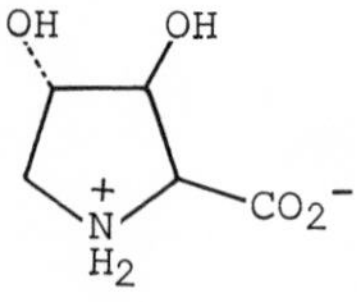

104

0.60 Å above the plane. Most examples, however, have atom 4 as
the most puckered center [(249); cf. discussion and references
in ref. 252].

TABLE 17
Torsional Angles (deg) in Proline Rings
of Some Crystalline Derivatives

Compound	$\phi_{1,2}$	$\phi_{2,3}$	$\phi_{3,4}$	$\phi_{4,5}$	$\phi_{5,1}$	Reference
101	5.2	14.0	-27.2	29.9	-22.1	249,19
102	33.7	-35.0	23.0	-3.8	-18.4	250
103	16.1	-31.5	60.0	-25.1	4.5	251

Solution studies have also been performed using NMR
techniques. Quite early, proline derivatives were subjected to
a Karplus-type treatment (251). Thus *101*, for example, was
assigned the envelope form *105a* in analogy with the result from
X-ray-diffraction analysis [(247) and Table 17)], but with a
puckering angle of 53°! With present-day knowledge of the pit-
falls of such a treatment, one should regard this result with
caution despite the elaborate effort involved (251b).
    NMR studies using Karplus techniques are also available
for L-proline (252) and its *cis*- and *trans*-4-fluoro derivatives
106 (253) in aqueous solutions. Highly puckered models are
generally accepted, and for *106* the dihedral angles (along with
other ring parameters) were calculated (253) by fitting coupling
constants ($^3J_{HH}$ and $^3J_{HF}$) to the calculated geometry through the
use of empirical Karplus-type equations. Since, however, the
geometry was obtained by assuming strictly tetrahedral ring
atoms, hence trigonal projection angles, one should again
regard the results with caution. Here also a slightly distorted
and strongly puckered (ϕ_{max} ca. 53°) envelope *105b* was deduced
for *trans*-4-fluoro-L-proline [cf. *105a* and ref. 251b] but a near
half-chair *107* (ϕ_{max} ca. 52°) for the cis isomer [and *not* an

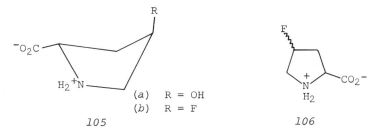

(a) R = OH
(b) R = F

105   *106*

107   108

envelope as mentioned in the original paper (253)]. These large
torsion angles are in contrast to the much smaller ones *measured* in the crystals (Table 17). One must, of course, keep in
mind that the NMR spectra were taken in water solution; whether
the discrepancies in ring puckering are real or caused by solvent vs. packing effects is hard to say at this stage.

The introduction of an sp^2 center into the ring in this
class of heterocycles also causes flattening, and for ethylene
thiourea (*108*) a nearly coplanar structure was found (254,19).

3.   *Sulfur Heterocycles*

Thiolane (*35c*) had been predicted to be a strongly inhibited pseudorotator [ΔV_C = 3 kcal/mol (9)] and experimental work
soon confirmed this prediction.

Thus thermodynamic data indicated a barrier to pseudorotation of 2.8 kcal/mol (255). Early IR and Raman spectra (54)
showed *35c* to be appreciably puckered, and later (256) far-IR
spectra clearly suggested restricted pseudorotation. A more
recent far-IR study (257,13) led to the value of a 2.21 kcal/mol
barrier with the C_2 form preferred.

An ED study of thiolane, in conjunction with energy-
geometry calculations (50b), indicated a C_2 form (2-3 kcal/mol
more stable than a C_S form at P = 90° away) with internal
angles θ_1 = 93.4°, θ_2 = θ_5 = 106.1°, and θ_3 = θ_4 = 105.0° (the
ring torsional angles are shown in formula *109*). Similarly
large puckering angles were found in various other sulfur -
containing five-membered rings, such as 1,2- and 1,3-dithio-
lanes, which have been reported in crystallographic (258,259)
or solution (260,261) studies. Thus 2,2'-bi-1,3-dithiolane
(*110*) (258) was found to exist in the crystal as half-chairs
(*111*) [the torsional angles (19) are given in the formula]
whereas crystalline 1,2-dithiolane-4-carboxylic acid was
found as a slightly distorted envelope (*112*) (259). A temperature-dependent CD study of (R)-α-lipoic acid (*113*) was only
qualitatively conclusive concerning its conformation (260).

Attempts to study the conformation of 2-substituted 1,3-
dithiolanes (261) and their 1,1,3,3-tetraoxides (*114*) (262) by
NMR spectroscopy using R values are subject to the same
criticism that this approach has met with in other five-member-

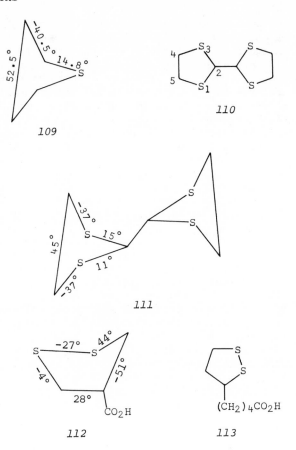

109

110

111

112

113

ed rings (see Sects. IV-A, B.). The measured NMR data were
apparently overinterpreted, but the qualitative correlations
with torsion angles are probably conclusive for the
existence of large puckering angles for 1,3-dithiolanes and
somewhat smaller ones for the tetroxides 114 (262).

Similar remarks apply to solution studies of 2-substituted
1,3-oxathiolanes (115) (261,263,264) and 1,3-thiazolidines
(116) (225). Although detailed NMR data have been presented for

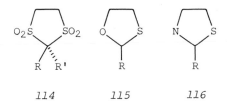

114   115   116

all of them, such data are useful mainly for configurational
assignments and for assessing qualitatively the puckering
tendency in the ring. Thus the 1,3-oxathiolanes appear to
have normally puckered rings with the sulfur in the flat part
of the molecule (263). This agrees well both with results of
an X-ray analysis of cholestan-4-one-3-spiro(2,5-oxathiolane)
(117) (266) and with force-field calculations (51). However,
the assertion that pseudorotation is absent in these systems
(263b) is not warranted, and calculations actually predict
considerable pseudolibration (51). A special caveat concerns
the calculation of conformational free energies for substituents
in such systems. Thus, for example, in 115 -ΔG for R = Me and
i-Pr was calculated (263b) to be 1.13 and 2.01 kcal/mol,
respectively. This was based on the assumption that 2-t-butyl-
1,3-oxathiolane (115, R = t-Bu) is conformationally homogeneous and that the chemical shifts of the 5-hydrogens are
extreme values. It was, however, shown that in the cyclopentane series even t-butyl-substituted compounds have
appreciable pseudolibration (18) and, moreover, that the
t-butyl group seriously deforms the ring (23) as it adopts a
more or less fixed conformation in relation to it. On the
other hand, in N-alkyl pyrrolidine radicals (100) the corre-
esponding hyperfine splittings are taken to indicate a
conformational free energy difference of 0.33 kcal/mol for
R = CH₃ and 0.62 kcal/mol for R = t-Bu. To be sure one can
hardly regard these values, as well as the above NMR-derived
ones, as more than educated guesses.

We end this section by mentioning an ED study of 7-thia-
bicyclo[2.2.1]heptane (118), which has been investigated
along with the [2.1.1] derivative and may be compared with the
parent compounds (Sect. IV-D).

117

4. Selenium Heterocycles

Selenolane was studied in the crystal as an iodine or
bromine complex (119) (269,270). For both a nearly perfect
and extremely puckered half-chair (120) was found.

A far-IR spectroscopic study yielded, accordingly, a
very high pseudorotational barrier of 5.3 kcal/mol (271). MW
spectra seem to confirm these features (272).

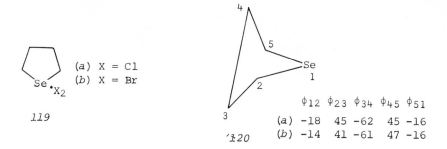

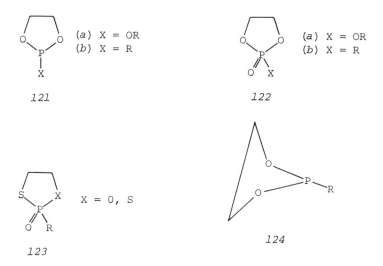

(a) X = Cl
(b) X = Br

119

	ϕ_{12}	ϕ_{23}	ϕ_{34}	ϕ_{45}	ϕ_{51}
(a)	-18	45	-62	45	-16
(b)	-14	41	-61	47	-16

'120

5. Phosphorus Heterocycles

In this class of compounds five-membered cyclic phosphites
121a, phosphates 122a, and the corresponding phosphinates 121b
and phosphonates 122b have been investigated (273-277) with NMR
techniques. In addition, some related sulfur-containing
derivatives 123 have been studied (278,279).

Recent consensus seems to favor distorted envelope or
half-chair ring conformations (124) with very slow (if any)
inversion at phosphorus.

(a) X = OR
(b) X = R

121

(a) X = OR
(b) X = R

122

X = O, S

123

124

A number of X-ray-diffraction analyses have also been
reported as, for example, that of methyl ethylene phosphate
(122a) (R = Me) (280), which was found as a slightly distorted
envelope having a puckering angle of ca. 11° (cf. also ref.
53).

An exotic molecule whose static and dynamic behavior has
been studied is (PCF$_3$)$_5$ (125). X-ray analysis provided qualita-

tive evidence that the ring is in a conformation with symmetry
intermediate between C_2 and C_S (281), and ^{19}F- and ^{31}P-NMR
temperature-dependent spectra indicated (282) a barrier to
pseudorotation of ca. 23 kcal/mol.

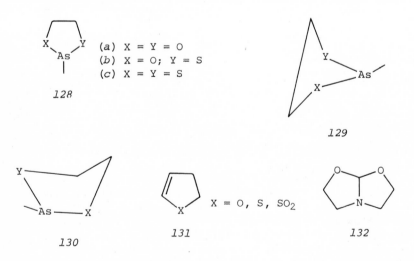

6. Miscellaneous Heterocyclic Systems

     Both sila- and germacyclopentane (*126,127*) have been
studied by far-IR spectroscopy. The IR as well as Raman spectra
of *126* and MW spectra of *127* indicate C_2 conformations with
high barriers to pseudorotation, ca. 4 kcal/mol for *126* and 6
kcal/mol for *127* (13).
     Organoarsenic compounds such as dioxaarsolanes (*128a*) and
some thia analogues (*128b,128c*) have been investigated (283,284)
by NMR spectroscopy, mainly variable-temperature techniques.
The conformation assignments vary between equilibrium mixtures
of half-chairs (*129*) (283) and of envelopes (*130*) (284). There
are many recent publications dealing with a variety of other
five-membered rings, from polysubstituted oxygen heterocycles
(285,286) through unsaturated oxygen and sulfur heterocycles
of type *131* (287-289) to bicyclic compounds such as *132* (289).

The investigative tool in these studies was NMR at various levels of sophistication. However, as in a number of previously mentioned cases, it is hard at this stage to evaluate the validity of the interpretations. In general the wealth of available coupling constants that have been treated (or mistreated) by simple-minded Karplus approaches should be amenable to unifying and consistent reinterpretation (279).

H. Sugars

This class of compounds certainly deserves its own section. The conformational aspects of six-membered carbohydrates have been extensively reviewed in 1965 (2). The five-membered-ring sugars had been relatively sparsely studied at that time (2) but have been extensively investigated during the last decade.

To start with, one should mention the five-membered cyclitols, that is, the polyhydroxycyclopentanes (291). Whereas the six-membered cyclitols behave according to well-understood conformational criteria (2), the physical and chemical behavior of the five-membered analogues is dictated almost exclusively by configurational criteria (291b). Nevertheless certain conformational aspects of various derivatives in the series have been described (291,157,149a).

The furanose ring is obviously the most important member of this category, both by itself (e.g., *133*) and as the central moiety (*134*) of the sugar phosphate chains of nucleic acids. Its detailed structural features are rather well known from

133   *134*

X-ray-diffraction studies of nucleosides and nucleotides, which have been exhaustively reviewed (292,17c). In solution the method of choice was, naturally, NMR spectroscopy, which has been extensively used in this area(293-305). After a number of pioneering studies (294-297) the techniques were gradually upgraded until it was found that reliable correlations between coupling constants and geometry could be made (193,305).

Altona and Sundaralingam (17c) calculated the puckering amplitude and its phase angle p of pseudorotation for the furanose ring in a large number of β-purine and β-pyrimidine

nucleosides and nucleotides using eq [4] (17) and the known
internal ring torsion angles. Performing a statistical classification of the compounds for which p is in a certain range,
they were then able to show (17c) that only two pseudorotational
ranges are preferred by β-sugars in solid form, each occupying
less than 10% of the total pseudorotational circuit. This
circuit is depicted in Figure 15, and the two "canonic" half-
chair (twist) forms of the furanose ring are shown in formulas
135 and 136. From the calculated p values a nomenclature
emerged, according to the range of phase angles, that is,
"type N" (i.e., North) in the 0° range or "type S" (i.e.,
South) in the 180° range (Fig. 16). This treatment is very
illuminating.

This approach was extended to solution studies by taking
external torsion angles ϕ_{HH} from X-ray studies and coupling
constants from literature data. A correlation was then performed using a Karplus expression and allowing for pseudorotation in the ranges N and S (293). The purine ribosides
showed a *small* conformational preference for the type S
conformer 136, and pyrimidine sugars a small preference for the

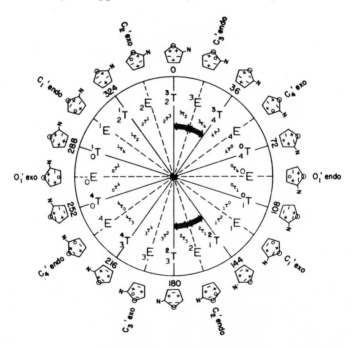

Fig. 15. Pseudorotational circuit of the furanose ring,
$P = 0$ to $360°$, E = envelope; T = twist (half-chair) forms. The
heavy arrows indicate the preferred pseudorotational range
(taken from reference 17c by permission).

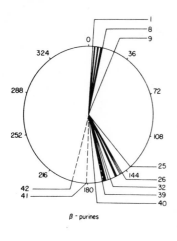 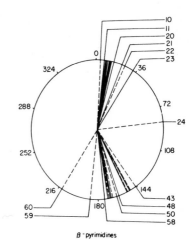

β - purines   β - pyrimidines

Fig. 16. Phase angle (*P*) values for β-purine and β-pyrimidine glycosides. Numbers refer to compounds tabulated in ref. 17c (taken from ref. 17c by permission).

type N conformer *135*. Again the evidence is overwhelming (293) and, with a few exceptions, recent studies appear largely to confirm these results (299-304).

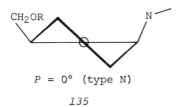 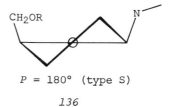

$P = 0°$ (type N) $P = 180°$ (type S)

135   *136*

I. Prostaglandins

We end this chapter with a discussion of a worthy member of the club--the prostaglandins. The tremendous development of this class of biologically important compounds during the last decade has included generation of conformational information obtained mainly by X-ray analyses. Thus the conformations of the five-membered ring in prostaglandins, $F_1\beta$ (306), A_1 (309), and E_2 (310), are depicted in formulas *137*, *138*, and *139*, respectively. Once again it is seen that there is hardly any ring-imposed conformational pattern but rather that the substituents have the last word in the matter.

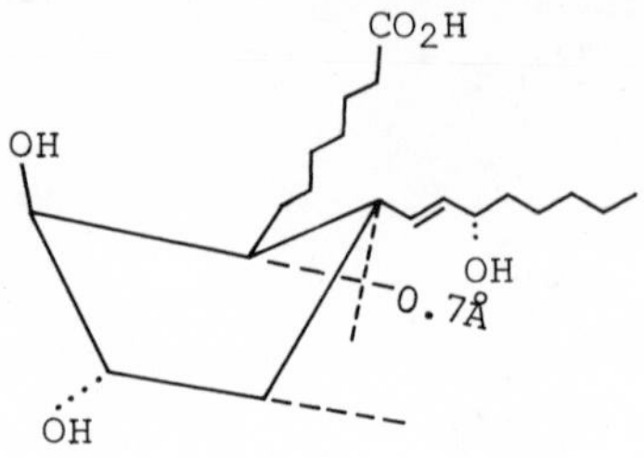

137

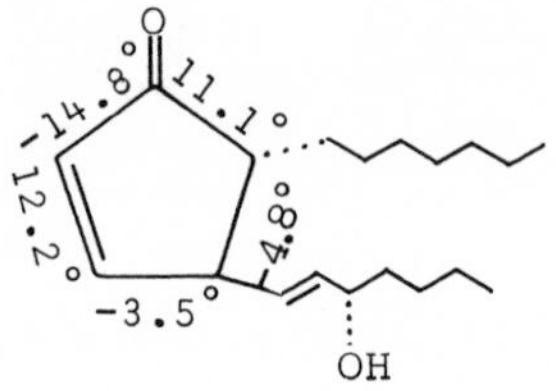

138

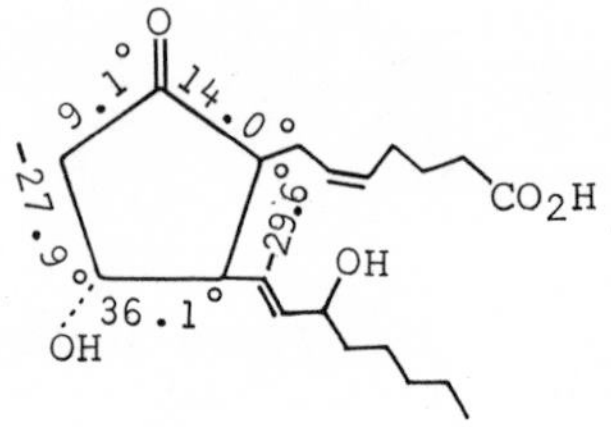

139

V. CONCLUDING REMARKS

Beyond conveying specific information this chapter was written in an attempt to convey a message. Most of those who have actively contributed to the knowledge of five-membered ring conformation have emphasized that, except for extremely rigid or heavily substituted systems, ring conformation is not well defined and one can speak only of more or less shallow potential wells in the pseudorotational circuit. Moreover the preferred forms are in most cases *not* symmetrical ring conformations. Consequently the notions "axial" and "equatorial" are not similar to their meanings for six-membered rings, mainly in that they are not well defined. It would be useful if educators would keep this in mind when teaching five-membered-ring conformational analysis.

Finally, one might have expected a special section of this chapter to be dedicated to conformation-reactivity relationships. There have been, indeed, a considerable number of attempts to correlate various reactions with conformational features. Most of them have failed to do so, and some have been mentioned throughout the discussion. To mention the rest would be prohibitive because of space demands and not very illuminating. As it is, we have tried to be as comprehensive as possible while keeping within the borders of conformational analysis of five-membered rings. It is hoped that not many papers in this area have escaped attention.

VI. ADDENDUM

Since the completion of this chapter a number of additional papers appeared. These are discussed here, the indexing being in accord with the section designations of the main text.

III-B. Methyl- and fluorocyclopentane were calculated by Pople and co-workers (311) within the framework of the *ab initio* MO theoretical treatment. Both substituents appear to favor the envelope form, leading to two separate potential minima for axial and equatorial substitution. However, while the equatorial form of methylcyclopentane is more stable, the reverse is true of fluorocyclopentane. The latter result is supported experimentally (see Sect. IV-B) but its origin remains rather obscure. It should be mentioned that, again, the relative energy differences are no more than 0.3-0.4 kcal/mol.

III-E. Cyclopentene as well as 1-pyrazolines were subjected to calculation by the LCMO method (312) with no essential new features.

IV-B. It is becoming apparent that the chiroptical properties of five-membered rings are of considerable interest (313). Thus chirality functions for some substituted cyclopentanes (with idealized D_{5h} symmetry!) were developed and compared with experimental values. As those initiated in cyclopentane behavior would have expected, 1,2-disubstituted derivatives show good agreement, but 1,3-derivatives do not.

IV-C. Still in the chiroptical context, very interesting results on cyclopentanone derivatives have become available (314-317). Thus CD data (n,π^*) for chiral cyclopentanones (315) were analyzed (314), and it was suggested that C_α-H bonds make a dominant "antioctant" contribution. In other system, namely (1S,5S)-bicyclo[3.2.0]heptan-3-one, the unusually large rotatory strength of the n,π^* transition (316) was calculated (by CNDO/S) to result mainly from the twisting of the five-membered ring (317).

Vibrational spectroscopy studies were performed on 2-halocyclopentanones (318) and also on *cis*-3,4-dimethylcyclopentanone, which was also investigated by NMR (319). The molecules were concluded to exist in the ubiquitous half-chair conformation.

Five-membered carbo- and heterocyclic radicals were investigated by ESR in an attempt to ascertain shape and interconversion, with modest success.

IV-E. In the bicyclo[3.1.0]hexane series a MW spectral study of the 6-thia derivative (321) led to the well-accepted

boat conformation, whereas an NMR study of the isomeric
thujanols (322) led to a challenge of this form but with very
little solid evidence.

IV-G. The conformational aspects of a considerable
number and variety of five-membered heterocycles continue to
receive attention. Examples are 1,3-dioxolanes (NMR) (323)
and trioxolanes (MW) (324), substituted pyrrolidines (NMR)(325),
2-pyrazoline (326), 3-germacyclopentanols (as compared to the
carbocyclic analogues) (NMR-LIS (327), 1,3-dithiolans (UV-PES
and MO calculations) (328), substitued 1,3-oxathiolans (epimer
equilibria, NMR) (329,330), 3-aryl-1,2,3-oxathiazolidin-2-
oxides (NMR) (331), and various arsolanes (NMR) (332, 333).
A noteworthy paper on proline conformation (molecular mechanics
and review of literature data on atomic coordinates) was
published very recently (334).

IV-H. A C-13 NMR study (chemical shifts) of furanosides as
compared with cyclopentanols has appeared (335).

IV-I. Finally, after a recent chapter on steroid
structures, Duax and collaborators (336) have reviewed prostaglandin conformations as they emerge from crystallographic
data (337).

ACKNOWLEDGMENTS

The first draft of this chapter was written during a
Visiting Professorship at the University of North Carolina at
Chapel Hill. The kind hospitality of the faculty of the
Department of Chemistry there is gratefully acknowledged.
It is a pleasure to express thankful appreciation to Dr.
Cornelis Altona for his careful reading of the manuscript and
his pertinent comments. Finally, it was a privilege to enjoy
the encouragement, help, and stimulation of Prof. Ernest L.
Eliel in this endeavor.

REFERENCES

1. W. G. Dauben and K. S. Pitzer, in *Steric Effects in
    Organic Chemistry,* M. S. Newman, Ed., Wiley, New York,
    1956, Chap. 1.
2. E. L. Eliel, N. L. Allinger, S. J. Angyal, and G. A.
    Morrison, *Conformational Analysis,* Interscience, New
    York, 1965, pp. 200-206.
3. M. Hanack, *Conformation Theory,* Academic Press, New York,
    1965, pp. 72-86.

4. (a) H. C. Brown, J. H. Brewster, and H. Shechter, *J. Am. Chem. Soc.*, 76, 467 (1954); cf. also H. C. Brown, *J. Org. Chem.*, 22, 439 (1957), and references cited there. (b) S. A. Barker and R. Stephens, *J. Chem. Soc.*, 1954, 4550. (c) E. Gil-Av and J. Shabtai, *Chem. Ind. (Lon.)*, 1959, 1630.

5. L. P. Kuhn, *J. Am. Chem. Soc.*, 74, 2492 (1952); 76, 4323 (1954).

6. (a) J. G. Aston, S. C. Schumann, H. L. Fink, and P. M. Doty, *J. Am. Chem. Soc.*, 63, 2029 (1941). (b) J. G. Aston, H. L. Fink and S. C. Schumann, *J. Am. Chem. Soc.*, 65, 341 (1943).

7. K. S. Pitzer, *Science,* 101, 672 (1945).

8. J. E. Kilpatrick, K. S. Pitzer, and R. Spitzer, *J. Am. Chem. Soc.*, 80, 6697 (1958).

9. K. Pitzer and W. E. Donath, *J. Am. Chem. Soc.*, 81, 3213 (1959).

10. F. V. Brutcher, Jr., T. Roberts, S. J. Barr, and N. Pearson, *J. Am. Chem. Soc.*, 81, 4915 (1959).

11. R. Bucourt, "The Torsion Angle Concept in Conformational Analysis," in *Topics in Stereochemistry,* Vol. 8, E. L. Eliel and N. L. Allinger, Eds., Interscience, New York, 1974, p. 159.

12. (a) J. D. Cox and G. Pilcher, *Thermochemistry of Organic and Organometallic Compounds,* Academic Press, New York, 1970. (b) J. W. Knowlton and F. D. Rossini, *J. Res. Natl. Bur. Stand.*, 43, 113 (1949). (c) F. D. Rossini, K. S. Pitzer, R. L. Arnett, R. M. Brown, and G. C. Pimentel, *Selected Values of Physical and Thermodynamic Properties of Hydrocarbons and Related Compounds,* Carnegie Press, Pittsburgh, 1953.

13. J. Laane, in *Vibrational Spectra and Structure,* Vol. 1, J. R. Durig, Ed., Marcel Dekker, New York, 1972.

14. J. B. Hendrickson, *J. Am. Chem. Soc.*, 83, 4537 (1961), and erratum, 85, 4059 (1963).

15. E. M. Engler, J. D. Andose, and P. v. R. Schleyer, *J. Am. Chem. Soc.*, 95, 8005 (1973).

16. (a) S. Lifson and A. Warshel, *J. Chem. Phys.*, 49, 5116 (1968). (b) A. Warshel, Ph.D. Thesis, Weizmann Institute of Science, 1969.

17. (a) C. Altona, H. J. Geise, and C. Romers, *Tetrahedron,* 24, 13 (1968). (b) H. J. Geise, C. Altona, and C. Romers, *Tetrahedron Lett.*, 1967, 1383. (c) C. Altona and M. Sundaralingam, *J. Am. Chem. Soc.*, 94, 8205 (1972).

18. C. Altona, in *Conformational Analysis: Scope and Present Limitations,* G. Chiurdoglu, Ed., Academic Press, New York, 1971, p. 1. and references cited there.

19. C. Romers, C. Altona, H. R. Buys, and E. Havinga, in *Topics in Stereochemistry,* Vol. 4, E. L. Eliel and N. L. Allinger, Eds., Interscience, New York, 1969.

20. (a) N. L. Allinger, J. A. Hirsch, M. A. Miller, I. J.
     Tyminski, and F. A. Van Catledge, *J. Am. Chem. Soc.*, <u>90</u>,
     1199 (1968). (b) N. L. Allinger, M. T. Tribble, M. A.
     Miller, and D. W. Wertz, *J. Am. Chem. Soc.*, <u>93</u>, 1637
     (1971).

21. J. D. Dunitz, *Tetrahedron*, <u>28</u>, 5459 (1972); J. D. Dunitz
     and J. Waser, *J Am. Chem. Soc.*, <u>94</u>, 5654 (1972); J. D.
     Dunitz and J. Waser, *Elem. Math.*, <u>27</u>, 25 (1972).

22. D. Cremer and J. A. Pople, *J. Am. Chem. Soc.*, <u>97</u>, 1354
     (1975). (b) D. Cremer and J. A. Pople, *J. Am. Chem. Soc.*,
     <u>97</u>, 1358 (1975).

23. (a) F. V. Brutcher, Jr., and R. C. Lugar, *Abstracts of the
     160th ACS Meeting*, 153, September, 1970. (b) R. C. Lugar,
     Ph.D. Thesis, University of Pennsylvania, 1970.

24. R. Hoffmann, *J. Chem. Phys.*, <u>39</u>, 1397 (1963).

25. G. Dallinga and P. Ross, *Rec. Trav. Chim.*, <u>87</u>, 906 (1968).

26. H. Cambron-Bruderlein and C. Sandorfy, *Theor. Chim. Acta*,
     <u>4</u>, 224 (1966).

27. N. C. Baird, *Tetrahedron*, <u>26</u>, 2185 (1970).

28. T. R. Ferguson and C. L. Beckel, *J. Chem. Phys.*, <u>59</u>,
     1905 (1973).

29. J. R. Hoyland, *J. Chem. Phys.*, <u>50</u>, 2775 (1969).

30. H. M. Pickett and H. L. Strauss, *J. Chem. Phys.*, <u>55</u>,
     324 (1971).

31   (a) D. O. Harris, G. G. Engerholm, C. A. Tolman, A. C.
     Luntz, R. A. Keller, H. Kim, and W. D. Gwinn, *J. Chem.
     Phys.*, <u>50</u>, 2438 (1969). (b) G. G. Engerholm, A. C.
     Luntz, W. D. Gwinn, and D. O. Harris, *J. Chem. Phys.*, <u>50</u>,
     2446 (1969).

32. J. Reisse, L. Nagels, and G. Chiurdoglu, *Bull. Soc. Chim.
     Belg.*, <u>74</u>, 162 (1965).

33. C. Altona, H. R. Buys, and E. Havinga, *Rec. Trav. Chim.*,
     <u>85</u>, 973 (1966).

34. L. A. Carreira, *J. Chem. Phys.*, <u>55</u>, 181 (1971).

35. B. Fuchs and P. S. Wechsler, *Tetrahedron*, <u>33</u>, 57 (1977).

36. R. Bucourt and N. C. Cohen, *Bull. Soc. Chim. Fr.*, <u>1969</u>,
     2015.

37. (a) R. H. Boyd, *J. Chem. Phys.*, <u>49</u>, 2574 (1968). (b) C.
     F. Shieh, D. McNally, and R. H. Boyd, *Tetrahedron*, <u>25</u>,
     3653 (1969). (c) S. Chang, D. McNally, S. Shary-Tehrany,
     M. J. Hickey, and R. H. Boyd, *J. Am. Chem. Soc.*, <u>92</u>,
     3109 (1970).

38. R. Bucourt and D. Hainaut, *Bull. Soc. Chim. Fr.*, <u>1965</u>,
     1366.

39. C. Ouannes and J. Jacques, *Bull. Soc. Chim. Fr.*, <u>1965</u>,
     3601.

40. (a) N. L. Allinger and M. T. Tribble, *Tetrahedron*, <u>28</u>,
     1191 (1972). (b) N. L. Allinger, M. T. Tribble, and M. A.
     Miller, *Tetrahedron*, <u>28</u>, 1173 (1972).

41. (a) F. V. Brutcher, Jr., and W. Bauer, Jr., *J. Am. Chem.
     Soc.*, <u>84</u>, 2233 (1962). (b) F. V. Brutcher, Jr., and W.

Bauer, Jr., *J. Am. Chem. Soc.*, 84, 2236 (1962). (c) F. V. Brutcher, Jr., and E. J. Leopold, *J. Am. Chem. Soc.*, 88, 3156 (1966).

42. H. Krieger, *Suom. Kemistil.*, B31, 348 (1958); B32, 109 (1959).

43. C. F. Wilcox, Jr., *J. Am. Chem. Soc.*, 82, 414 (1960).

44. G. A. Sim, *J. Chem. Soc.*, 1965, 5994.

45. C. Altona and M. Sundaralingam, *J. Am. Chem. Soc.*, 92, 1995 (1970).

46. (a) N. L. Allinger, J. A. Hirsch, M. A. Miller, and I. J. Tyminski, *J. Am. Chem. Soc.*, 90, 5773 (1968).
    (b) N. L. Allinger and J. T. Sprague, *J. Am. Chem. Soc.*, 94, 5734 (1972).

47. E. L. James, *Diss. Abstr.*, 24, 1398 (1963).

48. H. M. Seip, *Acta Chem. Scand.*, 23, 2741 (1969).

49. C. M. Venkatachalam, B. J. Price, and S. Krimm, *Macromolecules*, 7, 212 (1974).

50. (a) Z. Nahlovski, B. Nahlovski, and H. M. Seip, *Acta Chem. Scand.*, 24, 1903 (1970).
    (b) Z. Nahlovski, B. Nahlovski, and H. M. Seip, *Acta Chem. Scand.*, 23, 3534 (1969).

51. G. E. Wilson, *J. Am. Chem. Soc.*, 96, 2426 (1974).

52. (a) R. A. Rouse, *Int. J. Quant. Chem.*, 1974, 201.
    (b) R. A. Rouse, *J. Am. Chem. Soc.*, 95, 3640 (1973).

53. D. A. Usher, E. A. Dennis, and F. H. Westheimer, *J. Am. Chem. Soc.*, 87, 2320 (1965).

54. H. Tschamler and H. E. Voetter, *Monatsh.*, 83, 302, 835, 1228 (1952).

55. D. R. Douslin and H. M. Huffman, *J. Am. Chem. Soc.*, 68, 173 (1946).

56. R. S. Spitzer and H. M. Huffman, *J. Am. Chem. Soc.*, 69, 211 (1947).

57. R. S. Spitzer and K. S. Pitzer, *J. Am. Chem. Soc.*, 68, 2537 (1946).

58. J. P. McCullough, *J. Chem. Phys.*, 29, 966 (1958).

59. J. P. McCullough, R. E. Pennington, J. C. Smith, I. A. Hossenlop, and G. Waddington, *J. Am. Chem. Soc.*, 81, 5880 (1959).

60. F. A. Miller and R. G. Inskeep, *J. Chem. Phys.*, 18, 1519 (1950).

61. B. Curnutte and W. H. Shaffer, *J. Mol. Spectrosc.*, 1, 239 (1957).

62. C. G. LeFevre and R. J. W. LeFevre, *J. Chem. Soc.*, 1956, 3549; 1957, 3458; M. Low, *Tetrahedron Lett.*, 1960, 3.

63. E. J. Rosenbaum and H. F. Jacobson, *J. Am. Chem. Soc.*, 63, 2841 (1941); K. Tanner and A. Weber, *J. Mol. Spectrosc.*, 10, 381 (1963).

64. J. R. Durig and D. W. Wertz, *J. Chem. Phys.*, 49, 2118 (1968).

65. I. M. Mills, *Mol. Phys.*, 20, 127 (1971).

66. R. Davidson and P. A. Warsop, *J. Chem. Soc., Faraday Trans. II,* <u>68</u>, 1875 (1972).

67. L. A. Carreira, G. J. Tiang, W. B. Person, and J. N. Willis, *J. Chem. Phys.,* <u>56</u>, 1440 (1972).

68. T. Ikeda, R. C. Lord, T. B. Malloy, and T. Ueda, *J. Chem. Phys.,* <u>56</u>, 1434 (1972).

69. E. S. Glazer, R. Knorr, C. Ganter, and J. D. Roberts, *J. Am. Chem. Soc.,* <u>94</u>, 6026 (1972).

70. O. Hassel and H. Viervoll, *Acta Chem. Scand.,* <u>1</u>, 149 (1947).

71. O. Bastiansen, O. Hassel, and L. K. Lund, *Acta Chem. Scand.,* <u>3</u>, 297 (1949).

72. A. Almeningen, O. Bastiansen, and P. N. Skancke, *Acta Chem. Scand.,* <u>15</u>, 711 (1961).

73. W. J. Adams, H. J. Geise, and L. S. Bartell, *J. Am. Chem. Soc.,* <u>92</u>, 5013 (1970).

74. K. B. Wiberg and B. J. Nist, *J. Am. Chem. Soc.,* <u>83</u>, 1226 (1961).

75. (a) J. B. Lambert, J. J. Papay, S. A. Khan, K. A. Kappauf, and E. S. Magyar, *J. Am. Chem. Soc.,* <u>96</u>, 6112 (1974).
     (b) J. B. Lambert, J. J. Papay, E. S. Magyar, and M. K. Neuberg, *J. Am. Chem. Soc.,* <u>95</u>, 4458 (1973).

76. (a) J. B. Lambert, *Acc. Chem. Res.,* <u>4</u>, 87 (1971).
     (b) J. B. Lambert, *J. Am. Chem. Soc.,* <u>89</u>, 1836 (1967).

77. (a) H. R. Buys, *Rec. Trav. Chim.,* <u>88</u>, 1003 (1969).
     (b) H. R. Buys, *Rec. Trav. Chim.,* <u>89</u>, 1253 (1970).

78. (a) R. L. Lipnick, *J. Mol. Struct.,* <u>21</u>, 411 (1974).
     (b) R. L. Lipnick, *J. Mol. Struct.,* <u>21</u>, 423 (1974).
     (c) R. L. Lipnick, *J. Am. Chem. Soc.,* <u>96</u>, 2941 (1974).

79. H. R. Buys, C. Altona, and E. Havinga, *Rec. Trav. Chim.,* <u>87</u>, 53 (1968).

80. C. Altona, *Tetrahedron Lett.,* <u>1968</u>, 2325.

81. (a) I. O. C. Ekejiuba and H. E. Hallam, *J. Mol. Struct.,* <u>6</u>, 341 (1970).
     (b) I. O. C. Ekejiuba and H. E. Hallam, *Spectrochim. Acta,* <u>26A</u>, 59 (1970).
     (c) I. O. C. Ekejiuba and H. E. Hallam, *Spectrochim. Acta,* <u>26A</u>, 67 (1970).

82. (a) W. C. Harris, J. M. Karriker, and J. R. Durig, *J. Mol. Struct.,* <u>9</u>, 139, (1971).
     (b) J. R. Durig, J. M. Karriker, and D. W. Wertz, *J. Mol. Spectrosc.,* <u>31</u>, 237 (1969).

83. F. Vovelle, A. LeRoy, and S. Odiot, *J. Mol. Struct.,* <u>11</u>, 53 (1972).

84. D. W. Wertz and W. E. Shasky, *J. Chem. Phys.,* <u>55</u>, 2422 (1971).

85. D. W. Wertz, *J. Mol. Struct.,* <u>17</u>, 163 (1973).

86. W. Hückel and J. Kurz, *Ann. Chem.,* <u>645</u>, 162 (1961).

87. G. Chiurdoglu and W. Masschelein, *Bull. Soc. Chim. Belg.*, 71, 59 (1962).

88. B. Fuchs and R. G. Haber, *Tetrahedron Lett.*, 1966, 1323.

89. I. O. C. Ekejiuba and H. E. Hallam, *J. Chem. Soc. (B)*, 1970, 209.

90. G. Chiurdoglu, *Bull. Soc. Chim. Belg.*, 53, 45 (1944), and previous papers.

91. (a) G. Vavon and M. Barbier, *Bull. Soc. Chim. Fr.*, 49, 572 (1931).
    (b) G. Vavon and C. Zaremba, *Bull. Soc. Chim. Fr.*, 49, 1857 (1931).

92. W. Hückel and M. Hanack, *Ann. Chem.*, 616, 18 (1958).

93. W. Hückel and E. Mögle, *Ann. Chem.*, 649, 13 (1961); W. Hückel and H. D. Sauerland, *Ann. Chem.*, 592, 190 (1955).

94. W. Hückel and R. Bross, *Ann. Chem.*, 664, (1963).

95. W. Hückel, W. Egerer, and F. Mössner, *Ann. Chem.*, 645, 162 (1961).

96. (a) O. A. Aref'ev, V. A. Zakharenko, and A. A. Petrov, *Neftekhimiya*, 4, 854 (1964).
    (b) O. A. Aref'ev, V. A. Zakharenko, and A. A. Petrov, *Neftekhimiya*, 6, 505 (1966).

97. M. B. Epstein, G. M. Barrow, K. S. Pitzer, and F. D. Rossini, *J. Res. Natl. Bur. Stand.*, 43, 245 (1949).

98. A. L. Lieberman, O. V. Bragin, G. K. Gur'yanova, and B. A. Kazanskii, *Dokl. Chem.* (Engl. Transl.), 148, 70 (1963).

99. (a) G. L. Fonken and S. Shiengthong, *J. Org. Chem.*, 28, 3485 (1963).
    (b) D. S. Seigler and J. J. Bloomfield, *J. Org. Chem.*, 38, 1375 (1973).

100. D. Y. Curtin, H. Gruen, Y. G. Hendrickson, and H. E. Knipmeyer, *J. Am. Chem. Soc.*, 83, 4838 (1961).

101. F. Reynolds, Ph.D. Thesis, University of Pennsylvania, 1963.

102. (a) W. D. Kumler and A. C. Huitric, *J. Am. Chem. Soc.*, 78, 3369 (1956).
     (b) W. D. Kumler, A. C. Huitric, and H. K. Hall, Jr., *J. Am. Chem. Soc.*, 78, 4345 (1956).

103. C. Altona, H. R. Buys, and E. Havinga, *Rec., Trav. Chim.*, 85, 983 (1966).

104. C. Altona, H. R. Buys, H. G. Hageman, and E. Havinga, *Tetrahedron*, 23, 2265 (1967).

105. H. R. Buys, C. Altona, and E. Havinga, *Tetrahedron*, 85, 998 (1966).

106. C. Altona, H. J. Hageman, and E. Havinga, *Spectrochim. Acta*, 24A, 633 (1968).

107. (a) K. Kozima and W. Suetaka, *J. Chem. Phys.*, 35, 1516 (1961).
     (b) W. Suetaka and K. Kozima, *J. Chem. Phys.*, 41, 1519 (1964).

108. H. R. Buys, C. Altona, and E. Havinga, *Rec. Trav. Chim.*, 86, 1007 (1967).

109. E. Benedetti, P. Corradini, and C. Pedone, *J. Phys. Chem.*, 76, 790 (1972).

110. T. N. Margulis, L. R. Dalton, and A. L. Kwiram, *Nature (Phys.)*, 242, 82 (1973).

111. T. N. Margulis, *Acta Crystallogr.*, B31, 1049 (1975).

112. S. F. Birch and R. A. Dean, *J. Chem. Soc.*, 1953, 2477.

113. (a) K. von Auwers, *Ann. Chim.*, 84, 820 (1920).
     (b) H. van Bekkum, A. van Veen, P. E. Verkade, and B. M. Wepster, *Rec. Trav. Chim.*, 80, 1310 (1961).
     (c) N. L. Allinger, *J. Am. Chem. Soc.*, 79, 3443 (1957).

114. J. N. Haresnape, *Chem. Ind. (Lond.)*, 1953, 1091.

115. K. A. Saegebarth, *J. Org. Chem.*, 25, 2212 (1960).

116. H. C. Brown and G. Zweifel, *J. Org. Chem.*, 27, 4708 (1960).

117. C. Darby, H. B. Henbest, and I. McClenaghan, *Chem. Ind. (Lond.)*, 1962, 462.

118. H. Baumann, N. C. Franklin, and H. Möhrle, *Tetrahedron*, 23, 4331 (1967).

119. B. Fuchs and R. G. Haber, *Bull. Res. Counc. Isr.*, 11A, 30 (1962).

120. R. G. Haber and B. Fuchs, *Tetrahedron Lett.*, 1966, 1447.

121. (a) A. Warshawsky and B. Fuchs, *Tetrahedron*, 25, 2633 (1969).
     (b) A. Warshawsky, M.Sc. Thesis, Israel Institute of Technology, 1967.

122. J. C. Richer and G. Gilardeau, *Can. J. Chem.*, 43, 3419 (1963).

123. P. S. Wechsler and B. Fuchs, *J. Chem. Soc., Perkin II*, 1976, 943.

124. B. Fuchs and P. S. Wechsler, *J. Chem. Soc., Perkin II*, 1977, 75.

125. S. C. Temin and M. E. Baum, *Can. J. Chem.*, 45, 705 (1963).

126. I. Lillien and K. Khaleeluddin, *Chem. Ind. (Lond.)*, 1964, 1028.

127. A. P. Krapcho and D. E. Horn, *Tetrahedron Lett.*, 1966, 6107.

128. H. Fuhrer, V. B. Kartha, P. J. Krueger, H. H. Mantsch, and R. N. Jones, *Chem. Rev.*, 72, 439 (1972), and references cited there.

129. J. R. Durig, G. L. Coulter, and D. W. Wertz, *J. Mol. Spectrosc.*, 27, 285 (1968).

130. L. A. Carreira and R. C. Lord, *J. Chem. Phys.*, 57, 3225 (1969).

131. T. Ikeda and R. C. Lord, *J. Chem. Phys.*, 56, 4450 (1972).

132. H. E. Howard-Lock and G. W. King, *J. Mol. Spectrosc.*, 35, 393 (1970); 36, 53 (1970).

133. V. B. Kartha, H. H. Mantsch, and R. N. Jones, *Can. J. Chem.*, 51, 1749 (1973).

134. R. Cataliot and G. Paliani, *Chem. Phys. Lett.*, 20, 280 (1973).

135. C. Sablayrolles, L. Bardet, and G. Fleury, *J. Chem. Phys.*, 66, 1139 (1968).

136. G. Erlandson, *J. Chem. Phys.*, 22, 563 (1954).

137. J. H. Burkhalter, *J. Chem. Phys.*, 23, 1172 (1955).

138. H. Kim and W. D. Gwynn, *J. Chem. Phys.*, 51, 1815 (1969).

139. H. J. Geise and F. C. Mijlhoff, *Rec. Trav. Chim.*, 90, 577 (1971).

140. C. Altona, R. A. DeGraaf, C. H. Leeuwestein, and C. Romers, *J. Chem. Soc., Chem. Commun.*, 1971, 1305.

141. F. A. L. Anet, *Can. J. Chem.*, 39, 2316 (1961).

142. C. Ouannes and J. Jacques, *Bull. Soc. Chim. Fr.*, 1965, 3611.

143. C. Djerassi, R. Records, C. Ouannes, and J. Jacques, *Bull. Soc. Chim. Fr.*, 1966, 2878.

144. J. Jacques, *Bull. Soc. Chim. Fr.*, 1969, 3505.

145. L. Bardet, R. Granger, C. Sablayrolles, and J. P. Girard, *J. Mol. Struct.*, 13, 59 (1972).

146. C. Sablayrolles, R. Granger, J. P. Girard, H. Bodot, J.-P. Aycard, and L. Bardet, *Org. Magn. Reson.*, 6, 161 (1974).

147. F. S. Richardson, D. D. Shillady, and J. E. Bloor, *J. Phys. Chem.*, 75, 2466 (1971).

148. S. E. Pfeffer and S. F. Osman, *J. Org. Chem.*, 37, 2425 (1972).

149. (a) F. G. Coçu, G. Wolczunowicz, L. Bors, and Th. Pasternak, *Helv. Chim. Acta*, 53, 739 (1970); (b) G. Hajdukov and M. L. Martin, *J. Mol. Struct*, 20, 105 (1974).

150. D. Q. Quan, *C. R.*, 261, 2374 (1965).

151. L. C. G. Goaman and D. F. Grant, *Tetrahedron*, 19, 1531 (1963).

152. C. S. Blackwell and R. C. Lord, in *Vibrational Spectra and Structure*, Vol. 1, J. R. Durig, Ed., Dekker, New York, 1972, p. 1.

153. G. W. Rathjens, Jr., *J. Chem. Phys.*, 36, 2401 (1962).

154. S. S. Butcher and C. C. Costain, *J. Mol. Spectrosc.*, 15, 40 (1965).

155. M. I. Davis and T. W. Muecke, *J. Phys. Chem.*, 74, 1104 (1970); R. K. Bohn, H. Shintani, T. Fukuyama, and K. Kuchitsu, *Acta Crystallogr.*, 28A, 517 (1972).

156. (a) G. V. Smith and H. Kriloff, *J. Am. Chem. Soc.*, 85, 2016 (1963). (b) E. W. Garbisch, Jr., *J. Am. Chem. Soc.*, 86, 5561 (1964).

157. (a) R. S. Steyn and H. Z. Sable, *Tetrahedron*, 27, 4429
      (1971).
      (b) H. Z. Sable, M. W. Ritchey, and J. E. Nordlander,
      Carbohydr. Res., 1, 10 (1965); *J. Org. Chem.*, 31, 3771
      (1966).
158. H. J. Jakobsen, *Tetrahedron Lett.*, 1967, 1991
159. (a) H. R. Buys and E. Havinga, *Tetrahedron*, 24, 4967
      (1968).
      (b) R. A. Austin and C. P. Lillya, *J. Org. Chem.*, 34,
      1327 (1969).
      (c) M. Hiscock and G. B. Porter, *J. Chem. Soc.*, B, 1971
      1631.
160. R. L. Cook and T. B. Malloy, *J. Am. Chem. Soc.*, 96, 1703
      (1974).
161. W. J. Lafferty, *J. Mol. Spectrosc.*, 36, 84 (1970).
162. T. B. Malloy, *J. Mol. Spectrosc.*, 49, 432 (1974).
163. R. A. Creswell and W. J. Lafferty, *J. Mol. Spectrosc.*,
      46, 371 (1973).
164. R. C. Lord and T. B. Malloy, *J. Mol. Spectrosc.*, 46, 358
      (1973); L. A. Carreira and R. C. Lord, *J. Chem. Phys.*,
      51, 2735 (1969).
165. M. F. Grostic, D. J. Duchamp, and C. G. Chidester, *J. Org.
      Chem.*, 36, 2929 (1971).
166. (a) P. K. Freeman, M. F. Grostic, and F. A. Raymond,
      J. Org. Chem., 30, 771 (1965).
      (b) P. K. Freeman, F. A. Raymond, J. C. Sutton, and W. R.
      Kindley, *J. Org. Chem.*, 33, 1448 (1968), and other papers
      in the series.
167. S. Winstein, E. C. Frederick, R. Baker, and Y. Lin,
      Tetrahedron, 58, 621 (1966).
168. J. J. McCullough, H. B. Henbest, R. J. Bishop, G. M.
      Glover, and L. E. Sutton, *J. Chem. Soc.*, 1965, 5496.
169. K. Tori, *Chem. Pharm. Bull.*, 13, 1439 (1965).
170. M. S. Bergqvist and T. Norin, *Ark Kemi*, 22, 137 (1964);
      T. Norin, *Acta Chem. Scand.*, 17, 738 (1963).
171. T. Norin, S. Stromberg, and M. Weber, *Acta Chem. Scand.*,
      27, 1579 (1973).
172. R. J. Abraham, C. M. Holden, P. Loftus, and D. Whittaker,
      Org. Magn. Reson., 6, 184 (1974).
173. R. P. Linstead, *Ann. Rep. Chem. Soc.*, 1935, 305.
174. J. W. Barret and R. P. Linstead, *J. Chem. Soc.*, 1935, 436,
      1069; 1936, 611.
175. R. Granger, P. F. G. Nau, and J. Nau, *Bull. Soc. Chim.
      Fr.*, 1959, 1807, 1811.
176. R. Granger, P. F. G. Nau, and J. Nau, *Bull. Soc. Chim.
      Fr.*, 1960, 217, 1225, 1350.
177. N. L. Allinger and R. J. Curby, *J. Org. Chem.*, 26, 933
      (1961).

178. (a) N. L. Allinger, M. Nakazaki, and V. Zalkov, *J. Am. Chem. Soc.*, 81, 4074 (1959).
(b) E. L. Eliel, N. L. Allinger, S. J. Angyal, and G. A. Morrison, *Conformational Analysis*, Interscience, New York, 1965, pp. 176, 227.

179. (a) R. Granger, L. Bardet, C. Sablayrolles, and J. P. Girard, *Bull. Soc. Chim. Fr.*, 1971, 391.
(b) R. Granger, L. Bardet, C. Sablayrolles, and J. P. Girard, *Bull. Soc. Chim. Fr.*, 1971, 1771.
(c) R. Granger, L. Bardet, C. Sablayrolles, and J. P. Girard, *Bull. Soc. Chim. Fr.*, 1971, 4454.
(d) R. Granger, L. Bardet, C. Sablayrolles, and J. P. Girard, *Bull. Soc. Chim. Fr.*, 1971, 4458.

180. I. Tabushi, K. Fujita, and R. Oda, *J. Org. Chem.*, 35, 2383 (1970).

181. J. P. Vidal, R. Granger, J. P. Girard, J. C. Rossi, and C. Sablayrolles, *C. R. Acad. Sci. Paris*, 274C, 905 (1972).

182. L. N. Owen and A. G. Peto, *J. Chem. Soc.*, 1955, 2383.

183. E. L. Eliel, N. L. Allinger, S. J. Angyal, and G. A. Morrison, *Conformational Analysis*, Interscience, New York, 1965, pp. 228-230.

184. N. L. Allinger, *J. Org. Chem.*, 21, 915 (1966).

185. A. D. Cross and P. Crabbé, *J. Am. Chem. Soc.*, 86, 1221 (1964).

186. J. Fishman, (a) *J. Am. Chem. Soc.*, 87, 3455 (1965); (b) *J. Org. Chem.*, 31, 520 (1966).

187. T. Takemoto, Y. Kondo, and H. Mori, *Chem. Pharm. Bull.*, 13, 897 (1965).

188. D. A. Swann and J. H. Turnbull, *Tetrahedron*, 24, 1441 (1968).

189. J. C. Espie, A. M. Giroud, and A. Rassat, *Bull. Soc. Chim. Fr.*, 1967, 809.

190. J. P. Jennings, W. P. Mose, and P. M. Scopes, *J. Chem. Soc.*, C, 1967, 1103.

191. M. J. Brienne, A. Heymes, J. Jacques, G. Snatzke, W. Klyne, and S. R. Wallis, *J. Chem. Soc.*, C, 1970, 423.

192. Cf. references cited in refs. 193 and 17a.

193. (a) J. Geise, C. Altona, and C. Romers, *Tetrahedron*, 23, 439 (1967).
(b) H. J. Geise, C. Romers, and E. W. M. Rutten, *Acta Crystallogr.*, 20, 249 (1966).
(c) H. J. Geise and C. Romers, *Acta Crystallogr.*, 20, 257 (1966).

194. D. F. High and J. Kraut, *Acta Crystallogr.*, 21, 88 (1966).

195. D. A. Norton, G. Kartha, and C. T. Lu, *Acta Crystallogr.*, 16, 89 (1963).

196. D. C. Hodgkin, B. M. Rimmer, J. D. Dunitz, and K. N. Trueblood, *J. Chem. Soc.*, 1963, 4945.

197. R. Bucourt, "The Torsion Angle Concept in Conformational
      Analysis," in *Topics in Stereochemistry,* Vol. 8, E. L.
      Eliel and N. L. Allinger, Eds., Interscience, 1974, p.
      212.

198. Y. Morino, K. Kuchitsu, and A. Yokozeki, *Bull. Chem.
      Soc. Jpn.,* 40, 1552 (1967); A. Yokozeki and K. Kuchitsu,
      Bull. Chem. Soc. Jpn., 44, 2356 (1971).

199. J. F. Chiang, C. F. Wilcox, and S. H. Bauer, *J. Am.
      Chem. Soc.,* 90, 3149 (1968).

200. G. Dallinga and L. H. Toneman, *Rec. Trav. Chim.,* 87, 795
      (1968).

201. C. Altona and M. Sundaralingam, *Acta Crystallogr.,* B28,
      1806 (1972).

202. A. V. Fratini, K. Britts, and I. L. Karle, *J. Phys.
      Chem.,* 71, 2482 (1967).

203. R. A. Alden, J. Kraut, and T. G. Traylor, *J. Am. Chem.
      Soc.,* 90, 74 (1968).

204. See compilation of references in ref. 45.

205. E. Pretch, H. Immer, C. Pascual, K. Schaffner, and W.
      Simon, *Helv. Chim. Acta,* 50, 105 (1967).

206. J. M. Mellor and C. F. Webb, *Tetrahedron Lett.,* 1971,
      4025.

207. R. J. Ouellette, J. D. Rawn, and S. N. Jreissaty, *J. Am.
      Chem. Soc.,* 93, 7117 (1971).

208. G. Dallinga and L. H. Toneman, *Rec. Trav. Chim.,* 86, 171
      (1976); C. F. Wilcox, Jr., *J. Am. Chem. Soc.,* 82, 414
      (1971).

209. W. J. Lafferty, D. W. Robinson, R. V. St. Louis, J. W.
      Russel, and H. L. Strauss, *J. Chem. Phys.,* 42, 2915
      (1965).

210. J. A. Greenhouse and H. L. Strauss, *J. Chem. Phys.,* 50,
      124 (1969).

211. A. Almenningen, H. M. Seip, and T. Willadsen, *Acta
      Chem. Scand.,* 23, 2748 (1969).

212. H. J. Geise, W. J. Adams, and L. S. Bartell, *Tetrahedron,*
      25, 3045 (1969).

213. H. R. Buys, C. Altona, and E. Havinga, *Tetrahedron,* 24,
      3019 (1968).

214. L. Phillips and V. Wray, *J. Chem. Soc., Chem. Commun.,*
      1973, 90, and references cited therein.

215. M. Holik and I. Borkovco, *Chem. Zvesti,* 28, 374 (1974).

216. A. D. Mighell and R. A. Jacobsen, *Acta Crystallogr.,* 17,
      1554 (1964).

217. (a) J. Bogner, J. C. Duplan, Y. Infarnet, J. Delmau, and
      J. Huet, *Bull. Soc. Chim. Fr.,* 1972, 3616;
      (b) D. Gagnaire and P. Vottero, *Bull. Soc. Chim. Fr.,*
      1972, 873.

218. K. V. Sarkanen and A. F. A. Wallis, *J. Heterocycl. Chem.,*
      10, 1025 (1973).

219. G. A. Jeffrey, R. D. Rosenstein, and M. Vlasse, *Acta Crystallogr.*, 22, 725 (1967); A. McL. Mathieson, *Tetrahedron Lett.*, 1963, 81, and references cited there.

220. J. B. Lowry and N. V. Riggs, *Tetrahedron Lett.*, 1964, 2911.

221. M. Anteunis and M. Vanderwalle, *Spectrochim. Acta*, 27A, 2119 (1971).

222. M. Ehrenberg, *Acta Crystallogr.*, 19, 698 (1965).

223. J. R. Durig and D. W. Wertz, *J. Chem. Phys.*, 49, 675 (1968).

224. S. Furberg and O. Hassel, *Acta Chem. Scand.*, 4, 1584 (1950).

225. C. Altona and A. P. M. van der Veek, *Tetrahedron*, 24, 4377 (1968).

226. W. E. Willy, G. Binsch, and E. L. Eliel, *J. Am. Chem. Soc.*, 92, 5394 (1970).

227. F. Borremans, M. Anteunis, and F. Anteunis-DeKetelaere, *Org. Magn. Reson.*, 5, 299 (1973).

228. M. Anteunis, R. Van Cauwenberghe, and C. Becu, *Bull. Soc. Chim. Belg.*, 82, 591 (1973).

229. R. U. Lemieux, J. D. Stevens, and R. R. Fraser, *Can. J. Chem.*, 40, 1955 (1962).

230. B. Matthiason, *Acta Chem. Scand.*, 17, 2133 (1963).

231. F. A. L. Anet, *J. Am. Chem. Soc.*, 84, 747 (1962).

232. R. J. Abraham, *J. Chem. Soc.*, 1965, 256.

233. D. Gagnaire and J. B. Robert, *Bull. Soc. Chim. Fr.*, 1965, 3646.

234. F. Alderweireldt and M. Anteunis, *Bull. Soc. Chim. Belg.*, 74, 488 (1965).

235. M. Anteunis and F. Alderweireldt, *Bull. Soc. Chim. Belg.*, 73, 889, 903 (1964).

236. N. Baggett, J. M. Duxbury, A. B. Foster, and J. M. Webber, *J. Chem. Soc.*, C, 1966, 208, and previous papers.

237. M. Anteunis, F. Borremans, J. Gelan, L. Heyndrickx, and W. Vandenbroucke, *Bull. Soc. Chim. Belg.*, 76, 533 (1967).

238. Y. Rommelaere and M. Anteunis, *Bull. Soc. Chim. Belg.*, 79, 11 (1970).

239. G. Lemière and M. Anteunis, *Bull. Soc. Chim. Belg.*, 80, 215 (1971).

240. T. D. Inch and N. Williams, *J. Chem. Soc.*, C, 1970, 263.

241. F. Kametani and Y. Sumi, *Chem. Pharm. Bull.*, 20, 1479 (1972).

242. F. G. Riddell and M. J. T. Robinson, *Tetrahedron*, 27, 4163 (1973).

243. C. J. Brown, *Acta Crystallogr.*, 7, 92 (1954).

244. I. Wang, C. O. Britt, and J. E. Boggs, *J. Am. Chem. Soc.*, 87, 4950 (1965).

245. A. Almenningen, P. Kolsaker, H. M. Seip, and T. Willadsen, *Acta Chem. Scand.*, 23, 3398 (1969).

246. J. P. McCullough, D. R. Douslin, W. N. Hubbard, S. S.
      Todd, J. F. Messerly, I. A. Hossenlopp, F. R. Frow, J. P.
      Dawson, and G. Waddington, *J. Am. Chem. Soc.*, <u>81</u>, 5884
      (1959).

247. J. Donohue and K. N. Trueblood, *Acta Crystallogr.*, <u>5</u>, 414,
      419 (1952).

248. Y. Mitsui, M. Tsuboi, and Y. Iitaka, *Acta Crystallogr.*,
      <u>B25</u>, 2182 (1969).

249. I. L. Karle, *J. Am. Chem. Soc.*, <u>94</u>, 81 (1972), and
      references cited therein.

250. I. L. Karle, *Acta Crystallogr.*, <u>B26</u>, 765 (1970).

251. (a) R. J. Abraham and W. A. Thomas, *J. Chem. Soc.*, <u>1964</u>,
      3739.
      (b) R. J. Abraham and K. A. McLauchlan, *Mol. Phys.*, <u>5</u>,
      195,513 (1962).
      (c) R. J. Abraham, K. A. McLauchlan, S. Dalby, G. W.
      Kenner, R. C. Sheppard, and L. F. Burroughs, *Nature*, <u>192</u>,
      1150 (1961).

252. L. Pogliani, M. Ellenberg, and J. Valat, *Org. Magn.
      Reson.*, <u>7</u>, 61 (1975).

253. J. T. Gerig and R. S. McLeod, *J. Am. Chem. Soc.*, <u>95</u>, 5725
      (1973).

254. P. J. Wheatley, *Acta Crystallogr.*, <u>6</u>, 369 (1953).

255. W. N. Hubbard, H. L. Finke, D. W. Scott, J. P. McCullough,
      C. Katz, M. E. Gross, J. F. Messerly, R. E. Pennington,
      and G. Waddington, *J. Am. Chem. Soc.*, <u>74</u>, 6025 (1952).

256. G. A. Crowder and D. W. Scott, *J. Mol. Spectrosc.*, <u>16</u>,
      122 (1965).

257. D. W. Wertz, *J. Chem. Phys.*, <u>51</u>, 2133 (1969).

258. L. B. Brahde, *Acta Chem. Scand.*, <u>8</u>, 1145 (1954).

259. O. Foss and O. Tjomsland, *Acta Chem. Scand.*, <u>12</u>, 1810
      (1958).

260. (a) L. A. Neubert and M. Carmack, *Tetrahedron Lett.*,
      <u>1974</u>, 3543.
      (b) I. L. Karle, J. A. Estlin, and K. Britts, *Acta
      Crystallogr.*, <u>22</u>, 567 (1967).
      (c) R. M. Stroud and C. H. Carlisle, *Acta Crystallogr.*,
      <u>B28</u>, 304 (1972).
      (d) O. Foss, A. Hordvick, and J. Sletten, *Acta Chem.
      Scand.*, <u>20</u>, 1169 (1966).

261. L. A. Sternson, D. S. Coviello, and R. S. Egan, *J. Am.
      Chem. Soc.*, <u>93</u>, 6529 (1971).

262. L. A. Sternson, L. C. Martinelli, and R. S. Egan, *J.
      Heterocycl. Chem.*, <u>11</u>, 1117 (1974).

263. (a) K. Pihlaja, *Suom. Kemistil. B*, <u>43</u>, 143 (1970).
      (b) D. J. Pasto, F. M. Klein, and T. W. Doyle, *J. Am.
      Chem. Soc.*, <u>89</u>, 4368 (1967).

264. G. E. Wilson, Jr., M. G. Huang, and F. A. Bovey, *J. Am.
      Chem. Soc.*, <u>92</u>, 5907 (1970).

265. G. E. Wilson, Jr., and T. J. Bazzone, *J. Am. Chem. Soc.*, **96**, 1465 (1974).

266. A. Cooper and D. A. Norton, *J. Org. Chem.*, **33**, 3535 (1968).

267. B. C. Gilbert, R. O. C. Norman, and M. Trenwith, *J. Chem. Soc., Perkin II, 1974*, 1033.

268. T. Fukuyama, K. Kuchitsu, Y. Tamaru, Z. Yoshida, and I. Tabushi, *J. Am. Chem. Soc.*, **93**, 2799 (1971).

269. H. Hope and J. D. McCullough, *Acta Crystallogr.*, **17**, 712 (1964).

270. M. Barlow and I. Barlow, *Acta Crystallogr.*, **A21**, 103 (1966).

271. W. H. Green, A. B. Harvey, and J. A. Greenhouse, *J. Chem. Phys.*, **54**, 850 (1971).

272. A. K. Mamleev, N. N. Magdesiev, and N. M. Pozdeev, *J. Struct. Chim.*, **11**, 1053 (1970).

273. B. Fontal and H. Goldwhite, *Tetrahedron*, **22**, 3275 (1966).

274. P. Haake, J. P. McNeal, and E. J. Goldsmith, *J. Am. Chem. Soc.*, **90**, 715 (1968).

275. R. H. Cox, B. S. Campbell, and M. G. Newton, *J. Org. Chem.*, **37**, 1557 (1972).

276. R. H. Cox and M. G. Newton, *J. Am. Chem. Soc.*, **94**, 4212 (1972).

277. J. Deviller, M. Cornus, and J. Navech, *Org. Magn. Reson.*, **6**, 211 (1974), and preceding papers in the series.

278. K. Bergensen, M. Bjoroy, and T. Gramstad, *Acta Chem. Scand.*, **26**, 3037 (1972).

279. B. Fuchs, in preparation.

280. T. A. Steitz and W. N. Lipscomb, *J. Am. Chem. Soc.*, **87**, 2488 (1965).

281. (a) J. Donohue, *Acta Crystallogr.*, **15**, 708 (1962).
      (b) C. J. Spencer and W. N. Lipscomb, *Acta Crystallogr.*, **14**, 250 (1961); **15**, 509 (1962).

282. E. J. Wells, H. P. K. Lee, and L. K. Peterson, *Chem. Commun.*, **1967**, 894.

283. Y. Y. Samitov, N. K. Tazeeva, and N. A. Chadaeva, *J. Struct. Chem.*, **16**, 29 (1975).

284. D. W. Aksnes and O. Vikane, *Acta Chem. Scand.*, **27**, 1337 (1973); **26**, 835, 2535 (1972).

285. L. P. Glushko, T. M. Malinovskaya, M. S. Malinovskii, M. M. Kremlev, N. I. Pokhodenko, T. A. Zyablikova, and N. A. Zhikareva, *Khim. Geterotsikl. Soedin.*, **1974**, 593.

286. B. D. Lavrukhi, T. M. Filippov, and E. I. Fedin, *Org. Magn. Reson.*, **6**, 368 (1974).

287. E. Dradi and G. Gatti, *Org. Magn. Reson.*, **3**, 479 (1971).

288. E. Dradi, G. Tosolini, and G. Gatti, *J. Magn. Reson.*, **6**, 565 (1972).

289. J. T. Wrobel and K. Kabzinski, *Bull. Pol. Chim.*, **22**, 173 (1974).

290. M. Anteunis, J. Gelan, and R. Van Cauwenberghe, *Org. Magn. Res.*, 6, 362 (1974).

291. (a) R. Steyn and H. Z. Sable, *Tetrahedron*, 25, 3579 (1969), and other papers in the series.
      (b) B. Tolbert, R. Steyn, J. A. Franks, Jr., and H. Z. Sable, *Carbohydr. Res.*, 5, 62 (1967).

292. M. Sundaralingam, *Biopolymers*, 7, 821 (1969).

293. C. Altona and M. Sundaralingam, *J. Am. Chem. Soc.*, 95, 2333 (1973).

294. C. D. Jardetzki, *J. Am. Chem. Soc.*, (a) 82, 229 (1960); (b) 83, 2919 (1963); (c) 84, 62 (1962); (d) 82, 299 (1970).

295. (a) R. U. Lemieux, *Can. J. Chem.*, 39, 116 (1961).
      (b) R. U. Lemieux, J. D. Stevens, and R. R. Fraser, *Can. J. Chem.*, 40, 1955 (1962).

296. R. J. Abraham, L. D. Hall, L. Hough, and K. A. McLauchlan, *J. Chem. Soc.*, 1962, 3699.

297. L. D. Hall, *Chem. Ind. (Lond.)*, 1963, 950, and further papers.

298. R. C. Fahey, G. C. Graham, and R. L. Piccioni, *J. Am. Chem. Soc.*, 88, 193 (1966).

299. S. S. Danyluk and F. E. Hruska, *Biochemistry*, 7, 1038 (1968).

300. F. E. Hruska, A. A. Grey, and I. C. P. Smith, *J. Am. Chem. Soc.*, 92, 214, 4088 (1970).

301. (a) F. E. Hruska, K. K. Ogilvie, A. A. Smith, and H. Wayborn, *Can. J. Chem.*, 49, 2449 (1971).
      (b) F. E. Hruska, *Can. J. Chem.*, 49, 2111 (1971).

302. F. E. Hruska, A. A. Smith, and J. G. Dalton, *J. Am. Chem. Soc.*, 93, 4334 (1971).

303. R. H. Sarma, R. J. Mynott, D. J. Wood, and F. E. Hruska, *J. Am. Chem. Soc.*, 95, 6457 (1973).

304. (a) L. D. Hall, P. R. Steiner, and C. Pedersen, *Can. J. Chem.*, 48, 1155 (1970).
      (b) L. D. Hall, S. A. Black, K. N. Slessor, and A. S. Tracey, *Can. J. Chem.*, 50, 1912 (1972).

305. A. A. Akhrem, G. V. Zaitseva, A. S. Fridman, and I. Mikhailopulo, *Spectrosc. Lett.*, 7, 1 (1974).

306. S. Abrahamson, *Acta Crystallogr.*, 16, 409 (1962).

307. I. Rabinowitz, P. Ramwell, and P. Davison, *Nature (New Biol.)*, 233, 88 (1971).

308. J. R. Hoyland and L. B. Kier, *J. Med. Chem.*, 15, 84 (1972).

309. W. L. Duax and J. W. Edmonds, *Prostaglandins*, 4, 209 (1973).

310. J. W. Edmonds and W. L. Duax, *Prostaglandins*, 5, 275 (1974).

311. D. Cremer, J. S. Binkley, and J. A. Pople, *J. Am. Chem. Soc.*, 98, 6836 (1976).

312. G. Leroy, J. M. Mangen, M. Sana, and J. Weiler, *J. Chem. Phys.*, 74, 351 (1977).

313. W. J. Richter and B. Richter, *Isr. J. Chem.*, 15, 57 (1976/77).

314. D. N. Kirk, *J. Chem. Soc.*, *Perkin 1*, 1976, 2171.

315. D. N. Kirk and W. Klyne, *J. Chem. Soc.*, *Perkin 1*, 1976, 762.

316. C. A. Windhorst, *J. Chem. Soc.*, *Chem. Commun.*, 1976, 331.

317. T. D. Bouman, *J. Chem. Soc.*, *Chem. Commun.*, 1976, 665; F. S. Richardson, D. D. Shilady, and J. E. Bloor, *J. Phys. Chem.*, 75, 2466 (1971).

318. M. F. Guimon, O. Pfister-Guillouz, F. Metras, and J. Petrissans, *J. Mol. Struct.*, 33, 239 (1976).

319. C. Sablayrolles, R. Granger, and L. Bardet, *J. Chem. Phys.*, 73, 1000 (1976); *J. Raman Spectrosc.*, 5, 211 (1976).

320. B. C. Gilbert and M. Trenwith, *J. Chem. Soc.*, *Perkin 2*, 1975, 1083.

321. P. J. Mjoberg, W. M. Ralowski, S. O. Ljunggren, and J. E. Bäckwall, *J. Mol. Spectrosc.*, 60, 179 (1976).

322. V. Hach, *J. Org. Chem.*, 42, 1616 (1977).

323. J. Lukaszczyk, Z. Jedlinski, and M. Gibas, *Bull. Acad. Pol. Sci.*, *Ser. Chim.*, 23, 661 (1975), and previous papers in the series.

324. R. P. Lattimer, J. Mazur, and R. L. Kuczkowski, *J. Am. Chem. Soc.*, 98, 4012 (1976).

325. Y. Infarnet, J. Bogner, J. C. Duplan, J. Delmau, and J. Huet, *Bull. Soc. Chim. Fr.*, 1976, 137.

326. J. Elguero and A. Fruchiero, *An. Quim.*, 70, 141 (1974).

327. A. Lectard, C. Vaziri, J.-C. Richer, J.-C. Florence, and G. Manuel, *J. Organomet. Chem.*, 102, 153 (1975).

328. F. Bernardi, G. Distefano, A. Modelli, D. Pietropaolo, and A. Ricci, *J. Organomet. Chem.*, 128, 331 (1977).

329. R. Keskinen, A. Nikkila, and K. Pihlaja, *J. Chem. Soc.*, *Perkin 2*, 1977, 343.

330. K. Pihlaja, R. Keskinen, and A. Nikkila, *Bull. Soc. Chim. Belg.*, 85, 435 (1976).

331. F. Yamada, T. Nishiyama, and H. Samukawa, *Bull. Chem. Soc. Jpn.*, 48, 1878, 3313 (1975).

332. (a) D. W. Aksnes and M. Bjory, *Acta Chem. Scand.*, A, 29, 672 (1975).
      (b) D. W. Aksnes, F. A. Amer, and K. Bergesen, *Acta Chem. Scand.*, 30, 109 (1976).

333. J. Devillers, M. Cornus, J. Navech, and J. C. Wolf, *Org. Magn. Reson.*, 7, 411 (1975).

334. D. F. Detar and N. P. Luthra, *J. Am. Chem. Soc.*, 99, 1232 (1977).

335. R. G. S. Ritchie, N. Cyr, B. Korsch, H. J. Koch, and A. S. Perlin, *Can. J. Chem.*, 53, 1424 (1975).

336. W. L. Duax, C. M. Weeks, and D. C. Rohrer, in *Topics in Stereochemistry*, Vol. 9, N. L. Allinger and E. L. Eliel, Eds., Interscience, New York, 1976.

337. G. T. Detitta, J. W. Edmond, W. L. Duax, and H. Hauptman, *Adv. Prostagl. and Ther.*, 2, 869 (1976).

Absolute Stereochemistry
of Chelate Complexes

YOSHIHIKO SAITO

The Institute for Solid State Physics,
University of Tokyo,
Tokyo, Japan

I.   Introduction 96
     A. Designation of Absolute Configuration 97
     B. Abbreviation of the Ligands. 97
II. Tris(bidentate) Complexes 98
     A. Five-Membered Chelate Rings 98
     B. Six-Membered Chelate Rings 107
     C. Seven-Membered Chelate Rings 110
     D. Other Tris-bidentate Complexes 112
III. *cis*-Bis-bidentate Complexes 113
IV. *trans*-Bis-bidentate Complexes 114
V.   Complexes Involving Multidentate Ligands 116
     A. Terdentate 116
         1. *N*-(2'-Aminoethyl)-1,2-diaminoethane-
             (diethylenetriamine) 116
         2. (*S*)-Aspartic Acid 116
         3. (*S*)-2,3-Diaminopropionic Acid 119
         4. β-(*N*-Sarcosinyl)propionic Acid 121
         5. Tribenzo[*b.f.j*][1.5.9]triazacyclododeca-
             hexaene (TRI). 121
         6. (*R*)-2-Methyl-1,4,7-triazacyclononane
             (R-MeTACN) 122
         7. 1,1,1-Tris(aminoethyl)ethane (tame). 124
     B. Quadridentate 125
         1. 1,8-Diamino-3,6-diazaoctane(triethylene-
             tetramine, trien, or 2,2,2-tet) 125
         2. (+)$_{495}$-[Co(*S*-Glut)(en)$_2$]$^+$ 135
         3. (−)$_{589}$-[Co(sar)(en)$_2$]$^{2+}$ 135
         4. 5(*R*),7(*R*)-Dimethyl-1,4,8,11-tetraaza-
             undecane(*R,R*-2,3'',2-tet) 136
         5. 1,10-Diamino-4,7-diazadecane(3,2,3-tet). . . 137
         6. (1*R*,3*R*,8*R*,10*R*)-Tetramethyl-4,7-diazadecane-
             1,10-diamine(*R,R,R,R*-3'',2,3''-tet) 138

            7. Ethylenediamine-*N*,*N*'-diacetic Acid (edda). . 139
         C. Quinquedentate 141
            1. 1,11-Diamino-3,6,9-triazaundecane(tetraen) . 141
         D. Sexidentate 142
            1. 1,14-Diamino-3,6,9,12-tetraazatetradecane
                (linpen) 142
            2. *N*,*N*,*N*',*N*'-Tetrakis(2'-aminoethyl)-1,2-
                diaminoethane(penten) 147
            3. (-)-*N*,*N*,*N*',*N*'-Tetrakis(2'-aminoethyl)-1,2-
                diaminopropane(mepenten) 148
            4. Ethylenediaminetetraacetic Acid (edta) . . . 149
            5. Trimethylenediaminetetraacetic Acid (trdta). 150
            6. 1,3,6,8,10,13,16,19-Octaazabicyclo-
                [6.6.6]eicosane (Sepulchrate, Sep.). 151
     VI. Empirical Rules Relating Absolute Configuration
          and CD Spectrum 151
         A. cis-Bis-bidentate Complexes, *cis*-[CoX$_2$(en)$_2$]$^{n+}$. 156
         B. Multidentate Complexes 160
            1. Ion Pairing 163
            2. Exciton CD 163
    VII. Conclusions 166
          References 167

I. INTRODUCTION

This review comprises a survey of structural studies of
optically active complexes. The compounds included are
restricted to those chelate complexes whose absolute configuration has been established by means of X-rays. At present such
complexes amount to about 90 (not counting organometallic
compounds), and the number is still growing at an increasing
rate. The X-ray method is the ultimate means of establishing
the absolute configuration of a molecule. At this time no
other physical method is capable of determining the absolute
configuration unambiguously. One of the more important aims of
absolute configuration determination is to establish an empirical rule relating the absolute configuration and circular
dichroism (CD). This would then enable one to determine the
absolute configuration by simply measuring CD spectra, a
procedure much easier to carry out and less time consuming
than the tedious X-ray method. This topic is discussed
separately in Sect. VI.

Almost all the chelate complexes included in this survey
are six-coordinate octahedral complexes. They are discussed in
groups as follows:
    (A) tris(bidentate) complexes
    (B) *cis*-bis(bidentate) complexes
    (C) *trans*-bis(bidentate) complexes
    (D) complexes involving multidentate ligands

In each section some selected complexes of fundamental importance are described in some detail and the remaining ones are discussed briefly, with emphasis on the purpose of the study of each particular compound.

A. Designation of Absolute Configuration

The symbols used to denote the absolute configurations of metal complexes in this chapter follow IUPAC convention (1,2).

B. Abbreviations of the Ligands

acac	acetylacetonate
ala	alaninate
arg	argininate
asp	aspartate
atc	acetylcamphorate
bigua	biguanide
bpy	2,2'-bipyridyl
cat	1,2-benzenediolate
chxn	*trans*-1,2-diaminocyclohexane (*trans*-1,2-cyclohexanediamine)
cptn	*trans*-1,2-diaminocyclopentane (*trans*-1,2-cyclopentanediamine)
dien	diethylenetriamine
edda	ethylenediamine-*N*,*N*'-diacetate
edta	ethylenediamine-*N*,*N*,*N*',*N*'-tetraacetate
en	ethylenediamine
glut	glutamate
gly	glycinate
linpen	1,14-diamino-3,6,9,12-tetraazatetradecane (pentaethylenehexamine)
mal	malonate
N-meen	*N*-methylethylenediamine
mepenten	*N*,*N*,*N*',*N*'-tetrakis(2'-aminoethyl)-1,2-diaminopropane
MeTACN	2-methyl-1,4,7-triazacyclononane
ox	oxalate
phen	1,10-phenanthroline
penten	*N*,*N*,*N*',*N*'-tetrakis(2'-aminoethyl)-1,2-diaminoethane
pn	propylenediamine
pro	prolinate
ptn	2,4-diaminopentane
sar	sarcosinate
sarmp	β-(*N*-sarcosinato)propionate (sarcosinate-*N*-monopropionate)

sep	1,3,6,8,10,13,16,19-octaazabicyclo[6.6.6]-eicosane
tame	1,1,1-tris(aminoethyl)ethane
2',2,2'-tet	3,8-dimethyltriethylenetetramine (3,8-dimetrien)
2,3",2-tet	5,7-dimethyl-1,4,8,11-tetraazaundecane
3,2,3-tet	1,10-diamino-4,7-diazadecane
3",2,3"-tet	1,10-diamino-1,3,8,10-tetramethyl-4,7-diazadecane
tetraen	1,11-diamino-3,6,9-triazaundecane (tetramethylenepentamine)
tmd	1,4-diaminobutane (tetramethylenediamine)
tn	1,3-diaminopropane (trimethylenediamine)
trdta	trimethylenediaminetetraacetate
TRI	tribenzo[b.f.j][1.5.9]triazacyclododeca-hexaene
trien	N,N'-bis(2'-aminoethyl)-1,2-diaminoethane (trimethylenetetramine) (2,2,2-tet)

II. TRIS(BIDENTATE) COMPLEXES

A. Five-Membered Chelate Rings

The stability of complexes is known to be greatly enhanced if chelation involves five-membered chelate rings. The best known bidentate ligand is probably ethylenediamine. When three ethylenediamine molecules are coordinated octahedrally to a central metal atom, two optically active isomers can be formed (Fig. 1.). We call these isomers Δ and Λ. Since the five-mem-

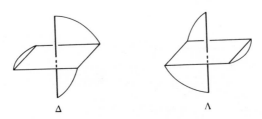

Δ   Λ

Fig. 1. Optical isomers of tris(bidentate) complexes.

bered chelate rings formed by ethylenediamine are not planar, there exist enantiomeric conformations, δ and λ, as shown in Figure 2. Accordingly there are eight possible configurations for tris(ethylenediamine)cobalt(III) ions, namely

$$1\begin{cases}\Lambda(\delta\delta\delta)\\\Delta(\lambda\lambda\lambda)\end{cases} \qquad 2\begin{cases}\Lambda(\delta\delta\lambda)\\\Delta(\lambda\lambda\delta)\end{cases} \qquad 3\begin{cases}\Lambda(\delta\lambda\lambda)\\\Delta(\lambda\delta\delta)\end{cases} \qquad 4\begin{cases}\Lambda(\lambda\lambda\lambda)\\\Delta(\delta\delta\delta)\end{cases}$$

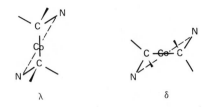

Fig. 2. Two possible conformations of a metal-ethylenediamine ring viewed along the twofold axis.

The difference in the combination of $\Lambda(\delta)$ and $\Lambda(\lambda)$ is as
follows: In Figure 3 the $\Lambda(\delta\delta\delta)$ and $\Delta(\lambda\lambda\lambda)$ isomers are shown as
examples. The C-C axes are eclipsed in the $\Lambda(\delta\delta\delta)$ and staggered
in the $\Lambda(\lambda\lambda\lambda)$ form. In the former the C-C axis is parallel to

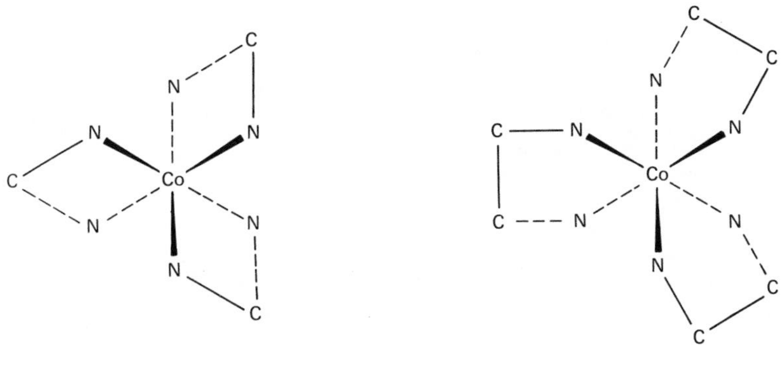

*lel*₃   *ob*₃

Fig. 3. $\Lambda(\delta\delta\delta)$ and $\Lambda(\lambda\lambda\lambda)$ isomers of $[Co(en)_3]^{3+}$.

the threefold axis, whereas it is slanted obliquely in the
latter. Thus they are called the *lel*₃ and *ob*₃ forms, respectively (3).
      Figure 4 shows a perspective drawing of the $(+)_{589}$-tris-
(ethylenediamine)cobalt(III)* complex ion $(+)_{589}$-$[Co(en)_3]^{3+}$.
This is the first chelate complex whose absolute configuration
was determined by means of X-rays (4,5). The tris(ethylenediamine)cobalt(III) ion has the *lel*₃ form, which agrees with the
result of calculation that the *lel*₃ form is more stable by about

      *Because of the frequently large and anomalous optical
rotatory dispersion of complex ions in the visible region of
the spectrum, the sign of their rotation is not meaningful
unless indexed with the wavelength at which the rotation is
measured.

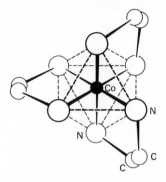

Fig. 4. $(+)_{589}$-$[Co(en)_3]^{3+}$ (4).

7.6 kJ/mol than the ob_3 form. The absolute configuration of $(+)_{589}$-$[Co(en)_3]^{3+}$ can be designated as $\Lambda(\delta\delta\delta)$. The complex has D_3 symmetry within the limits of the experiments. The shape and size of the cobalt-ethylenediamine ring can be summarized as follows (6):

Co-N = 1.978 ± 0.004 Å	N-Co-N = 85.4 ± 0.3°
N-C = 1.497 ± 0.010 Å	Co-N-C = 108.4 ± 0.5°
C-C = 1.510 ± 0.010 Å	N-C-C = 105.8 ± 0.7°
	N-C-C-N = 55.0°
	(dihedral angle)

The octahedron formed by the six nitrogen atoms is slightly distorted: the upper triangle formed by the three nitrogen atoms is rotated counterclockwise from the position expected for a regular octahedron by about 5° with respect to the lower triangle formed by the remaining three nitrogen atoms.

The carbon atoms of the chelate ring show thermal anisotropy best described as an oscillatory motion perpendicular to the C-C bond. A similar type of anisotropic vibration has also been reported for other complexes with metal-ethylenediamine rings {for example, $[Co(SO_3)(NCS)(en_2)]$ (7), $[Cu(en)_2(H_2O)]^{2+}$ (8), $[Co(NO_2)_2(en)_2]^+$ (9)}. This feature of anisotropic vibrations of the carbon atoms seems to support the existence of a puckering motion of the chelate ring in solution. Actually Mason and Norman (10,11) measured the CD spectra of dissolved $[Co(en)_3]^{3+}$ ions and suggested that different conformations of $[Co(en)_3]^{3+}$ coexist in solution (12). Beattie examined the nmr spectra of $[Co(en)_3]^{3+}$ and showed that the ligands undergo rapid inversion between δ and λ conformations in solution. Together with the statistical effect he suggested that the most abundant conformation in solution may be $\Lambda(\delta\delta\lambda)$ and not $\Lambda(\delta\delta\delta)$ (13).

Before 1968, all crystal structures containing [Me(en)₃] complex ions were found to exist as the *lel₃* conformers Λ(δδδ) or their enantiomers Δ(λλλ). In 1968, however, for [Cr(en)₃]³⁺, examples were reported for each of the other three possible configurations, Λ(δδλ), Λ(δλλ), and Λ(λλλ). In the structure of [Cr(en)₃][Ni(CN)₅]·1.5H₂O the complex cations assume *lelob₂* and *lel₂ob* conformations, whereas in crystals of [Cr(en)₃][Co(CN)₆]-·6H₂O the complex cations have the *ob₃* conformation (14,15). These results led the authors to suggest that hydrogen bonding specifically stabilizes these conformations, since their crystal structure permits more hydrogen bonds than the *lel₃* form. The *lel₃* form is the most compact and probably leads to better packing in the lattice. Crystals of racemic [Cr(en)₃]-Cl₃·3H₂O contain the *lel₃* form, which is isostructural with its cobalt(III) analogue (16).

Recently similar isomerism was observed for the first time in the [Co(en)₃]³⁺ ion. Crystals of [Co(en)₃][SnCl₃]Cl₂ contain *lel₂ob* isomers (17). Such isomers are again favored by N-H···Cl hydrogen bond formation in crystals.

When the bidentate ligand is propylenediamine, the number of isomers of the tris(bidentate) complex [Co(±pn)₃]³⁺ increases to 24. They constitute two enantiomeric series with absolute configurations Λ and Δ, respectively. In each of the series there exist two types of geometric isomers, *fac* and *mer*, with respect to the position of the substituted methyl groups for *lel₃* and *ob₃* isomers, and four with *lel₂ob* and *lelob₂* conformers (18).

Early workers concluded that an optically active ligand like (-)-propylenediamine favored the formation of one isomer to the complete exclusion of the other. Thus it was believed that (-)-pn gives only the (-)₅₈₉-[Co(-pn)₃]³⁺ isomer. However, in 1959 Dwyer and his collaborators succeeded in isolating (+)₅₈₉- and (-)₅₈₉-[Co(-pn)₃]I₃ (19). Figure 5 shows

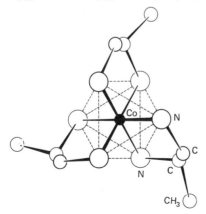

Fig. 5.   (-)₅₈₉-[Co(-pn)₃]³⁺ (20).

the structure of the complex ion $(-)_{589}$-[Co(-pn)$_3$]$^{3+}$ (20), the
most stable isomer. The absolute configuration of the complex
ion may be designated as $\Delta(\lambda\lambda\lambda)$, the *lel*$_3$ form. The three
methyl groups are attached in facial positions. The geometry
of the five-membered chelate ring is very similar to that of
the cobalt-ethylenediamine ring. Methyl substitution on the
chelate ring does not seem to disturb the overall features of
the rings. Each C-CH$_3$ bond lies in an equatorial position
relative to the plane of the five-membered chelate ring. For
[Co{(-)-*R*-propylenediamine}$_3$]$^{3+}$ complex ions there are four
possible isomers:

Δ(*lel*$_3$)(*fac*) Δ(*lel*$_3$)(*mer*)
Λ(*ob*$_3$)(*fac*) Λ(*ob*$_3$)(*mer*)

MacDermott succeeded in separating the Δ(*lel*$_3$)(*mer*) isomer
from the Δ(*lel*$_3$)(*fac*) isomer by fractional crystallization
(21). Yamasaki and his collaborators isolated the Λ(*ob*$_3$)(*fac*)
and Λ(*ob*$_3$)(*mer*) isomers in pure states by column chromatography on an ion exchange SP Sephadex column (22). Figure 6

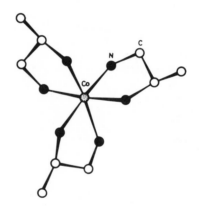

Fig. 6. $(+)_{589}$-[Co(-pn)$_3$]$^{3+}$ (23).

is a perspective drawing of the complex ion $(+)_{589}$-[Co(-pn)$_3$]$^{3+}$
as viewed down the pseudothreefold axis (23). This is the
facial isomer, in which the three chelate rings are puckered,
with the methyl groups in equatorial positions. The absolute
configuration of the complex ion may be designated as $\Lambda(\lambda\lambda\lambda)$.
The central C-C bond in the chelate ring is inclined by 61°
with respect to the threefold axis, whereas it is inclined by
only 5.7° in the case of the Δ(*lel*$_3$)(*fac*) isomer. The geometry
of the chelate ring is not very much different from that of the
(*lel*$_3$)(*fac*) isomer. The only difference is that the N-Co-N
angle is compressed by about 1.7°.

Unlike facial isomers, it was not possible to determine
the structures of the *mer* isomers because the salts containing
these isomers are amorphous glasses or because the complex
ions exhibit orientational disorder in the crystal lattice.
For example, $(-)_{589}$-[Co$(R$-pn$)_3$]$(+)_{589}$-[Cr$(mal)_3$]·3H$_2$O crystal-
lizes in a rhombohedral space group $R32$ and the complex
cations are on a set of special positions with D_3 site symmetry. The *mer*-[Co$(R$-pn$)_3$]$^{3+}$ ion has no strict overall
symmetry, but the nonmethylated fragment does have D_3 symmetry. The electron-density distribution of the complex
cation in the crystal looks like that of tris$(R,R$-2,3-diamino-
butane)cobalt(III), with methyl groups of half weight (47). The
hexacyanocobaltate(III) salt of the $\Lambda(ob_3)$(*mer*) isomers is
cubic. Again the complex ions exhibit orientational disorder,
and no conclusion can be drawn about the structures of the *mer*
isomer (24).

It appears that there are packing problems for the *mer*
isomer which lead to a disordered structure that grossly
resembles that of the facial isomer. The free energy difference at 25°C of the *lel*$_3$ and *ob*$_3$ isomers was calculated from the
equilibrium concentrations of these isomers. The *lel*$_3$ isomer is
more stable by about 6.7 kJ/mol (25). Recently Schäffer and his
collaborators determined the following free energy differences
between the isomers at 100°C (18):

$$\Delta G° (ob_3 \rightarrow lel_3) = -6.73 \text{ kJ/mol}$$
$$\Delta G° (ob_2lel \rightarrow lel_2ob) = -2.56 \text{ kJ/mol}$$
$$\Delta G° (lel_2ob \rightarrow lel_3) = +0.50 \text{ kJ/mol}$$

The isomers of [Co$(\pm chxn)_3$]$^{3+}$ comprise two enantiomeric
series with absolute configurations Λ and Δ around the cobalt
atom. The chelate rings formed by the ligands $(-)$-R,R-chxn and
$(+)$-S,S-chxn have the absolute configurations λ and δ, respectively. For each configurational series the possible ligand
conformations give rise to four diastereomers:

$$lel_3, \; lel_2ob, \; ob_2lel, \; \text{and} \; ob_3.$$

The structures and absolute configurations of all the
isomers have been determined. Figure 7 represents the complex
ion in the $(-)_{589}$-[Co$(+chxn)_3$]Cl$_3$·5H$_2$O crystal viewed along the
normal to the N(1)-N(2)-N(3) plane (26). The complex cation
has approximate D_3 symmetry. Each ligand molecule in the complex
is coordinated to the central cobalt atom by its nitrogen atoms.
All the C-C bonds in the chelate ring are nearly parallel to the
threefold axis of rotation; that is, the complex ion has *lel*$_3$
conformation. The geometry of the three-chelate-ring system is
very much like that of [Co(en)$_3$]$^{3+}$. The mean value of the
dihedral angle between the planes N(1)-C(1)-C(6) and N(4)-C(6)-
C(1) and the corresponding angles in the other two chelate rings

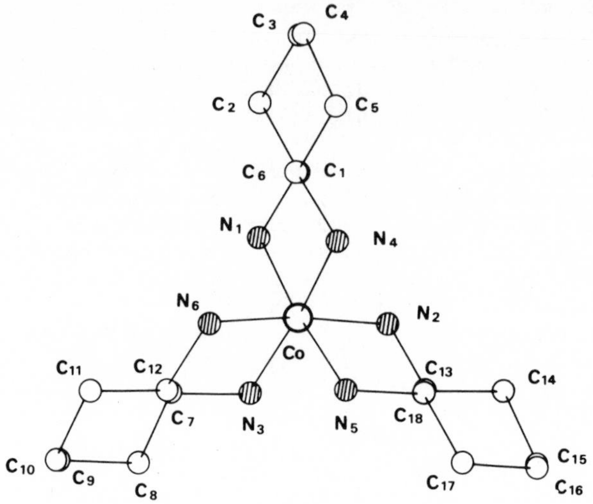

Fig. 7. $(-)_{589}$-[Co(+chxn)$_3$]$^{3+}$ (26).

is about 59.3°, which is almost identical with the value
expected in a free *trans*-1,2-diaminocyclohexane molecule. Thus
the ligand molecule seems to be only slightly strained by the
formation of the chelate rings. The cyclohexane ring has the
chair conformation. All the bond distances and angles within the
six-membered chelate ring are quite normal and agree well with
those observed for other related compounds.

In the ob_3 isomer, $(+)_{589}$-[Co(-chxn)$_3$]$^{3+}$, shown in Figure
8, the central C-C bond in the chelate ring is inclined at an
angle of 66° with respect to the threefold axis. The cyclohexane
ring assumes a chair conformation. The absolute configuration
may be designated as $\Lambda(\lambda\lambda\lambda)$. The N-Co-N angle in the chelate
ring is 84.2°, smaller by about 2.4° than that of the lel_3 isomer
(27).

The crystal structure of $(-)_{589}$-[Co(+chxn)$_2$(-chxn)]Cl$_3 \cdot$5H$_2$O
has been recently determined (28). Figure 9 shows a perspective
drawing of the complex ion. This is the lel_2ob isomer and has
an approximate twofold axis of symmetry. Its absolute configura-
tion may be designated as $\Lambda(\delta\delta\lambda)$ or in full as Λ-$(-)_{589}$-
[Co{(*S*,*S*)-(+)chxn}$_2${(*R*,*R*)-(-)chxn}$\delta\delta\lambda$]$^{3+}$. Each chelate ring has
an unsymmetrical skew conformation. The dihedral angles about
the C-C bond of the chelate rings are 53° on the average. The
nonbonded short hydrogen-hydrogen contacts occur between NH$_2$
and CH groups in adjacent chelate rings. The average H···H
distances are 2.46 Å between the lel rings and the ob rings.
The inclination angle of the coordination plane (formed by
Co and two N atoms) of the ob ring with respect to the pseudo-

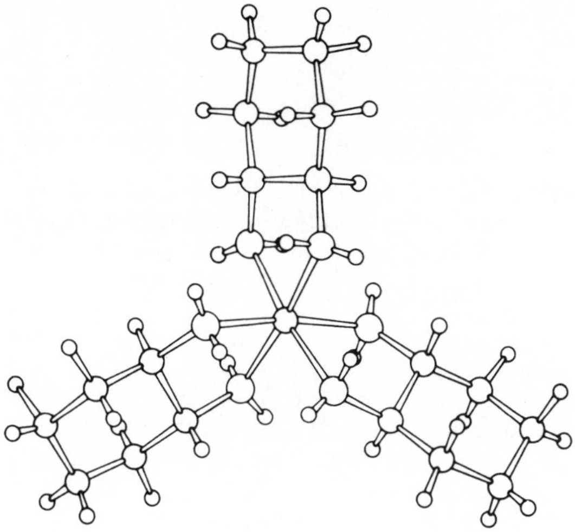

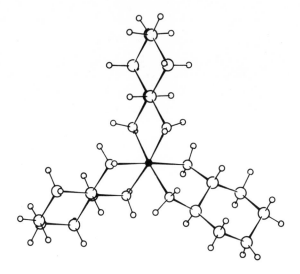

Fig. 8. $(+)_{589}$-[Co(-chxn)$_3$]$^{3+}$ (27).

Fig. 9. $(-)_{589}$-[Co(+chxn)$_2$(-chxn)]$^{3+}$ (28).

threefold axis of the complex ion, 35.7° is significantly
larger than that in the lel_3 or ob_3-isomer (31.8 and 31.5°,
respectively). This difference in inclination angle serves
to alleviate the nonbonded hydrogen interactions.

Schäffer and his collaborators established the equilibrium between the [Co(±chxn)$_3$]$^{3+}$ isomers on charcoal at 100°C

and pH = 7.0. The $\Delta G°$ values are as follows (29):

$$\Delta G° (lel_2 ob \rightarrow lel_3) \qquad -0.93 \text{ kJ/mol}$$
$$\Delta G° (ob_2 lel \rightarrow lel_3) \qquad -3.72 \text{ kJ/mol}$$
$$\Delta G° (ob_3 \rightarrow lel_3) \qquad -8.20 \text{ kJ/mol}$$

The different stabilities of the isomers may be rationalized in terms of the mutual interactions between the ligands. It is generally accepted that nonbonded hydrogen interactions are among the important factors determining the stability of such isomers.

When a molecule of *trans*-1,2-diaminocyclopentane is coordinated to a cobalt atom, the chelate ring system involves much strain. Jaeger was the first to describe this complex (30), but it was suggested later that its existence was doubtful (31). Figure 10 shows a perspective drawing of the complex ion

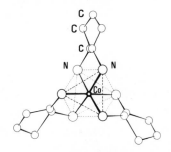

Fig. 10. $(-)_{589}-[Co(+cptn)_3]^{3+}$ (32).

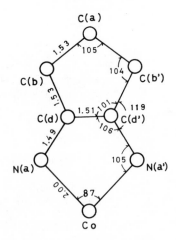

Fig. 11.   Bond lengths and angles of a chelate ring, averaged by assuming D_3 symmetry (32).

$(-)_{589}-[Co(+cptn)_3]^{3+}$ (32). The cobalt atom has a slightly
distorted octahedral coordination of six nitrogen atoms with
an average distance of 2.00 Å. The N-Co-N angles in the five-
membered chelate rings are 86.7° on the average. The upper
N-N-N triangle is rotated by about 5.5° with respect to the
lower N-N-N triangle from the position expected for a
regular octahedron. All the C-C bonds in the chelate ring point
along the threefold axis of the octahedron, hence the conformation of the complex ion is lel_3 and the absolute configuration
of the complex ion may be designated as $\Lambda(\delta\delta\delta)$. The conformation
of the cyclopentane ring is half-chair. The bond lengths are
normal, as shown in Figure 11; however, angles $C(b)-C(d)-C(d')$
of 101° and $N(a)-C(d)-C(b)$ of 119° deviate substantially from
the normal tetrahedral angle. Corresponding angles in the
trans-1,2-diaminocyclohexane analogue are 112 and 110°,
respectively. The observed conformation of the chelate ring
system agrees well with the result given by strain energy
minimization for a free ligand except for the two bond angles
mentioned above (32).

B. Six-Membered Chelate Rings

The six-membered chelate ring is more flexible than the
five-membered chelate ring, and the conformational problem
presented by its complexes is similar to that posed by cyclohexane, except that the metal ligand angle is nearly 90°.
There are three possible conformations for the cobalt-trimethyl-
enediamine ring: a chair form, a skewboat form, and a boat form.
Even simple molecular models, however, show at once that, owing
to steric hindrance, a boat form cannot accommodate the tris-
bidentate complex. The crystal structures of $(-)_{589}-[Co(tn)_3]-$
$Br_3 \cdot H_2O$ and $(-)_{589}-[Co(tn)_3]Cl_3 \cdot H_2O$ have been determined
(33,34). Figure 12 shows a perspective drawing of the complex
ion $(-)_{589}-[Co(tn)_3]^{3+}$, which has an approximately threefold
axis through the cobalt atom. The three six-membered chelate
rings are nearly but not exactly identical and assume chair
forms. The chelate ring is substantially flattened due to
nonbonded hydrogen interactions. The Co-N-C bond angles are
much larger than the normal tetrahedral angle, the average
value being 122.0°. The mean N-Co-N angle in the chelate ring
is 91.0°. Ellipsoids of thermal motion of that chelate ring
which is most loosely packed in the crystal indicate that the
largest amplitude of thermal vibration of the carbon atoms is
primarily perpendicular to the plane formed by the two C-C
or C-N bonds for each atom. This large thermal motion of the
carbon atoms is consistent with the results of conformational
analysis of the six-membered chelate rings (35,36) and suggests
a conformational equilibrium involving significant amounts of
two or more conformers in solution at room temperature (37).

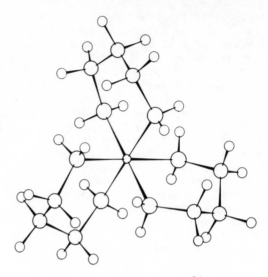

Fig. 12. $(-)_{589}-[Co(tn)_3]^{3+}$ (34).

The chair form of a six-membered chelate ring is not chiral but has a mirror plane. The skew-boat form is chiral and its helicity can be defined by the line joining the two coordinating atoms and the line joining the two adjacent atoms of the ligand. When three (or two) 1,3-diaminopropane molecules are coordinated to a metal atom to form three chelate rings with chair conformations, chirality is generated. The arrangement of the three chelate rings in the complex ion $\Lambda-[Co(tn)_3]^{3+}$ defines a clockwise rotation, as indicated by the arrow in Figure 13. The chair may then fold in such a way that the outer edges of the rings (C-C-C) define a direction parallel or antiparallel to the direction of rotation determined by the three chelate rings. Raymond (38) designated the two resulting conformations as *p* and *a*, respectively. The fold direction is determined by

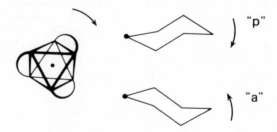

Fig. 13.   Two chair conformers of Co-tn rings. The upper ring is the *p* conformer; the lower is the *a* form.

the orientation of the C-C-C plane. The complex ion [Cr(tn)₃]³⁺
in crystals of [Cr(tn)₃][Ni(CN)₅]·2H₂O has a different conformation (39), one chelate ring taking a skew-boat form and the
other two chair conformations. The directions of fold of the
two chair rings are different. The complex cation may be
designated as Λ(p δ a) or its antipode.

2,4-Diaminopentane acts as a bidentate ligand and forms a
six-membered chelate ring. Three isomers exist for this ligand:
R,R, S,S, and R,S. The equatorial preference of the C-CH₃ bond
will fix the conformation of the chelate rings. Two isomers of
tris(R,R-2,4-diaminopentane)cobalt(III) were synthesized and
characterized (40). Figure 14 shows a perspective drawing of

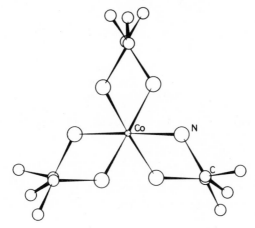

Fig. 14.   (-)₅₄₆-[Co(R,R-ptn)₃]³⁺ (41).

the complex ion (-)₅₄₆-[Co(R,R-ptn)₃]³⁺ (41). The complex ion
has rigorous C_2 symmetry, combined with approximate D_3 symmetry.
The mean Co-N distance is 1.984(5) Å. All other C-C or C-N
distances are normal. The N-C-N angles are 90°, the mean value
being 89.1(3)°. The equatorial preference of the two methyl
groups fixes the conformation of the six-membered chelate ring
in a twist-boat form. The six-membered chelate ring is thus
chiral and its absolute configuration may be designated as λ.
The line joining the two asymmetric carbon atoms is inclined
by about 2° with repsect to the threefold axis. Accordingly this
is the lel₃ isomer.

Figure 15 shows a projection of the complex ion (+)₅₄₆-
[Co(R,R-ptn)₃]³⁺ along the threefold axis (41). The complex
ion has approximate D_3 symmetry, the average Co-N distance is
1.99 Å, and the N-Co-N angle in the chelate ring is 89°. The
chelate ring again assumes a twist-boat conformation with
C-CH₃ bonds in equatorial positions. The chelate ring has the

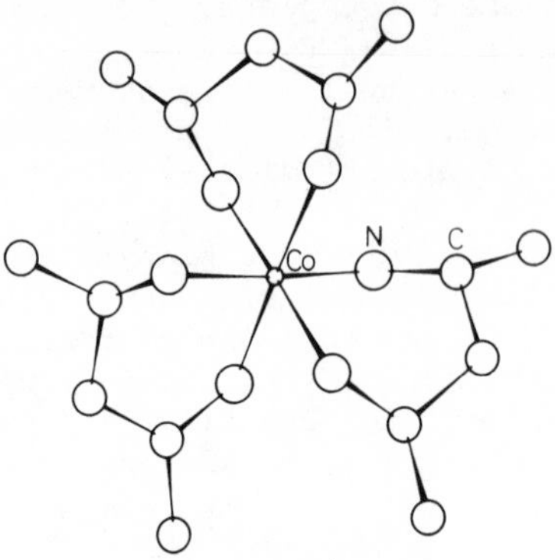

Fig. 15. (+)$_{546}$-[Co(R,R-ptn)$_3$]$^{3+}$ (42).

absolute configuration λ. This is the ob$_3$ isomer, and the line joining the two asymmetric carbon atoms is inclined by about 70° with respect to the threefold axis of the complex ion.

A metal-(R,S-2,4-diaminopentane) ring is expected to have a chair conformation if the equatorial preference of the two substituted methyl groups is taken into account. This was indeed verified in the structure of (+)$_{510}$-oxalatobis(R,S-2,4-diaminopentane)cobalt(III) perchlorate monohydrate, (+)$_{510}$-[Co(ox)(R,S-ptn)$_2$]ClO$_4$·H$_2$O (43). The complex ion has the absolute configuration Λ-trans(R,R)-cis(S,S) and the six-membered chelate ring takes on a chair form. As for the tris complex involving this ligand, it was believed that the facial isomer was obtained exclusively. Recently the mer and fac isomers of [Co(R,S-ptn)$_3$]$^{3+}$ were successfully separated and resolved by column chromatography on SP Sephadex (44). The absolute configuration of the (+)$_{589}$-fac isomer was determined to be Λ (45).

The chelated malonato-metal ring has a high degree of conformational flexibility. In the structure of Δ-[Co(R-pn)$_3$]Δ-[Cr(mal)$_3$]·3H$_2$O the three malonato-Cr(III) rings are equivalent by symmetry and possess an envelope conformation in which only the methylene carbon atom is significantly displaced from the plane of the chelate ring. In the diastereomer (+)$_{546}$-[Co(mal)$_2$(en)](−)$_{589}$-[Co(NO$_2$)$_2$(en)$_2$] both rings are reported to have an approximately planar conformation. The largest deviation from the mean plane of the Co-mal ring is 0.23 Å for a ligating oxygen atom, and the pattern for both rings suggests a distortion toward a skew-boat conformation (46–48).

C.   Seven-Membered Chelate Rings

The basic series of structures with three five- and six-membered chelate rings has recently been supplemented by a new important member, the tris(1,4-diaminobutane)cobalt(III) ion, containing three seven-membered chelate rings. The crystal structure and absolute configuration of (+)$_{589}$-tris(1,4-diaminobutane)cobalt(III) bromide has been determined (49). The compound was synthesized by Fujita and Ogino (50) from cobalt(II) nitrate and tetramethylenediamine in dimethyl sulfoxide by air oxidation. It was resolved into optical isomers by conversion to diastereomeric di-μ-(+)-tartrato-(4)-bis[antimonates(III)]. The resolution was also achieved by SP Sephadex column chromatography. Figure 16 shows a perspective drawing of the complex ion (+)$_{589}$-[Co(tmd)$_3$]$^{3+}$, which has D_3 symmetry. The average C–N distance is 1.991 Å and the

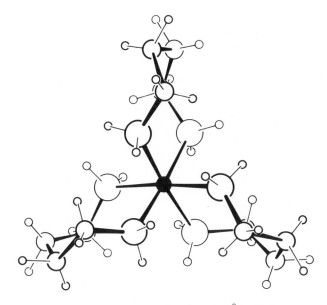

Fig. 16.   (+)$_{589}$-[Co(tmd)$_3$]$^{3+}$ (49).

average N–Co–N angle is 89.2°. The seven-membered chelate ring is much more flexible than are six-membered chelate rings. The chelate ring is strained; the Co–N–C, N–C–C, and C–C–C angles are 123.2, 113.6, and 116.1°, respectively, all greater than the normal tetrahedral angle. Figure 17 shows a projection of the chelate ring along the twofold axis. This chelate ring is chiral and may be designated as λ provided that the helicity is defined by the line joining nitrogen atoms and the line joining the two carbon atoms bonded to the nitrogen atoms. The line joining the two carbon atoms in the chelate ring is inclined by about 0.6° with respect to the threefold axis of the complex ion. Accordingly this is the *lel*$_3$ isomer and may be designated as Δ(λλλ) as a whole. It is

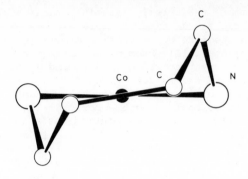

Fig. 17. Projection of the seven-membered chelate ring along the twofold axis (49).

to be noted here that the conformation of the segment Co-N-C-C is δ and that of N-C-C-C is λ.

D. Other Tris-bidentate Complexes

The lel_3 and ob_3 isomers of tris(trans-1,2-diaminocyclo-hexane)rhodium(III) salts have been isolated as chlorides and nitrates (51). The absolute structures of $(+)_{589}$-[Rh(-chxn)$_3$]-(NO$_3$)$_3\cdot$3H$_2$O, ob_3 isomer, and $(+)_{589}$-[Rh(-chxn)$_3$](NO$_3$)$_3\cdot$3H$_2$O, lel_3 isomer, have been determined (52,53). The complex ion in the former may be designated as $\Lambda(\lambda\lambda\lambda)$, that in the latter as $\Delta(\lambda\lambda\lambda)$. The geometries of the complex ions are very much like those of the analogous Co(III) complexes, the only difference being that the average Rh-N distance of 2.082 Å is greater than the Co-N distance of 1.978 (4) Å.

The absolute configurations of the following four D_3 complexes, as determined by X-ray analysis correlated with the CD bands in the ultraviolet region:

$(-)_{589}$-[Fe(phen)$_3$]$^{2+}$: Λ (54)
$(+)_{589}$-[Ni(phen)$_3$]$^{2+}$: Λ (55)
$(+)_{589}$-[Ni(bpy)$_3$]$^{2+}$: Λ (56)
$(-)_{589}$-[As(cat)$_3$]$^{-}$: Δ (57)

All the ligand molecules are practically planar and the complex ions possess approximate D_3 symmetry. The CD spectra of these complex ions are discussed in Sect. VI.

The absolute configurations of $(-)_{589}$-[Co(ox)$_3$]$^{3-}$ (55) and $(+)_{589}$-[Cr(ox)$_3$]$^{3-}$ (46) are Λ and Δ, respectively. Both complex ions possess effective D_3 symmetry. The ligand is planar and has no significant deformation on forming a chelate ring. When thiooxalate ion is introduced in place of the oxalato group, $(+)_{589}$-[Co(thiox)$_3$]$^{3-}$ has the absolute configuration Λ (58).

The three ligands coordinate to the cobalt atom with six sulfur atoms, with the Co-S bonds ranging from 2.230 to 2.259 Å and the S-Co-S angles in the chelate ring from 89.3 to 89.9°. In this complex ion the CoS_6 chromophore is elongated along the threefold axis. The Co-S bond is inclined by 53.8° with respect to the trigonal axis. (This angle would be 54.75° for a regular octahedral arrangement of the six S atoms.)

There are four possible isomers of $[M(S\text{-am})_3]$, where am is an amino acid, such as the $[Co(S\text{-pn})_3]^{3+}$ ion. Tris(S-alaninato)-cobalt(III) may be regarded as the parent compound. The $(+)_{589}$-α-isomer, the more easily obtainable violet isomer, was found to have the absolute configuraiton Λ (59). This is the mer isomer and it has no threefold axis. The chelate ring is nearly planar and the three methyl groups are attached in equatorial positions. The absolute configurations of a number of tris-amino acid complexes have been deduced by comparison of CD spectra.

The absolute configuration of $(-)_{546}$-tris(acetylacetonato)-cobalt(III), Λ, was successfully determined by means of quasiracemic crystals containing the partially resolved cobalt(III) complex and the racemic aluminum(III) analogue (60). The absolute configurations of $(-)_{589}$-$[Cr(bigua)_3]^{3+}$ (61) and $(+)_{589}$-$[Cr(atc)_3]^{3+}$ (62) have also been determined to be Λ.

Some octahedral complexes have also been reported with three chelate rings containing two different ligands. In $(+)_{589}$-$[Co(en)_2(tn)]^{3+}$ the Co-tn ring takes on a chair form and the two Co-en rings assume the lel conformation. The absolute configuration of the complex ion may be designated as Λ(δδ) (63). $(-)_{589}$-$[Co(ox)(en)_2]^+$ has the absolute configuration Δ(δλ); i.e., the two Co-en rings take lel ob conformation (64). $(-)_{589}$-$[Co(acac)(tn)_2]^{2+}$ has the absolute configuration Δ, with the two Co-tn rings in chair conformation (65). Other complexes of similar type containing amino acids or related ligands are discussed in Sect. V-B.

III. cis-BIS-BIDENTATE COMPLEXES

The cis-bis-bidentate complexes are structurally related to the tris-bidentate complexes, and their CD spectra can also be interpreted with reference to the parent tris-complexes. Thus the structures of cis-bis-bidentate complexes are only briefly reviewed here.

The $(+)_{589}$-cis-dichlorobis(ethylenediamine)cobalt(III) ion takes Λ absolute configuration, and the two Co-en rings have δ conformation; thus the ion is the lel_2 isomer (66). The analogous platinum(IV) complex $(+)_{450}$-cis-$[PtCl_2(en)_2]^{2+}$ is Δ(λλ), i.e., the lel_2 isomer (67). On the other hand, in $(-)_{589}$-$[Co(NO_2)_2(en)_2]^+$ the two chelate rings adopt the ob_2 conformation. The absolute configuration of the complex ion may be designated as Δ(δδ) (48). In crystals of $(-)_{589}$-

$[Co(NO_2)_2(en)_2](-)_{589}-[Co(NO_2)_2(ox)(NH_3)_2]$ the carbon atoms of the Co-en ring exhibit the greatest thermal motion perpendicular to the plane of the chelate ring (9). $(+)_{589}-cis-$ $[Co(CN)_2(en)_2]^+$ is found to have the absolute configuration $\Lambda(\lambda\lambda)$ (156).

The absolute configuration of $(+)_{589}-cis-[Co(NO_2)_2(R-pn)_2]^+$ was determined to confirm the assignment of the absolute configuration predicted from a comparison of its CD spectrum with that of $(+)_{589}-[Co(en)_3]^{3+}$. The absolute configuration was found to be $\Delta(\lambda\lambda)$, in agreement with the prediction (68). The C-CH$_3$ bond is in an equatorial position, and the methyl groups of the two chelate rings occupy trans positions. The complex ion $(+)_{589}-cis-[Co(NO_2)_2(R-ala)_2]^-$ has a rigorous twofold axis. The cobalt-amino acidato chelate ring is not planar but assumes the form of an asymmetric envelope, with the methyl group in an axial position. The absolute configuration may be designated as Λ (69). This conformation of the metal-amino acidato chelate ring is similar to one predicted by Freeman to be of minimum energy (70). The corresponding argininato complex $(-)_{589}-cis-$ $[Co(NO_2)_2(S-arg)_2]^+$ has the absolute configuration Λ (71).

The absolute configuration of the complex ion $(-)_{589}-$ $[Co(NCS)_2(tn)_2]^+$ may be designated as $\Lambda(pa)$. This complex has approximate twofold symmetry and the two six-membered chelate rings take on chair conformations (72).

IV. *trans*-BIS-BIDENTATE COMPLEXES

The main source of optical activity of the *trans*-bis-bidentate complexes $[MeX_2(bidentate)_2]^{n+}$ lies in the dissymmetry of the ligands. Only two absolute configurations of this type of complex have been determined. In this section some results obtained for square-planar bis-bidentate complexes are also described.

The $(-)_{589}-trans$-dichlorobis(R-propylenediamine)cobalt(III) ion has an approximate twofold symmetry along the Cl-Co-Cl bond. The chelate ring adopts the λ conformation with equatorial C-CH$_3$ bonds (73). The $(-)_{589}-trans$-dichlorobis(N-methylethylenediamine)cobalt(III) ion $(-)_{589}-[CoCl_2(N-meen)_2]^+$ also has an approximate twofold symmetry. The two chelate rings assume the δ conformation and the N-CH$_3$ bonds are equatorial with respect to the chelate ring. The absolute configuration of both asymmetric nitrogen atoms is R (74).

The absolute configuration of $(-)_{280}-cis$-dichloro(1-methylamino-2(S)-aminopropane)platinum(II) was determined (75). The complex is square planar, with an average Pt-Cl distance of 2.303 Å and an average Pt-N distance of 2.021 Å. The five-membered chelate ring adopts the gauche conformation with δ

absolute configuration. The methyl group on the carbon atom is
in the equatorial conformation with respect to the plane of
the chelate ring, while the methyl group attached to the nitrogen atom occupies the axial position.

In crystals of $(+)_{350}$-[Pt$(R$-pn)(Me$_2$en)][Sb$_2$(+)-tart$_2$]·2H$_2$O
the complex ion is planar (76). The five-membered chelate ring
formed by R-pn assumes λ absolute configuration, the C-CH$_3$ bond
being in equatorial position. The other five-membered ring composed of the Me$_2$en takes on the gauche conformation and the
absolute configuration is λ too. The two substituted N-methyl
groups are both in equatorial positions. Thus the absolute
configurations of the asymmetric nitrogen atoms are both S.

The structure of bis(S-prolinato)palladium(II) is also
known (77). The central palladium atom has square planar
coordination, with the two S-proline residues coordinated to it
in cis positions with respect to each other through imino nitrogen atoms and the carboxyl oxygen atoms. The two proline molecules are coordinated to the metal atom, with two pyrrolidine
rings disposed on opposite sides of the coordination plane when
they form the square complex (Figure 18).

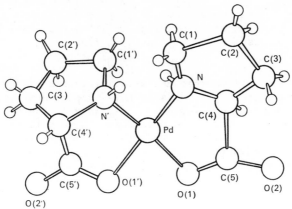

Fig. 18.   [Pd(S-pro)$_2$] (77).

The crystals of [PtCl$_2$(-chxn)$_2$][Pt(-chxn)$_2$]Cl$_4$ consist of
a stacking of chains along which square-planar [Pt(-chxn)$_2$]$^{2+}$
and tetragonal trans-[PtCl$_2$(-chxn)$_2$]$^{2+}$ alternate. The Pt-Cl
and Pt-N distances are 2.354 and 2.056 Å, respectively, and the
conformation of the chelate ring is all $(\lambda\lambda)$. The geometry of
the 1,2-diaminocyclohexane molecule is the same as that of the
isomers of [Co(-chxn)$_3$]$^{3+}$ (26,27). The analogous bromide
[PtBr$_2$(-chxn)$_2$][Pt(-chxn)$_2$]Br$_4$ is isostructural (180).

V. COMPLEXES INVOLVING MULTIDENTATE LIGANDS

A. Terdentate

1. N-(2'-Aminoethyl)-1,2-diaminoethane(diethylenetriamine)

There are three different ways of coordinating two diethylenetriamine molecules to a cobalt(III) ion, as illustrated in
Figure 19 (78). Of the three isomers the u-facial and mer

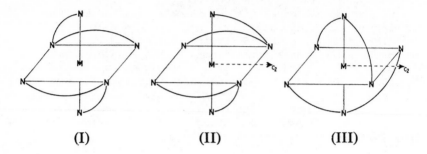

(I)   (II)   (III)

Fig. 19. Schematic drawings of (I) s-fac, (II) u-fac,
and (III) mer-isomer of [Co(dien)$_2$]$^{3+}$ (79).

isomers are chiral whereas the s-facial isomer is achiral. All
diastereomers and enantiomers in this system were isolated and
the diastereomeric configurations assigned to the optically
active isomers based on the difference in racemization behavior of the optically active u-facial and mer isomers (78).
The crystal structures of s-fac-[Co(dien)$_2$]Br$_3$ and (-)$_{589}$-
u-fac-[Co(dien)$_2$][Co(CN)$_6$]·2H$_2$O are known (79,80). This
latter crystal contains two different conformers. Figure 20
represents the two conformers of (-)$_{589}$-u-fac-[Co(dien)$_2$]$^{3+}$.
Two ligand molecules are coordinated to the cobalt atom by six
nitrogen atoms, adopting u-facial configurations. Both cations
may be designated as skew chelate pairs ΔΛΔ, and both have
twofold axes of rotation. In one complex ion (illustrated on
the left of Fig. 20) the conformations of the two chelate rings
formed by a dien molecule are δ and λ, whereas in the other
(right) they are both λ. In the crystal these complex ions are
packed in such a way that their conformations permit more
hydrogen bondings than would a single conformer.

2. (S)-Aspartic Acid

Three possible geometric isomers of the bis(S-aspartato)-
cobaltate(III) ion exist when the ligand acts as a terdentate
(Fig. 21). Two of these three isomers were obtained and charac-

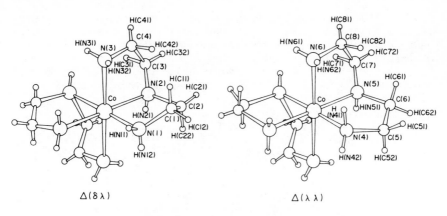

$\Delta(\delta\lambda)$ $\Delta(\lambda\lambda)$

Fig. 20. Perspective drawings of the two conformers of
$(-)_{589}$-u-fac-[Co(dien)$_2$]$^{3+}$ (80).

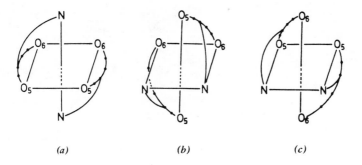

(a)   (b)   (c)

Fig. 21. Schematic drawings of (a) the $trans(N)$, (b) the
$cis(N)$-$trans(O_5)$, and (c) the $cis(N)$-$trans(O_6)$ isomers of the
[Co(S-asp)$_2$]$^-$ ion. Dots represent C atoms (83).

terized by Hosaka, Nishikawa, and Shibata (81). A few years
later all three isomers were isolated (82). O_5 and O_6 refer to
those oxygen atoms that form five- and six-membered chelate
rings with amino nitrogen atoms, respectively. The stereochemistry of aspartic acid as a terdentate ligand is different
from that of other terdentates. The three rings join on the
face of an octahedron at the asymmetric carbon atom in such a
manner that not all the rings define the edges of the octahedron. The presence of this type of chelate ring leads to
complexities in the CD spectra which make it difficult to
correlate the absolute configurations on the basis of CD (see
Sect. VI).

The IUPAC scheme for the designation of the absolute configuration (1,2) cannot be straightforwardly applied for
assigning a label to the isomers shown in Fig. 21*b* and *c*. As
described above, three chelate rings join in a vertex on a
face of the octahedron at noncoordinating atoms. Even if the
number of skew chelate pairs could be counted, it would be
impossible to obtain net chirality. This type of difficulty
generally arises when a multidentate ligand has a branch with
a ligating atom at its end and this branch is bonded to a
nonligating atom. On the other hand multidentate ligands like
dien, edta, trdta, and penten do not give rise to such difficulties. Figure 22 shows the absolute configuration of *cis*(N)-

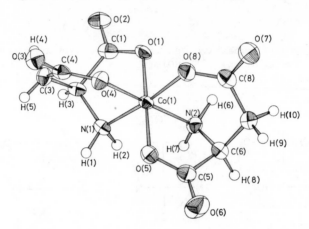

Fig. 22.   *cis*(*N*)-*trans*(O_5)-[Co(*S*-asp)$_2$]$^-$ (83).

trans(O_5)-[Co(*S*-asp)$_2$]$^-$ (83). Two aspartic acid residues are
octahedrally coordinated to a cobalt atom through two amino-
nitrogen atoms and four carboxylic oxygen atoms. Five-membered,
six-membered, and seven-membered chelate rings are formed.
Two oxygen atoms of the five-membered chelate rings are in
trans positions, and those of the six-membered chelate rings are
in cis positions. The conformations of the two five-membered
chelate rings assume the symmetric envelope form, whereas those
of the six-membered chelate rings are all asymmetric skew-
boat. Figure 23 is a perspective drawing of the *cis*(N)-
trans(O_6) isomer (84), in which the oxygen atoms of the
five-membered chelate rings and those of the six-membered
chelate rings are in cis and trans positions, respectively. The
five-membered chelate rings assume the symmetric envelope form
and the six-membered rings take on an asymmetric skew-boat
conformation.

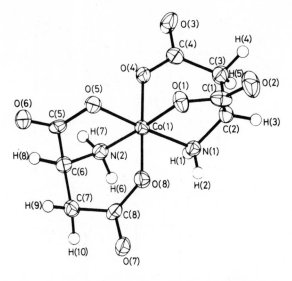

Fig. 23.   *cis*(*N*)-*trans*(*O*₆)-[Co(*S*-asp)₂]⁻ (84).

3.   (S)-2,3-*Diaminopropionic Acid*

The complexes of cobalt(III) with 2,3-diaminopropionic acid (*1,2*) were prepared by the reaction of tris(carbonato)-cobaltate(III) ion and (*R,S*)-2,3-diaminopropionic acid (85).

$$
\begin{array}{cc}
COOH & COOH \\
| & | \\
H_2N-C-H & H-C-NH_2 \\
| & | \\
CH_2-NH_2 & CH_2-NH_2 \\
\\
(S) & (R) \\
1 & 2
\end{array}
$$

The five possible diastereomers are illustrated in Figure 24. All are labeled according to the convention that the carboxyl groups are designated first, then the α-amino group, and final-ly the β-amino group. Thus the five isomers may be designated as (*a*) trans, cis, cis; (*b*) cis, cis, trans; (*c*) cis, trans, cis; (*d*) cis, cis, cis; and (*e*) trans, trans, trans. Isomers, (*a*), (*b*), and (*c*) require two ligands of the same absolute configurations (*S* in the above representations); (*d*) and (*e*) require that the two ligands be enantiomeric. All the isomers were isolated and, where possible, their configurations assigned on the basis of absorption, ORD, and CD spectra as

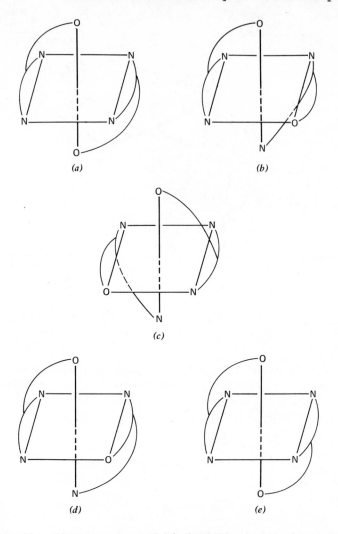

Fig. 24. Diastereomers of bis(2,3-diaminopropionato)-
cobalt (III) ion: (a) trans,cis,cis: (b) cis,cis,trans; (c)
cis,trans,cis: (d) cis,cis,cis; (e) trans,trans,trans.

well as by chemical methods. However, unambiguous differentiation of isomers (b) and (c) was impossible. The crystal
structure of the red crystals of $(-)_{546}-[Co(C_3H_7N_2O_2)_2]Br$ was
determined (86). This isomer proved to be the *S-cis, trans,
cis* isomer (Fig. 24c). The geometric arrangement of the donor
groups is shown in Figure 25. Two ligand molecules coordinate
to the central metal atom through four amino nitrogen atoms and

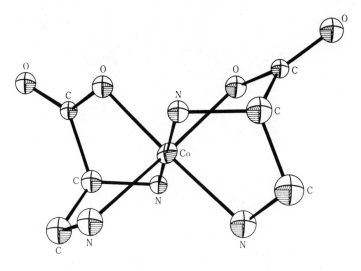

Fig. 25. $(-)_{546}-[Co(C_3H_7N_2O_2)_2]^+$ (86)

two carboxyl oxygen atoms to produce a distorted octahedral
coordination. Thus the ligand molecule acts as a terdentate
and forms a rigid fused-ring chelate system in which the absolute configuration of the ligand uniquely determines the
absolute configuration of the complex.

4.   β-(N-Sarcosinyl)propionic Acid (3)

The O, N, O-terdentate ligand coordinates facially to the

$$
\begin{array}{l}
CH_3 \\
| \\
N-CH_2-CH_2-COOH \\
| \\
CH_2-COOH
\end{array}
$$

3

metal atom. In $(+)_{546}-cis(O)-[Co(sarmp)(NH_3)_3]^+$ the resulting
six-membered chelate ring has the skew-boat form with δ conformation, the $N-CH_3$ and $C=O$ bonds both being in equatorial
positions. The five-membered chelate ring assumes an asymmetric
envelope form with λ conformation (87).

5.   Tribenzo[b.f.j][1.5.9]triazacyclododecahexaene (TRI) (4)

This terdentate ligand is the trimer formed by the self-
condensation of o-aminobenzaldehyde in the presence of metal
ions. This is a stereochemically rigid molecule. A cobalt(III)

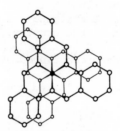

4

complex, $(+)_{546}$-$[Co(TRI)_2]^{3+}$, is shown in Figure 26 (88). The cobalt atom is coordinated to six nitrogen atoms, three from each ligand. The three nitrogen atoms from each ligand molecule define planes that are parallel to one another and that are 2.36 Å apart. The three nitrogen atoms of each ligand form an equilateral triangle, with the upper triangle rotated counter-clockwise by about 8° from an octahedral arrangement. This distortion of the CoN_6 chromophore occurs in the same sense as that observed for Λ-$[Co(en)_3]^{3+}$. Each TRI ligand is propeller shaped. The mean pitch (to the left) of the planar benzene is 14° with respect to the nitrogen planes.

Fig. 26. $(+)_{546}$-$[Co(TRI)_2]^{3+}$.

The CD spectrum of this complex ion in an aqueous solution in the first absorption region (i.e. the absorption region of longest wavelength) is very similar to that of Λ-$(+)_{589}$-$[Co(en)_3]^{3+}$, both showing a prominent positive peak at the longer-wavelength side and a weak negative peak at the shorter-wave-length side. This indicates that the distortion of the chrom-ophore plays an important role in the optical activity of these complexes.

    6.   (R)-2-Methyl-1,4,7-triazacyclononane (R-MeTACN)

This cyclic terdentate (5) and its cobalt(III) complex

were synthesized by Mason and Peacock (89). The complex was
found to exhibit the largest yet recorded ring-conformation-

$$
\begin{array}{c}
\text{H} \\
\mid \\
\text{N} \\
\diagup \quad \diagdown \\
(CH_2)_2 \qquad (CH_2)_2 \\
\mid \qquad\qquad \mid \\
\text{HN} \qquad\qquad \text{NH} \\
\diagdown\quad CH-CH_2 \quad\diagup \\
\mid \\
CH_3
\end{array}
$$

5

based optical activity for the [CoN$_6$] chromophore.* The crystal
structure of $(-)_{589}$-[Co$(R$-MeTACN)$_2$]I$_3$·5H$_2$O was determined (90).
The electron-density distribution of the complex ion looks as
if the complex had D_3 symmetry owing to the orientational
disorder. Three diastereomers are possible for [Co$(R$-MeTACN)$_2$]$^{3+}$
with regard to the positions of the three substituted methyl
groups. Figure 27 illustrates the structure of the complex.
Two cyclic ligands, spanned on a pair of opposite faces of an
octahedron, are coordinated to the cobalt atom with six
nitrogen atoms. The conformation of all the five-membered
chelate rings is λ. The substituted methyl group is attached
in an equatorial position with respect to the average plane

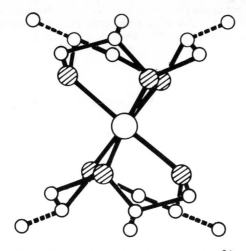

Fig. 27. $(-)_{589}$-[Co$(R$-MeTACN)$_2$]$^{3+}$.

*This activity is caused by the chiral conformation of
the chelate ring, which, in turn, induces optically active
transitions between the metal d-electron levels by the ligand
field.

of the chelate ring, and the [CoN$_6$] chromophore is twisted around threefold axis in the same way as in Δ-[Co(en)$_3$]$^{3+}$. Unlike the [Co(en)$_3$]$^{3+}$ ion [CoN$_6$] is elongated along the threefold axis, the Co-N bond being inclined at an angle of 51.28° with respect to the threefold axis. This angle is 54.75° for a regular octahedron. The distortion is of the same type as that reported for (+)$_{546}$-[Co(TRI)$_2$]$^{3+}$ (88). The existence of the six five-membered chelate rings with λ conformation may give rise to a large ring-conformation-based optical activity.*

7. 1,1,1-Tris(aminoethyl)ethane (tame)

This terdentate ligand, $CH_3C(CH_2NH_2)_3$, can coordinate to the cobalt atom to form the bis-complex [Co(tame)$_2$]$^{3+}$. It has its nonligating atoms above and below the trigonal planes of the ligating nitrogen atoms as does [Co(R-TeTACN)$_2$]$^{3+}$. The crystal structure of [(+)$_{589}$-[Co(tame)$_2$]]Cl[(+)$_{589}$-R,R-tart]·xH$_2$O was determined (181). The complex ion has approximate D_3 symmetry. The conformation of the six-membered chelate ring is intermediate between that of a regular skew boat and that of a true boat, and may be described as an asymmetric skew boat. The absolute configuration may be designated as λ, with the [CoN$_6$] chromophore twisted around the threefold axis. The distortion is equivalent to that in Δ-[Co(en)$_3$]$^{3+}$. Accordingly the absolute configuration can be fully designated as $\Delta\lambda\lambda$, where the two λ's refer to the conformations of the six-membered rings formed by the two different ligands, respectively. There

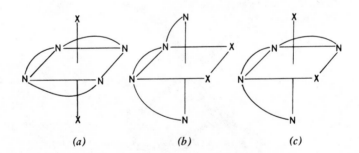

(a)   (b)   (c)

Fig. 28. Possible cobalt-trien coordination (44): (a) trans, (m_2); (b) cis-α (f_2); (c) cis-β (mf). For the notation in parentheses see Sect. V-D.

*This activity is caused by the chiral conformation of the chelate ring, which, in turn, induces optically active transitions between the metal d-electron levels by the ligand field.

are three possible isomers: λλ, δδ, and λδ (or δλ). The former
two are enantiomers, the last a meso isomer. Strain energy
calculation showed that the racemic isomer is more stable by
about 6.70 kJ/mol. The isomers may interconvert by a trigonal
twist of the ligand, whereby the methylene groups move from
one side to the other of the plane formed by the cobalt,
nitrogen, and the quaternary carbon atom. In fact, the
$[Co(tame)_2]^{3+}$ conformers equilibrate in solution. Thus the CD
spectrum was recorded for finely ground and dispersed powders
in polystyrene.

B. Quadridentate

1.   1,8-Diamino-3,6-diazaoctane(triethylenetetramine,
trien, or 2,2,2-tet)

The linear tetramine trien can act as a quadridentate
ligand and form complexes with Co(III). The three possible ways
of coordinating a trien molecule to a cobalt(III) ion are shown
in Figure 28. Further, two conformations of the trien ligand
for each cis-β isomer are possible, controlled by the symmetry
about the two secondary nitrogen atoms, which are shown in
Figure 29. Both conformations were detected in the Δ-cis-β-

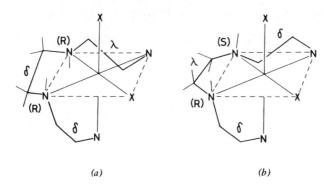

(a)   *(b)*

Fig. 29.   (a) Δ-R,R-cis-β and (b) Δ-R,S-cis-β isomers (94).

$[Co(trien)(H_2O)_2]^{3+}$ ion: Δ-cis-β-(RR) and Δ-cis-β-(RS), which
mutarotates to the thermodynamically more stable Δ-cis-β-(RR)
conformer (ΔG° 12.6 kJ/mol) (91). The Δ-cis-β-(RS) isomer could
not be obtained in a stable crystalline form. Freeman and
Maxwell determined the structure of racemic cis-β-(chloroaqua-
triethylenetetramine)cobalt(III) perchlorate. The structure
contains Λ-cis-β-(SS) and Δ-cis-β-(RR) isomers (92). When a
substituted trien ligand such as 3(S),8(S)-2',2,2'-tet was
used, the cis-β-(RS) form was obtained in the crystalline

form together with the *cis*-α isomer (93). Figure 30 shows the
complex ion $(-)_{546}$-*cis*-β-$[Co(NO_2)_2(3(S),8(S)-2',2,2'-tet)]^+$ (94).
The absolute configuration of the complex is Δ. The two

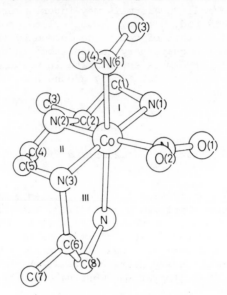

Fig. 30. $(-)_{546}$-$[Co(NO_2)_2(3(S),8(S)-2',2,2'-tet)]^+$ (94).

terminal chelate rings assume δ conformations, with the C–CH₃
bonds in equatorial positions. The central ring assumes the λ
conformation. The absolute configurations of the asymmetric
nitrogen atoms are enantiomeric, N(2) being (R), and N(3) (S).
One of the terminal five-membered chelate rings, I, is in the
eclipsed envelope conformation, while in the other terminal
chelate ring, III, the two carbon atoms are on the same side
of the plane formed by the two nitrogen atoms and the central
cobalt atom. The central chelate ring has an unsymmetrical skew
conformation. In the crystals of *cis*-β-$[CoCl(H_2O)(trien)](ClO_4)_2$,
the two outer chelate rings have unsymmetrical skew conformation
and the central one is of the envelope type (92). The C–C
distance in the central chelate ring of the present complex
ion, 1.57 Å, is somewhat greater and those in the chelate rings
at the two ends, 1.49 and 1.46 Å, are shorter than the normal
C–C bond distance of 1.52 Å in such chelate rings. This is in
contrast to the distances in *cis*-β-$[CoCl(H_2O)(trien)]^{2+}$; here
the C–C bond in the central ring is shorter than those in the
outer rings. The steric strain that arises from cis coordination of a substituted trien ligand appears to be partly accom-
modated by such distortion of the ligand.

The structure of $(-)_{589}$-*cis*-α-[Co(NO$_2$)$_2$(3(S),8(S)-2',2,2'-tet)]$^+$ is also known (95). Figure 31 shows a perspective drawing of the complex ion, which has approximate C_2 symmetry

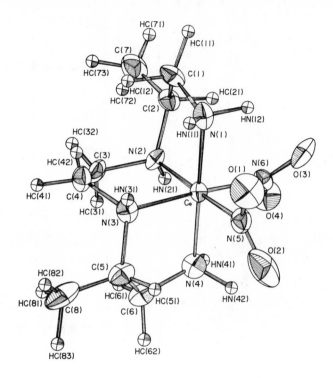

Fig. 31. $(-)_{589}$-*cis*-α-[Co(NO$_2$)$_2$(3(S),8(S)-2',2,2'-tet)]$^+$ (95).

through the central cobalt atom and the midpoint between N(2) and N(3). The absolute configuration can be described as a skew chelate pair Λ. The combination of the conformations for three chelate rings is δ, λ, and δ, and the two methyl groups lie in equatorial positions with respect to the planes of the chelate rings. As a result the absolute configurations about the two asymmetric nitrogen atoms are S. The angles subtended at the cobalt by the outer two chelate rings are 86.2° on the average, whereas the angle subtended by the inner chelate ring is slightly larger (87.6°). The two outer chelate rings have un-symmetrical skew conformations. In the central chelate ring, on the other hand, the skew conformation is nearly symmetrical.

Figure 32 represents a perspective drawing of the complex ion $(-)_{589}$-*trans*-[Co(NO$_2$)$_2$(3(S),8(S)-2',2,2'-tet)]$^+$ (96). The

TABLE 1

Comparison of Bond Lengths and Angles, Calculated by Energy Minimization, with Those Determined from X-Ray Crystal Structure Analysis

	cis-α Isomer		cis-β Isomer		trans Isomer	
	Minimi-zation	Crystal	Minimi-zation	Crystal	Minimi-zation	Crystal
Bond lengths (Å)						
Co···N(1)	1.96	1.94	1.97	1.98	1.99	1.97
Co···N(2)	1.97	1.96	1.99	1.99	1.95	1.96
Co···N(3)	1.99	1.96	1.97	1.94	1.96	1.95
Co···N(4)	1.99	1.95	1.99	1.97	2.01	2.03
Co···N(5)	1.96	1.90	1.97	1.94	1.99	1.93
Co···N(6)	1.94	1.88	1.97	1.92	2.01	1.98
Angles (°)						
N(1)CoN(2)	86°	86°	87°	85°	87°	85°
N(2)CoN(3)	87	88	87	87	86	86
N(3)CoN(4)	86	87	86	86	88	87

CoN(1)C(1)	110	114	107	108	106	108
CoN(2)C(2)	107	108	103	110	104	109
CoN(2)C(3)	111	110	108	109	110	110
CoN(3)C(4)	109	108	109	110	107	109
CoN(3)C(5)	108	108	106	108	106	107
CoN(4)C(6)	111	112	111	113	103	106
N(1)C(1)C(2)	109	108	109	113	110	111
C(1)C(2)N(2)	108	107	106	104	104	101
C(1)C(2)C(7)	111	114	110	109	109	115
C(7)C(2)N(2)	112	116	110	113	111	115
N(2)C(3)C(4)	110	111	107	107	104	105
C(3)C(4)N(3)	111	111	110	109	103	106
N(3)C(5)C(6)	108	107	109	110	102	106
C(5)C(6)N(4)	110	108	109	108	108	110
N(3)C(5)C(8)	111	114	110	110	112	113
C(8)C(5)C(6)	113	114	111	116	111	109
C(2)N(2)C(3)	112	115	117	118	119	122
C(4)N(3)C(5)	113	117	113	114	117	121

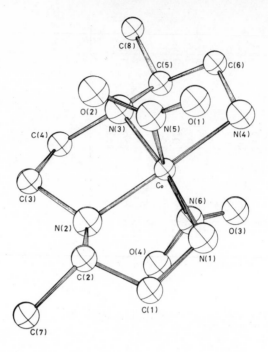

Fig. 32. $(-)_{589}$-*trans*-$[Co(NO_2)_2(3(S),8(S)-2',2,2'-tet)]^+$ (96).

complex ion has an approximate twofold axis of rotation through cobalt, bisecting the C-C bond in the central chelate ring. The two nitro groups are in trans positions. The quadridentate ligand forms a girdle about the cobalt atom, coordinating to the cobalt atom with its four nitrogen atoms. The conformations of the two chelate rings with methyl groups are δ; that of the other ring is λ. The two methyl groups lie in equatorial positions with respect to the plane of the chelate rings. As a result the absolute configurations of the two asymmetric nitrogen atoms are both R. The N-Co-N angles in the outer chelate rings are 85° and that in the central one is 88°. The angular strain of the ligand is further in evidence at the two asymmetric nitrogen atoms and at the two carbon atoms in the central chelate rings. The bond angles involving these atoms deviate considerably from the regular tetrahedral angle of 109.5°. The two outer chelate rings have unsymmetrical envelope conformations, whereas in the central chelate ring the two carbon atoms are on the same side of the plane formed by the central cobalt and the two nitrogen atoms. The strain energies of the three isomers were calculated according to Boyd's procedure (97,98). The bond lengths and angles in the three

isomers, reproduced within twice the standard deviation of the
values obtained by crystal structure analysis, are shown in
Table 1. The major angular distortions observed in the crystals
were accurately predicted from the minimization calculations.
For example, a remarkable angular strain was found in the
trans isomer at the secondary nitrogen atoms and the asymmetric
carbon atoms, as mentioned earlier. The calculated bond angles
involving these atoms agree well with observation. The final
energy terms obtained from the minimization for the three
isomers are tabulated in Table 2. The calculations indicate

TABLE 2
Distribution of Conformational Strain Energy (kJ/mol)

	cis-α	cis-β	trans
Bond length deformation	2.5	5.4	2.9
Bond angle deformation	5.9	8.8	24.3
Torsional strain	22.2	17.6	10.0
Nonbonded interaction	23.0	14.6	14.2
Total conformational energy	53.6	46.4	51.4
Energy difference (relative to cis-β)	7.2	0.0	5.0

that each isomer has a different prominent energy term. Of the
total strain energy, torsional strain and nonbonded interactions
are most prominent. The relative energy differences between the
cis-α and cis-β isomers and between the trans and cis-β
isomers are 7.2 kJ/mol and 5.0 kJ/mol, respectively. The result
indicates that the cis-β form is the most stable of the three
isomers. This is supported by the fact that the trans isomer
is easily isomerized to the cis-β form by recrystallization
from water. The calculation indicates that angular distortions
are important in deriving the relative stabilities of the complex ions. In fact, bond angles deform with a comparatively
small expenditure of energy for changes as large as several
degrees. Similarly, torsional angle distortion occurs easily.
These angular distortions can largely alleviate otherwise
serious nonbonded hydrogen interactions.

    In 1974 Yoshikawa and his collaborators (99) isolated
(+)$_{589}$-cis-β-carbonato(3(S),8(S)-2',2,2'-tet)cobalt(III) salts,
and the crystal structure of the perchlorate was determined
(100). It turned out to be the Λ-cis-β-R,R isomer, whose
structure is shown in Figure 33. In place of the two nitro
groups a planar carbonato group is coordinated to the cobalt

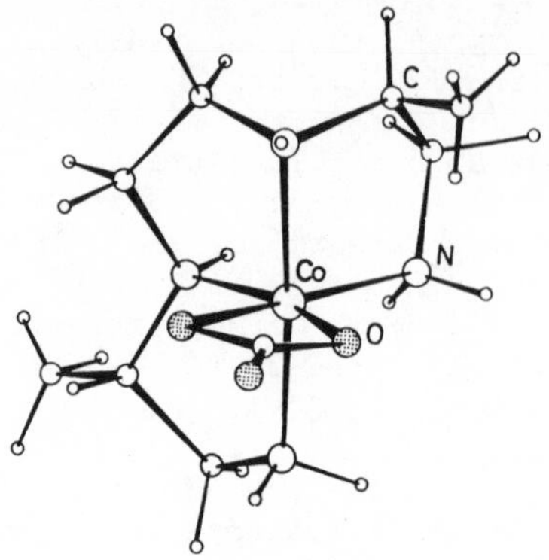

Fig. 33. $(+)_{589}$-cis-β-[Co(CO$_3$)(3(S),8(S)-2',2,2'-tet)]$^+$ (100).

atom. The absolute configurations of the three chelate rings are λ, λ, and δ, respectively. One of the two substituted methyl groups--the one in the apical chelate ring--is bonded axial with respect to the chelate ring, whereas the other one--bonded to the in-plane chelate ring--is in an equatorial position. (There is space to accommodate an axial methyl group, since instead of the two nitro groups a planar carbonato group is coordinated to the fifth and sixth sites and the O-Co-O angle is compressed to 68.6°.) The absolute configuration about both asymmetric nitrogen atoms is R. One of the two secondary nitrogen atoms connecting the two in-plane chelate rings is bonded to the cobalt atom with a significantly shorter distance (1.935 Å) than other Co-N distances (1.951 Å on the average). It is also to be noted here that the bond angles including this particular nitrogen atom are considerably distorted from the normal strainfree) angle. The chelate ring with an axial methyl group assumes an unsymmetrical skew conformation. The other two rings both assume an eclipsed envelope conformation.

The cis-α and cis-β isomers of [CoX$_2$(trien)]$^{n+}$ allow cis coordination of a bidentate. If the bidentate is unsymmetrical like glycine, two cis-β isomers are possible: cis-β$_1$, with the amino group of glycine in a trans position to a terminal NH$_2$ group of the trien ligand; and cis-β$_2$, with the amino group in a trans position to the secondary nitrogen atom of the quadridentate (101). The two isomers are shown in Figure 34. The structures of Λ-(-)$_{589}$-(R,R)- and Λ-(R,S)-β$_1$-[Co(gly)(trien)]$^{2+}$ were determined and the geometries were also adequately reproduced by strain energy minimization calculations (102). Table 3 shows the final energy terms for the minimized structures. The total strain energy difference between the Δ-β$_1$-(R,R) and Δ-β$_1$-(R,S) isomers is calculated as 3.4 kJ/mol in favor of the R,R

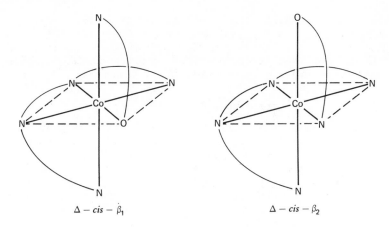

Fig. 34. Δ-*cis*-β_1 and Δ-*cis*-β_2 configurations of the [Co(gly)(trien)]$^{2+}$ ion.

TABLE 3
Final Energy Terms (kJ/mol) for Minimized Structures

	Δ-β_1-(R,R)	Δ-β_1-(R,S)
Bond length deformation	2.1	2.1
Bond angle deformation	16.3	18.4
Torsional strain	23.9	26.0
Nonbonded interaction	-13.0	-13.8
Total strain energy	29.3	32.7

isomer. Bond angle and torsional terms make the most significant contribution to the difference in the total strain energy. The measured difference in ΔH, obtained from the temperature dependence of the equilibrium constant, is less than 1.3 kJ/mol, which is in reasonable agreement with the calculated value at this stage of the development of force-field calculations. The strain energy calculated for Δ-β_2 isomers showed that the Δ-β_2-(R,R) isomer is favored by about 14.6 kJ/mol, which may be compared with the calorimetrically measured ΔH°_{298} of 4.6 kJ/mol. Thus the calculations correctly predict that the R,R isomer is more stable in both the β_1 and β_2 systems but by smaller a margin for the β_1 isomers. This is caused by the different orientations of the glycine ligand in the *cis*-β_1 and *cis*-β_2 coordination.

The structures of Δ-β_2-(R,R,S)-[Co(S-pro)(trien)]$^{2+}$ and Λ-β_2-(S,S,S)-[Co(S-pro)(trien)]$^{2+}$ were also determined (103).

Crystals were obtained from the reaction products of β-[Co(OH)-(trien)(H$_2$O)]$^{2+}$ and (S)-proline. The coordination of (S)-proline, with the O trans to the N of trien(β_2-coordination) agrees with the prediction that large nonbonded interactions would occur in the alternative β_1 configuration. Rough measurements using Dreiding stereomodels and a conservative H$\cdots$H nonbonded potential function indicated that the formation of the Λ-β_2-(S,S,S) form might be impossible owing to the nonbonded repulsions between the amino acid and the chelate ring. Unlike in the case of β_2-[Co(sar)(trien)]$^{2+}$ (104), no stereoselectivity was observed, however, in spite of the fixed configuration at the nitrogen atoms of (S)-proline by virtue of the five-membered pyrrolidine ring; viz. two [Co(S-pro)-(trien)]$^{2+}$ species Δ-β_2(R,R,S) and Λ-β_2(S,S,S), were shown to be formed from the reaction mixture in approximately equal amounts (105). The result of X-ray structure analysis showed that the formation of the Λ-β_2-(S,S,S) isomer is much more reasonable than originally expected. The relative ease of bond angle bending can be an important factor in determining the stability of isomers. The major geometrical difference between the Δ-β_2-(R,R,S) isomer and Λ-β_2-(S,S,S) isomer consists of the relative orientations of the proline moieties. In the Λ-β_2-(S,S,S) isomer the pyrrolidine ring is oriented toward the apical trien rings, whereas in the Δ-β_2-(R,R,S) isomer it is remote from the apical chelate ring. The measured free energy difference between these two isomers was only 5.4 kJ/mol in favor of the Δ-β_2-(R,R,S) isomer. This small energy difference is consistent with the conclusions based on the isomer geometries determined by X-ray analysis. In fact, when hydrogen atoms were placed at calculated positions on the final crystal structure model of Λ-β_2-(S,S,S) the resulting interactions were relatively small. The expansion of N(proline)-Co-N(trien) and Co-N(proline)-C angles clearly alleviates close nonbonded interactions between the amino acid and trien moieties, where N(trien) stands for a secondary nitrogen atom of the trien ligand in cis position with respect to N(proline).

In a recent study of the related complex ion, N-methyl-(S)-alaninatobis(ethylenediamine)cobalt(III) (106), it was shown that at pH > 12 mutarotation about the C center of the alaninato chelate occurred, and that the equilibrium Δ-R : Δ-S ratio was about 4; no mutarotation was observed at pH < 7. The crystal structure of Δ-R-[Co(en)$_2$(N-Me-(S)-ala)]I$_2$ was determined (106). In this complex ion the two ethylenediamine cobalt rings adopt δ and λ conformations. The amino acid contains trans methyl groups with the (R) and (S) configurations around the secondary nitrogen atom and the asymmetric carbon atom, respectively. Both methyl groups are attached in equatorial positions with respect to the chelate ring. The methyl nitrogen bond is largely distorted owing to the steric repulsion

between the methyl group and the cobalt ethylenediamine ring,
the Co-N-C(methyl) angle being 120.2°. The observed geometry
of the complex ion agrees well with the results of strain
energy minimization. The calculated strain energy increases in
the order $N(S)$-$C(R)$ < $N(R)$-$C(S)$ < $N(S)$-$C(S)$ < $N(R)$-$C(R)$. This
relationship results in large part from the nonbonded interactions between the methyl groups bonded to the carbon atom of
the amino acid ligand and the neighboring ethylenediamine ring.
The number and severity of the nonbonded H•••H, C•••H, and
C•••C contacts increase in the order $S(R)$ < $R(S)$ ~ $S(S)$ < $R(R)$
for the Δ configuration.

2. $(+)_{495}$-$[(S$-$Glut)(en)_2]^+$

The absolute configuration of $(+)_{495}$-$[Co(S$-$glut)(en)_2]^+$
was determined to be $\Lambda(\delta\delta)$, where δ refers to the two five-
membered chelate rings (107). The γ-carboxylate group forms
an intramolecular hydrogen bond with the amino nitrogen atom
of S-glut, the N-H•••O distance being 2.8 Å. The tendency for
stereoselective reaction of (S)-glutamic acid is relatively
small, since the chelated amino acid ring is nearly planar.
The reaction of (S)-glutamic acid with racemic $[Co(CO_3)(en)_2]^+$
results in $(+)$- and $(-)$-$[Co(S$-$glut)(en)_2]^+$ in equal amounts,
though kinetically stereoselective formation of first the
$(+)$-isomer and then, on further reaction, the $(-)$-isomer
was observed (108,109).

3. $(-)_{589}$-$[Co(sar)(en)_2]^{2+}$

The stereochemistry of the sarcosinatobis(ethylenediamine)cobalt(III) complex ion was first studied as early as in
1924 (110). This complex ion contains two chiral centers, one
around Co(III) and the other around the asymmetric nitrogen atom
of the coordinated sarcosine. Isolation of all four possible
isomers was claimed (110), but when the work was repeated
very carefully, only two forms were obtained (111). The newer
study produced evidence from a number of sources that the
sarcosinato ion was coordinated stereospecifically about one
configuration of the Co(en)₂ moiety. The stereospecificity
involved in this system was verified by the crystal structure
analysis of $(-)_{589}$-$[Co(sar)(en)_2]I_2 \cdot 2H_2O$ (112). Figure 35 shows
the absolute configuration of the complex ion. The sarcosinato-
cobalt(III) ring is slightly puckered, the conformation of the
five-membered chelate ring being λ. The two cobalt-ethylenediamine chelate rings assume δ and λ conformations, respectively. These combinations of the chelate ring conformations
presumably minimizes the H(methyl)•••H(amino) interactions.
The absolute configuration of the whole complex may be designat-
ed as $\Delta(\lambda_{sar}\lambda_{en}\delta_{en})$. The absolute configuration of the asymmetric nitrogen atom is S. In this stable Δ-S form the hydrogen

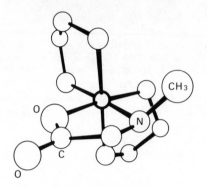

Fig. 35. $(-)_{589}-[Co(sar)(en)_2]^{2+}$.

atom is balanced over the adjacent Co(en) ring and the CH_3
group lies in the space between the two ethylenediamine rings.
    In 1976 Yamatera and his collaborators (113) succeeded in
separating all four possible isomers by chromatography on an SP
Sephadex column. The crystal structure of the less stable isomer
is not yet known. From the formation ratio the $\Delta-[Co(R-sar)(en)_2]^{2+}$
isomer appears to be only 3.8 kJ/mol less stable than the
$\Delta-[Co(S-sar)(en)_2]^{2+}$ isomer. This experimental difference in
$\Delta G°$ is about half the value of 7.1 kJ/mol calculated by strain
energy minimization (106).

4.   5(R),7(R)-Dimethyl-1,4,8,11-tetraazaundecane(R,R-2,3",2-tet)

    The structure of $(-)_{546}-cis-\beta-[Co(ox)(R,R-2,3",2-tet)]ClO_4$
was determined to establish the conformation of the central
six-membered chelate ring (114). Figure 36 shows a perspective
drawing of the complex ion. The quadridentate ligand coordinated to the cobalt atom with four nitrogen atoms has cis-β
configuration. The absolute configuration of the complex ion
may be designated as Λ; those of the two terminal five-membered chelate rings are δ. The central six-membered chelate ring
takes on a chair conformation, with one methyl group axial and
the other equatorial, unlike the cobalt-(R,R)-2,4-diaminopentane
chelate ring in $(-)_{546}-[Co(R,R-ptn)_3]^{3+}$ (41) and in $(+)_{546}-$
$[Co(R,R-ptn)_3]^{3+}$ (42). The two secondary nitrogen atoms have the
absolute configuration S. It is to be noted here that the
Co-N(2) distance of 2.007 Å is longer than the three other
Co-N bonds, the average being 1.950 Å. This particular nitrogen
atom [N(2)] is bonded to the carbon atom to which the methyl
group is attached in an axial position. The N-Co-N angle in the
six-membered chelate ring is 93.5°, while that in $(-)_{546}-$
$[Co(R,R-ptn)_3]^{3+}$ is 89.5°. The other two N-Co-N angles in

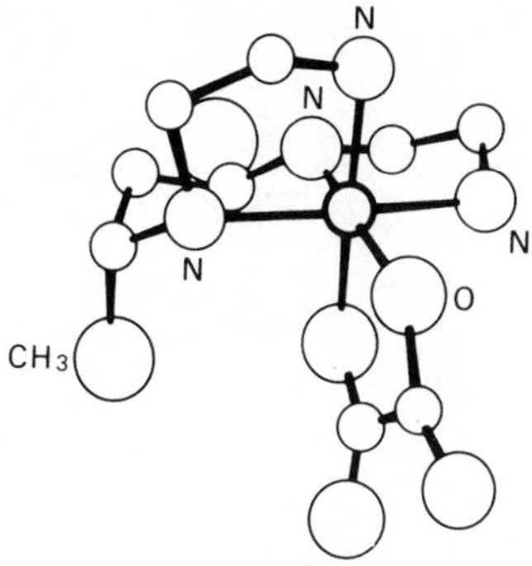

Fig. 36. $(-)_{546}$-cis-β-[Co(ox)(R,R-2,3',2-tet)]$^+$.

the five-membered chelate ring are both less than 90 deg., as
is the O-Co-O angle in the oxalato-cobalt ring. Thus the
distortion of the CoN$_6$ chromophore is similar to that in Λ-
[Co(en)$_3$]$^{3+}$ and in Λ-[Co(ox)(en)$_2$]$^+$. In fact, the CD spectra of
these compounds are very similar in the first absorption
region: they display a prominent positive peak on the longer
wavelength side.

5.   1,10-Diamino-4,7-diazadecane(3,2,3-tet)

trans-Diacidato complexes of cobalt(III) with 3,2,3-tet
were studied to establish a rule that would allow the prediction of absolute configurations from CD spectra associated with
d-d transitions. A new source of dissymmetry arises in the
trans structure from the secondary amines. There are three
possible isomers depending on the configurations at the two
secondary nitrogen atoms: the enantiomeric R,R and S,S forms,
and the internally compensated (R,S) (meso) form. A study of
a molecular model shows that for the (R,S) configuration the
ethylene fragment in 3,2,3-tet will assume an eclipsed form,
whereas the R,R or S,S configurations result in a gauche conformation. The trans-[Co(NO$_2$)$_2$(3,2,3-tet)]$^+$ cation was resolved
and the absolute configuration of $(+)_{546}$-[Co(NO$_2$)$_2$(3,2,3-tet)]$^+$
determined (115). The result is shown in Figure 37. The five-
membered chelate ring adopts the gauche conformation, with
absolute configuration δ. The two six-membered chelate rings
assume the chair conformation. The absolute configurations of
the two asymmetric nitrogen atoms are both (R). Two additional
sources of dissymmetry, apparently contributing to the observed
optical activity in the solid state, were recognized. First,
the four ligating nitrogen atoms of the tetramine ligand in
the equatorial plane of the coordination octahedron showed

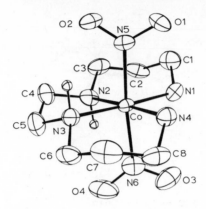

Fig. 37. $(+)_{546}$-[Co(NO$_2$)$_2$(3,2,3-tet)]$^+$ (115).

alternate deviations from the plane. This kind of deviation
generates two nonorthogonal skew lines that define a chiral
system at the cobalt atom. Second, the two trans ligands are
not axially symmetric. The plane of one nitro group is rotated
by about 10° around the NO$_2$-Co-NO$_2$ axis relative to the plane
of the other nitro group. In the case of a cation containing
the same tetramine ligand, but with axially symmetric chloride
ion, $(-)_{589}$-*trans*-[CoCl$_2$(3,2,3-tet)]$^+$, the deviations of the
four nitrogen atoms are similar in magnitude but opposite in
sign to those observed in $(+)_{546}$-[Co(NO$_2$)$_2$(3,2,3-tet)]$^+$ (116).
The coordination geometry of 3,2,3-tet is the same as that in
the *trans*-dinitro analogue.

6.   *(1R,3R,8R,10R)-Tetramethyl-4,7-diazadecane-1,10-diamine (R,R,R,R-3",2,3"-tet)*

The complex ion *trans*-[Co(NO$_2$)$_2$(R,R,R,R-3",2,3"-tet)]$^+$ has
a rigorous twofold axis of rotation (117). The absolute configurations of the two secondary nitrogen atoms are both S. The
central five-membered chelate ring assumes the gauche conformation with δ absolute configuration. The two six-membered
chelate rings take on the skew-boat conformation with methyl
groups in equatorial positions. The absolute configuration of
these chelate rings is λ. The geometry of the complex ion
agrees with the prediction based on the strain energy minimization technique (118). $(-)_{546}$-[Co(ox)(R,R,R,R-3",2,3"-tet)]$^+$
assumes Δ absolute configuration (119). The quadridentate
ligand has the *cis*-β configuration, and the absolute configuration of the two asymmetric secondary nitrogen atoms is S.
Both six-membered chelate rings assume chair conformations and
in each ring one methyl group is axial and the other equatorial.

Of the axial methyl groups one is attached to the carbon atom
next to the terminal nitrogen atom of the in-plane six-membered
chelate ring and the other to the carbon atom next to the
secondary nitrogen atom in the out-of-plane six-membered
chelate ring. The five-membered chelate ring is normal and its
conformation is δ.

7. *Ethylenediamine*-N,N′-*diacetic Acid* (*edda*)

For complexes of the type $[Co(edda)(L)]^n$, where L represents a bidentate ligand, two geometric isomers were possible:
symmetrical cis (*s-cis* or *cis-α*) and unsymmetrical cis (*u-cis*
or *cis-β*) (Figure 38). The cis-β coordination is most similar
to that of chelated edta, possessing both an in-plane girdling
ring and an out-of-plane ring. If the ligand L is *R*-pn, there
are four possible isomers of *cis-β*-$[Co(edda)(R-pn)]^+$, as shown
in Figure 39.

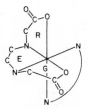

u-cis-$[Co(EDDA)(R-pn)]^+$

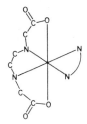

s-cis-$[Co(EDDA)(R-pn)]^+$

Fig. 38.   Two possible isomers of $[Co(edda)(L)]^n$.

Figure 40 illustrates the molecular structure of one of
the four isomers, $\Delta\Delta\Delta\Lambda$-*cis,trans*-(N-O)-*cis-β*-$[Co(edda)(R-pn)]^+$,
as revealed by X-ray structure analysis (120). The quadridentate
edda-Co(III) chelate is essentially a fragment of the larger
sexidentate edta-Co(III) chelate (Fig. 49). The absolute con-

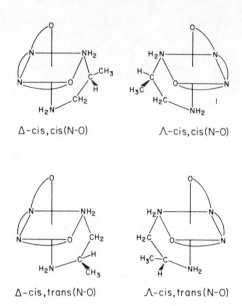

Δ-cis,cis(N-O)   Λ-cis,cis(N-O)

Δ-cis,trans(N-O)   Λ-cis,trans(N-O)

Fig. 39.   Isomers of *cis*-β-[Co(edda)(*R*-pn)]$^+$ (120).

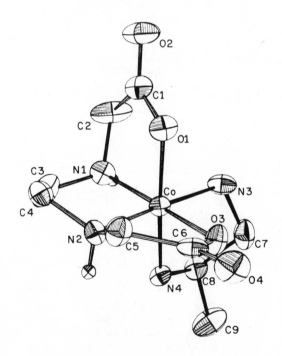

Fig. 40.   ΔΔΔΛ-*cis,trans*-u-*cis*-β-[Co(edda)(*R*-pn)]$^+$ (120).

figurations of the secondary nitrogen atoms are found to be
R,S. The R-pn ring conformation is λ with the methyl group in
the stable equatorial position. A structural study of [Co(edta)]$^-$
has shown that the in-plane girdling glycinate rings are
quite strained compared with the backbone ethylenediamine ring
and an out-of-plane glycinate ring (121). In the present complex, however, the ring strain is distributed over all three
rings.

C. Quinquedentate

1. 1,11-Diamino-3,6,9-triazaundecane (tetraen)

There exist four modes of coordination of 1,4,7,10,13-
pentaazatridecane to the cobalt(III) atom [Fig. 41, (I)-(IV)]

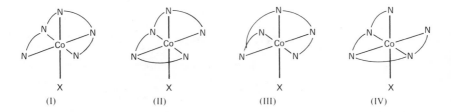

Fig. 41. Four possible geometrical isomers of [Co(tetraen)-
X]$^{2+}$ (122): (I) f_3; (II) $f_2m(mf_2)$; (III) fmf; (IV) $fm_2(m_2f)$.

(122). Form (I) has a plane of symmetry, if the conformations
of the chelate rings are ignored; (II), (III), and (IV) will
always be dissymmetric. In addition to the stereochemistry
generated by the topology of the coordinated ligand, the
coordinated secondary nitrogen atoms become chiral. This leads
to two diastereomeric forms for the f_2m structure and four
such forms for the fm_2 structure. The two orientations of the
central NH proton in the fmf array lead to identical structures,
since they are equivalent by rotation around the vertical
N-Co-X axis. Several isomers of cobalt(III) complexes of
[CoCl(tetraen)]$^{2+}$ have been isolated (122,123). The structures
of f_3-[CoCl(tetraen)]$^{2+}$, (+)$_{540}$-f_2mS-, and (+)$_{540}$-f_2mR-
[CoCl(tetraen)]$^{2+}$ are known (124,125). These structures were
submitted to a full and quantitative energy minimization
procedure to determine whether the conformation found in the
crystal corresponds to the calculated "gas state" geometry
(125,126). Figure 42 presents the f_2mS- and f_2mR-isomers. The
absolute configurations are both $\Lambda\Delta\Lambda$, and the chelate rings in
sequence I ~ IV have the conformations $\lambda\lambda\delta\delta$ for f_2mS and $\lambda\delta\lambda\delta$
for f_2mR isomers, respectively. The fusion of the $\lambda\lambda$ rings in

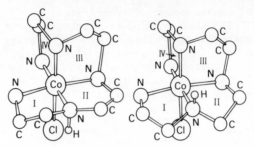

Fig. 42. f_2mS- and f_2mR-isomers of [CoCl(tetraen)]$^{2+}$ (122).

the f_2mS isomer is energetically less favorable in terms of the torsional energy, though not in overall energy.

D. Sexidentate

1. 1,14-Diamino-3,6,9,12-tetraazatetradecane(linpen)

Linear pentaethylenehexamine (6) can act as a sexidentate ligand. It consists of two dien parts linked by an ethylene

$$H_2N-(CH_2)_2-NH-(CH_2)_2-NH$$
$$\qquad\qquad\qquad\qquad\qquad\quad |$$
$$\qquad\qquad\qquad\qquad\qquad (CH_2)_2$$
$$\qquad\qquad\qquad\qquad\qquad\quad |$$
$$H_2N-(CH_2)_2-NH-(CH_2)_2-NH_2$$

6

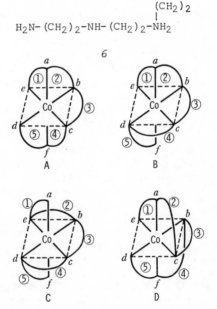

Fig. 43. Four possible geometric isomers of [Co(linpen)]$^{3+}$ (127).

group. The four possible geometric isomers of $[Co(linpen)]^{3+}$
are illustrated in Figure 43.

In isomer A the six nitrogen atoms are involved in forming
the facial structure exclusively. Accordingly, this isomer may
be designated as f_4. Isomer B, containing one NH group exhibiting meridional structure, may be represented by f_2mf or by fmf_2
starting from the other end of the chain molecule. In the same
way we can designate C as fm_2f and D as mf_2m. The number of
isomers increases to eight if the absolute configurations
of the dissymmetric secondary nitrogen atoms are taken into
account:

			Symmetry
A	f_4-R,S,S,R = I		C_2
B	f_2mf-R,S,S,S = II-1		C_1
	f_2mf-R,S,S,S = II-1		C_1
C	fm_2f-S,R,R,S = III		C_2
	fm_2f-S,S,S,S = IV		C_2
D	mf_2m-R,R,R,R = V		C_2
	mf_2m-R,R,R,S = VI		C_1
	mf_2m-S,R,R,S = VII		C_2

Among these eight diastereomers I, III, IV, V, and VII have
twofold axes of rotation. All eight have corresponding enantiomeric forms. Yoshikawa and Yamasaki (127) obtained the complex
by the reaction of the ligand and $[CoBr(NH_3)_5]Br_2$ in the presence of active charcoal. They separated all the eight diastereomers by means of column chromatography on SP Sephadex and
characterized them by their electronic and IR absorption, CD,
and PMR spectra. Isomers II-1 and II-2 were labile, isomeriz-
ing into a mixture of the two during isolation; they could not
be obtained in pure states. Unfortunately only one isomer, I,
gave good crystals for X-ray structure analysis: $(+)_{589}$-
$[Co(linpen)][Co(CN)_6]\cdot 3H_2O$. The crystal structure and
absolute configuration were determined to verify the character-
ization and to gain conformational details of the complex cation
(128). Figure 44 shows a perspective drawing of the complex ion
$(+)_{589}$-$[Co(linpen)]^{3+}$, isomer I. The ligand molecule is
coordinated to the cobalt atom with the six nitrogen atoms to
form an octahedral complex. Any three consecutive nitrogen
atoms are in the facial positions. The complex ion has an
approximate twofold axis of rotation. None of the five-membered
chelate rings is planar. The conformations of the chelate rings
are δ, λ, δ, λ, and δ, respectively. The deviations of the two
carbon atoms from the plane defined by the cobalt and the two
nitrogen atoms in each ring are not symmetrical with respect to
the plane. In Figure 45 the chelate rings A, B, and C have an
unsymmetrical skew conformation, ring D is of an eclipsed
envelope type, and in ring E the two carbon atoms are both on
the same side of the N-Co-N plane. The average value of the

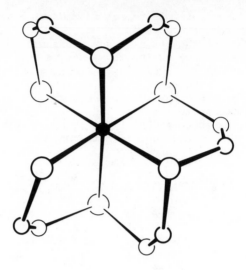

Fig. 44.   (+)$_{589}$-[Co(linpen)]$^{3+}$ (128).

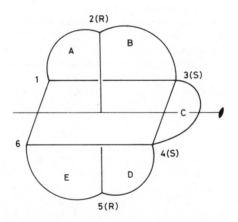

Fig. 45.   Labeling of the chelate rings and absolute configuration of the secondary nitrogen atoms in (+)$_{589}$-[Co(linpen)]$^{3+}$
(128).

dihedral angles about the C-C bonds in the ligand is 42.7°,
much smaller than that in a typical gauche structure. The
conformations of the chelate rings may be largely determined by
the nonbonded hydrogen-hydrogen interactions (129). The
absolute configuration of the complex ion may be designated as
ΛΛΛΔ, and the absolute configurations of the secondary nitrogen
atoms are (R) for N(2) and N(5), (S) for N(3) and N(4).

TABLE 4

Final Energy Terms for the Isomers of $[Co(linpen)]^{3+}$ [a]

Isomer	Absolute configuration	Proposed prefix	Bond Length deformation	Bond angle deformation	Torsional strain	Nonbonded interaction	Total Conformational energy	Formation percentage
I	ΛΛΛΔ	f_4-R,S,S,R	5.8	15.1	42.9	38.9	102.6	9
II-1	ΛΛΛΔ	f_2mf-R,S,R,S	5.6	23.3	41.0	35.7	105.6	15
II-2	ΛΛΛΔ	f_2mf-R,S,S,S	6.1	22.8	39.7	36.4	105.0	15
III	ΛΛΛΔ	fm_2f-S,R,R,S	5.6	39.9	34.5	29.9	109.9	1
IV	ΛΛΛΔ	fm_2f-S,S,S,S	5.6	34.5	32.7	31.9	104.7	2
V	ΛΛΛΔΔ	mf_2m-R,R,R,R	5.8	22.8	36.1	31.3	95.9	47
VI	ΛΛΛΔΔ	mf_2m-R,R,R,S	6.7	33.1	33.1	32.2	105.0	23
VII	ΛΛΛΔΔ	mf_2m-S,R,R,S	7.4	33.3	33.3	36.1	115.7	3

[a] In kJ/mol.

Since the crystals of the other isomers were not suitable
for X-ray work, the strain energy minimization was carried out
for all the possible diastereomers. Table 4 lists the final
energy terms and formation percentages of the isomers (129).
Isomer I contains only the facial arrangement of coordinating
nitrogen atoms. Accordingly the strain for the bond angle in it
is smaller than that in the other isomers, which contain meridional arrangement of nitrogen atoms. By the same token, isomers
III to VII, containing two meridional arrangements, have quite
large strain for angle bending with the exception of the isomer
V. The angle strain of this latter isomer seems to be alleviated
by increasing the torsional deformation. Figure 46 shows a plot

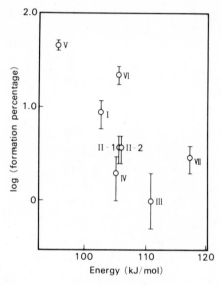

Fig. 46. Plot of log of formation percentage vs. minimized strain energy.

of formation percentage vs. minimized strain energy. As can be
seen from the figure, a very roughly linear relationship
exist, suggesting that formation percentages are largely
thermodynamically controlled. In constructing this plot, the
formation percentage of II-1 and II-2 has been divided by 2 to
take into account the statistical factor. The points for VI and
VII deviate appreciably, indicating higher than expected
abundance of these isomers. This suggests a mechanism of formation and/or interconversion in which the very high stability of
isomer V seems to result in excess formation of the VI and VII
isomers.

2.   N,N,N',N'-*Tetrakis(2'-aminoethyl)-*
     1,2-diaminoethane(penten)

The ligand 7 can act as a sexidentate one, giving com-

$$H_2N-CH_2-CH_2 \diagdown$$
$$\qquad\qquad\qquad N-CH_2-CH_2-N \diagup ^{CH_2-CH_2-NH_2}_{CH_2-CH_2-NH_2}$$
$$H_2N-CH_2-CH_2 \diagup$$

7

plexes related to those derived from ethylenediaminetetraacetic
acid (130,131). The complex is dissymmetric and can be resolved
into optical isomers. However, the absolute configuration proposed on the basis of CD spectra by Yoshikawa, Fujii and
Yamasaki (132) and by Gollogly and Hawkins (133) is enantiomeric
to that proposed by other workers (131,11). The crystal structure of $(+)_{589}$-[Co(penten)][Co(CN)$_6$]·2H$_2$O was determined (134).
Figure 47 shows a perspective drawing of the complex ion
$(+)_{589}$[Co(penten)]$^{3+}$. The six nitrogen atoms surrounding the

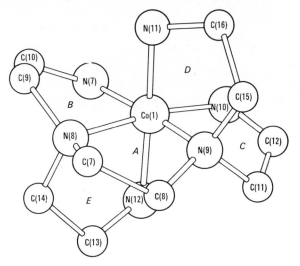

Fig. 47. $(+)_{589}$-[Co(penten)]$^{3+}$ (134).

cobalt atom form a distorted octahedron. The angles N(7)-Co(1)-
N(10) and N(9)-Co(1)-N(12) are 102.2 and 97.6°, respectively.
Five five-membered chelate rings are formed in the complex
cation. Roughly speaking, three of the five chelate rings form
a girdle about the cobalt atom. Approximately at right angles
to the girdle and to one another are two other five-membered

chelate rings. The N-Co-N angles in the chelate rings are 86°
except for N(8)-Co(1)-N(9), which is 89.5°. The deviations of
the two carbon atoms from the plane formed by the cobalt and
the two nitrogen atoms are not symmetrical, unlike those of
the chelate rings in [Co(en)$_3$]$^{3+}$. The chelate rings A and D
take the δ conformation, whereas B, C, and E are λ. The distortion of the chelate rings B and C is most noticeable: the
two carbon atoms are both on the same side of the coordination
plane. This is clearly a result of the constraints attending
multiple, as well as fused, ring formation. Nevertheless, the
disposition of bonds around the tertiary nitrogen atoms is
nearly regular tetrahedral. The absolute configuration of the
complex ion is ΛΔΛ.

3.   *(-)-N,N,N',N'-Tetrakis(2'-aminoethyl)-1,2-diaminopropane*
(mepenten)

This ligand is a methyl substituted penten *(8)*. Optically

$$H_2N-CH_2-CH_2 \diagdown_{\textstyle N-CH-CH_2-N} \diagup^{\textstyle CH_2-CH_2-NH_2}$$

H$_2$N-CH$_2$-CH$_2$⟍
⟍N-CH-CH$_2$-N⟋
H$_2$N-CH$_2$-CH$_2$⟋ | ⟍CH$_2$-CH$_2$-NH$_2$
                 CH$_3$
                      ⟋CH$_2$-CH$_2$-NH$_2$

8

active (-)-mepenten obtained from (-)-propylenediamine coordinates with cobalt to give (-)$_{589}$-[Co(-mepenten)]$^{3+}$. No
evidence for the second isomer could be found. The sexidentate
ligand complexes in completely stereospecific fashion (131).
Figure 48 illustrates the absolute configuration of (-)$_{589}$-
[Co(-mepenten)]$^{3+}$, as determined by means of X-rays (135). The
absolute configuration of the unique chelate ring A is λ, with

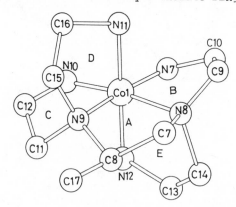

Fig. 48.   (-)$_{589}$-[Co(-metpenten)]$^{3+}$ (135).

the substituted methyl group in an equatorial position. The configuration of rings B and C is δ, whereas it is λ in D and E. In $(-)_{589}$-[Co(-mepenten)]$^{3+}$ the conformations of the chelate rings are enantiomeric with those in $(+)_{589}$-[Co(penten)]$^{3+}$ except for ring E. The δ conformation of the ring E is impossible owing to steric repulsion. The conformation of the complex cation agrees with the prediction by Gollogly and Hawkins (136) based on the strain energy minimization technique.

4. Ethylenediaminetetraacetic Acid (edta)

$$\begin{array}{l} HOOC-CH_2 \\ \diagdown \\ N-CH_2-CH_2-N \\ \diagup \\ HOOC-CH_2 \end{array} \begin{array}{l} CH_2-COOH \\ \diagup \\ \\ \diagdown \\ CH_2-COOH \end{array}$$

9

The structure of the complex ion [Co(edta)]$^-$ was determined in 1959 (121); however, its absolute configuration was established only recently by means of X-rays (87). Figure 49 shows the complex anion $(+)_{546}$-[Co(edta)]$^-$, whose absolute configuration is $\Delta\Lambda\Delta$. The central five-membered chelate ring has the asymmetric gauche conformation λ. The two glycinato cobalt rings that lie in the same plane as that of the central diamine ring are strained and nonplanar, assuming an asymmetric envelope form with δ conformation, whereas the other two glycinato chelate rings, which are nearly perpendicular to the first two, also take envelope forms but are less strained than

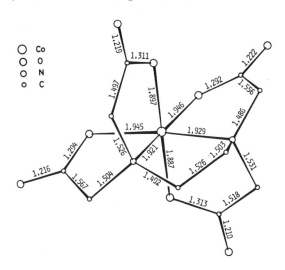

Fig. 49. $(+)_{546}$-[Co(edta)]$^-$ (87).

the former (121). That this strain can influence chemical behavior has been demonstrated in a structural study of the penta-
coordinate edta complex of the larger cation, Ni(II), in
which one of the in-plane acetate arms fails to coordinate
(137). Also the α-carbon protons of the out-of-plane glycinate
ring of [Co(edta)]⁻ and similar metal chelates exhibit a much
more rapid rate of H-D exchange than those of the corresponding in-plane chelate ring (138-141).

5. Trimethylenediaminetetraacetic Acid (trdta)

Cobalt(III) and chromium(III) complexes with trimethylene-
diaminetetraacetic acid, H_4trdta (10), were synthesized and the

$$HOOC-CH_2 \diagdown \diagup CH_2-COOH$$
$$N-CH_2-CH_2-CH_2-N$$
$$HOOC-CH_2 \diagup \diagdown CH_2-COOH$$

10

ligand was shown to act as a sexidentate ligand in both cases
(142,143). The potassium salt K[Co(trdta)]·$2H_2O$ was found to
undergo spontaneous resolution at room temperature. Figure 50
presents a perspective drawing of the complex ion $(-)_{546}$-
[Co(trdta)]⁻ (144).

The shape of the complex ion is broadly similar to that
of [Co(edta)]⁻ (87,121). It has a twofold axis of rotation

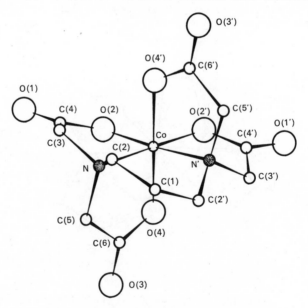

Fig. 50. $(-)_{546}$-[Co(trdta)]⁻ (144).

through the cobalt atom and the carbon atom of the central methylene group. A glycinic chelate ring that lies in the plane of the six-membered chelate ring, forming a girdle around the cobalt atom, shows significant departure from planarity, and O(1) is 0.18 Å out of the plane of the chelate ring. This strain is similar to but not as great as that observed in [Co(edta)]⁻ (121). Another crystallographically independent ring that lies in a plane nearly perpendicular to the girdle is almost planar, and O(3) deviates by about 0.11 Å from the average plane of that chelate ring. The six-membered chelate ring assumes a twist-boat form with the absolute configuration δ. The whole absolute configuration of the complex anion is ΛΔΛ.

6. *1,3,6,8,10,13,16,19-Octaazabicyclo[6.6.6]eicosane (Sepulchrate, sep)*

The ligand *11* acts as a sexidentate one, and the metal ions

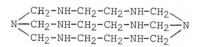

11

are encapsulated in the cage-shaped ligand. The cobalt(III) complex [Co(sep)]³⁺ was prepared by condensation of [Co(en)₃]³⁺ with formaldehyde and ammonia. The crystal structure of (-)₅₈₉-[Co(sep)]Cl₃·H₂O was determined (182). The complex ion has approximate overall D_3 symmetry. Its geometry can be most conveniently described as a Λ(δδδ)-*lel*₃-[Co(en)₃]³⁺ ion to which the two tris(methylene)amino caps are added at both ends. The absolute configuration of the six ligating nitrogen atoms is S. The encapsulation of metal ions in the cage ligand inhibits ligand substitution but offers interesting prospects for the study of intramolecular rearrangements, electron transfer, and spectroscopic properties. The complex ion [Co(sep)]³⁺ can be reduced to [Co(sep)]²⁺ with retention of absolute configuration. [Co(sep)]²⁺ can be easily oxidized to [Co(sep)]³⁺. This is essentially a reversible redox phenomenon. The measured electron transfer rate was about 10⁵-fold greater than that for [Co(en)₃]²⁺/[Co(en)₃]³⁺. The reason for this pronounced difference is not yet fully understood.

VI. EMPIRICAL RULES RELATING ABSOLUTE CONFIGURATION AND CD SPECTRUM

The absolute configuration of transition metal complexes can be determined by using the X-ray anomalous scattering method. The absolute configurations of about 90 complexes have now been established by means of X-rays. The known absolute configurations of the complexes thus determined taken in conjunction with CD studies allow the configurations of the enantiomers of a number of complexes to be settled with reasonable certainty. The alternative method for determining absolute configuration is the nonempirical calculation of exciton CD. There are, however, some necessary premises for this method to be applied successfully. In this section some empirical rules are reviewed that are mainly applicable to cobalt(III) complexes containing nitrogen atoms as ligating atoms.

In 1955 it was shown that the tris(ethylenediamine)cobalt (III) isomer, $(+)_{589}-[Co(en)_3]^{3+}$, which is dextrorotatory at the sodium D line, has the Λ configuration (4). An absolute basis was thus provided for the empirical relation of Mathieu, who had proposed that tris-chelated complexes having the same configuration as $(+)_{589}-[Co(en)_3]^{3+}$ give a predominantly positive CD in the longest-wavelength absorption (145). The absorption spectra of hexaammine complexes of cobalt(III) containing the octahedral $[CoN_6]$ chromophore consist of two weak ligand-field bands, one in the visible (the first absorption band), the other in the near UV (the second absorption band), plus a strong ligand-to-metal charge-transfer band in the far UV (CT band). In solution the optical activity associated with the first absorption band is generally more pronounced than that associated with the second absorption band. In the case of $[Co(en)_3]^{3+}$ it consists of a major positive and a minor negative CD band. These bands are ascribed to $^1A_1 \rightarrow {}^1E$ and $^1A_1 \rightarrow {}^1A_2$ transitions of O_h parentage in a D_3 environment. The crystal measurements showed that the observed solution CD spectra are the residual wing absorptions resulting from extensive cancellation of the large rotatory strengths of the two transitions E and A_2. The longer-wavelength band was assigned to that of E and the shorter one to that of A_2 symmetry (146). The CD associated with the first absorption band is diagnostic of the absolute configuration of the complex ion. Table 5 shows the CD spectra of some tris-bidentate cobalt(III) complexes having five-membered chelate rings. With the sole exception of $(+)_{589}-[Co(cptn)_3]^{3+}$ those complexes that show prominent positive CD in the first absorption region possess Λ absolute configuration. The symmetry of the longer-wavelength bands is E (147).

Table 6 lists the CD bands of tris-diamine complexes containing six-membered chelate rings. Except for $[Co(tn)_3]^{3+}$ the same empirical rule described earlier holds for these complexes. The CD spectrum of $\Delta-[Co(tn)_3]^{3+}$ in solution changes with temperature. It consists of two bands of opposite sign at room temperature, the longer-wavelength band being negative. On lowering the temperature the longer-wavelength negative peak

TABLE 5

CD Spectra of Tris-diamine Complexes of Co(III) with Five-Membered Chelate Rings

Complex	CD		Reference	Absolute Configuration (X-ray method)	Reference
	$10^3\tilde{\nu}$ cm^{-1}	$\Delta\varepsilon$			
$(+)_{589}$-[Co(en)$_3$]$^{3+}$	20.28	+1.89	148	$\Lambda(\delta\delta\delta)$ lel_3	4
	23.37	-0.17			
$(+)_{589}$-[Co(S-pn)$_3$]$^{3+}$	20.28	+1.95	148	$\Lambda(\delta\delta\delta)$ lel_3	20
	22.78	-0.58			
$(+)_{589}$-[Co(R-pn)$_3$]$^{3+}$	21.0	+2.47	149	$\Lambda(\lambda\lambda\lambda)$ ob_3	23
$(-)_{589}$-[Co(S,S-chxn)$_3$]$^{3+}$	20.0	+2.28	150	$\Lambda(\delta\delta\delta)$ lel_3	26
	22.5	-0.69			
$(+)_{589}$-[Co(R,R-chxn)$_3$]$^{3+}$	20.8	+3.9	150	$\Lambda(\lambda\lambda\lambda)$ ob_3	27
$(+)_{589}$-[Co(S,S-cptn)$_3$]$^{3+}$	18.9	+0.59	32	$\Lambda(\delta\delta\delta)$ lel_3	32
	21.1	-1.91			

TABLE 6

CD Spectra of Tris-diamine Cobalt(III) Complexes with Six-Membered Chelate Rings

Complex	CD		Reference	Absolute Configuration	Reference
	$10^3 \bar{\nu}$ cm^{-1}	$\Delta\varepsilon$			
$(-)_{589}-[\text{Co(tn)}_3]^{3+}$	18.69 30.0	+0.08 −0.17	151	$\Lambda(\underline{ppp})$	33
$(-)_{546}-[\text{Co}(meso-\text{ptn})_3]^{3+}$	20.0	−6.1	40	$\Delta(\underline{ppp})$	45
$(-)_{546}-[\text{Co}(R,R-\text{ptn})_3]^{3+}$	19.6	−6.2	40	$\Delta(\lambda\lambda\lambda)$, lel_3	41
$(+)_{546}-[\text{Co}(R,R-\text{ptn})_3]^{3+}$	20.9	+26.8	40	$\Lambda(\lambda\lambda\lambda)$, ob_3	42

progressively increases in strength, with concomitant
diminution of the positive peak. This observation may be
related to the flexibility of the unsubstituted six-membered
chelate rings: there may be conformational equilibrium in
solution between a tris-skew boat and a tris-chair form (37).
   The absorption and CD spectra of Λ-$[Co\{H_2N(CH_2)_nNH_2\}_3]^{3+}$
[n = 2, 3 (151), 4 (50)] are compared in Figure 51. The CD

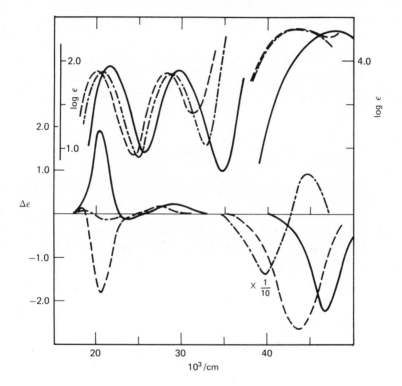

Fig. 51.   Absorption and CD spectra of Λ forms of $[Co(en)_3]^{3+}$
(———), $[Co(tn)_3]^{3+}$ (–•–•–), and $[Co(tmd)_3]^{3+}$ (————) (50).

spectrum in the first absorption region changes markedly with
n; however, the sign of the longer-wavelength band is always
positive and that of the first CD band in the UV region is
negative throughout the series. The longer-wavelength positive
CD band of $[Co(tmd)_3]^{3+}$ (n = 4) was found to have E symmetry by
single-crystal CD measurement (147).
   In $[Co(en)_3]^{3+}$ the transition of E symmetry dominates the
sign of the CD for the first absorption region. The sign will
be carried over to the two $A(E)$ levels and A_2 in the complex
ion $[Co(amino\ acid)(en)_2]^{2+}$ ([CoN$_5$O] chromophore, C_{4v}). These

suggestions were supported by the structural analyses of
$[Co(sar)(en)_2]^{2+}$ (112) and $(+)_{495}-[Co(S-glut)(en)_2]^+$ (107).
Thus all complexes in this series having a positive CD peak at
500 nm may be safely assigned the same absolute configuration
(see Table 7). Furthermore $(+)_{589}-\Lambda-\alpha-mer-[Co(S-ala)_3]$ (59)
shows a dominant positive CD band on the longer-wavelength side
of the first absorption region (Table 7). This suggests that the
series of complexes $mer-[Co(amino\ acid)_3]$ may all be assigned
the Λ configuration if a predominant positive CD band is
observed on the longer-wavelength side of the first absorption
region.

Table 8 lists the CD data of four tris-bidentate complexes
with Λ absolute configuration involving $[Mo_6]$ or $[MS_6]$ chromophore. It may be said that those complexes exhibiting prominent
positive CD in the first absorption region have Λ absolute configuration.

A. cis-Bis-bidentate Complexes, $cis-[CoX_2(en)_2]^{n+}$

The replacement of one chelate ligand in a tris-bidentate
complex (of D_3 symmetry) by X_2 converts it into a $[CoN_4X_2]$ type
chromophore of C_2 symmetry. In the C_2 ligand field the first
absorption arises from two transitions with B symmetry and one
with A symmetry. The A transition arises from desymmetrization
of the E transition in the C_2 field. This lowering of symmetry
results in the shifting and splitting of absorption bands. The
corresponding CD spectra vary in appearance with the nature
of X. Table 9 lists the CD spectra of some cis-bis-bidentate
complexes whose absolute configurations were established by
means of X-ray analysis. The observed CD spectrum is interpreted
as an unresolved composite of $A_2 + B_2$ (E) and B_1 (A_2). The sign
of the former is that of the parent (E) transition, namely
positive for Λ absolute configuration (155). The shifting and
splitting of the energy levels depend on the nature of the
ligand. The spectrochemical series provides a convenient
arrangement of the various ligands in order of increasing
ligand field splitting of the energy levels for a fixed metal
ion. Thus the ligand CN or NO_2 is behind en in the spectrochemical series, and the prominent $A_2 + B_2$ band lies on the
longer-wavelength side in the first absorption region. In the
case of $[CoCl_2(en)_2]^+$, Cl comes prior to en in the series, and
the positive $A_2 + B_2$ band is shifted to the shorter-wavelength
side of the first absorption band. This means that the same
empirical rule may be applied to these complexes as that which
correlates the absolute configuration of $[Co(en)_3]^{3+}$ and its CD
spectra.

TABLE 7

CD Spectra of Amino Acid Complexes

| Complex | CD | | | Absolute Configuration | Reference |
	$10^3\bar{\nu}$ cm^{-1}	$\Delta\varepsilon$	Reference		
$(-)_{589}-[Co(sar)(en)_2]^{2+}$	19.4	-1.8^a	152	Δ	112
$(+)_{495}-[Co(S-glut)(en)_2]^{2+}$	19.6	$+2.5^a$	109	Λ	107
	22.7	-0.1			
$(+)_{589}-[Co(S-ala)_3]$	18.5	$+1.3$	153	Λ	59
	21.0	-0.2			

^{a}Taken from figure.

157

TABLE 8

CD Spectra of Tris-bidentate Complexes Containing O or S as Ligating Atoms

Complex	CD			Absolute Configuration	Reference
	$10^3 \bar{\nu}$ cm^{-1}	$\Delta\varepsilon$	Reference		
$(-)_{589}-[Co(ox)_3]^{3-}$	16.2	+3.3	146	Λ	55
$(+)_{546}-[Co(thiox)_3]^{3-}$	15.8	-0.2	154	Λ	58
	18.9	+2.8			
$(+)_{589}-[Cr(ox)_3]^{3-}$	15.9	-0.6	146	Λ	46
	18.1	+2.8			
$(+)_{589}-[Cr(mal)_3]^{3-}$	16.1	-0.07	146	Λ	46
	18.0	+0.20			

TABLE 9

CD Spectra of *cis*-Bis-bidentate Cobalt(III) Complexes

Complex	CD			Absolute configuration	Reference
	$10^3 \bar{\nu}$ cm^{-1}	$\Delta\varepsilon$	Reference		
(+)$_{589}$-[CoCl$_2$(en)$_2$]$^+$	16.3	-0.6	155	Λ	66
	18.6	+0.7			
(-)$_{589}$-[Co(NO$_2$)$_2$(en)$_2$]$^+$	21.7	-1.4	155	Δ	48
	25.0	-0.65			
(+)$_{589}$-[Co(CN)$_2$(en)$_2$]$^+$	22.7	+0.30	155	Λ	156
	27.3	+0.17			
(+)$_{589}$-[Co(NO$_2$)$_2$(R-pn)$_2$]$^+$	21.7	-1.1a	157	Δ	68
	24.5	+0.6			

aTaken from figure.

B. Multidentate Complexes

Correlation of the absolute configuration of multidentate complexes with their CD spectra may be achieved using two methods. One is Hawkins and Larsen's octant rule (158); the other is the "ring pairing method" of Legg and Douglas (159). The two methods are essentially equivalent and both are empirical. Thus only the ring pairing method, which is more convenient to apply, is briefly described here. For a given complex all possible combinations of the two chelate rings are written down and the chirality (Λ or Δ) according to IUPAC convention (1,2) of each set is determined. The net (or dominant) chirality should be governed by the chirality that occurs the greatest number of times. If the net chirality is Λ, the CD spectrum of a cobalt(III) complex will show a positive Cotton effect in the longer-wavelength band in the region of the octahedral T_{1g} absorption ($^1A_{1g} \rightarrow {}^1T_{1g}$); if it is Δ, the longer-wavelength CD band will have a negative sign. In Legg and Douglas' original paper the use of Δ and Λ is opposite to that of IUPAC nomenclature used here. For example, $(+)_{589}$-[Co(penten)]$^{3+}$ has the absolute configuration $\Lambda\Delta\Lambda$, which gives the net chirality Λ. The complex shows a positive CD at the longer-wavelength side of the first absorption region, as shown in Table 10.

Table 10 lists the CD data in the first absorption region of complexes containing multidentate ligands. As shown in the table, the ring-pairing method covers the chromophores [CoN$_6$], [CoN$_4$O$_2$], and [CoN$_2$O$_4$], and the signs of the Cotton effect in the first absorption region can be correlated with the net chirality. In addition to these rules, more refined regional rules correlating the position of a substituent to tetragonal or octahedral chromophores (160-162) have been devised to predict the optical activity of the d-d transitions of these complexes. They require, however, somewhat more detailed geometrical information about the complex, hence, are less useful in predicting the absolute configuration on the basis of CD.

The ligand 2,3,2-tet and related linear quadridentate ligands are of interest in explaining the origins of optical activity in *trans*-diacidato cobalt(III) complexes. The optical properties of these complexes arise from the conformations of the chelate rings, since no net chirality is associated with these materials (163). A series of optically active linear quadridentate complexes of cobalt(III) containing different ring sizes have been described (163) and their CD spectra interpreted on the basis of chelate ring conformations (164-166). No simple empirical rule has yet been obtained.

An example is presented here to illustrate the case in which the assignment of the absolute configuration on the basis of CD is ambiguous. Figure 52 shows the CD spectra for

TABLE 10

CD Data for Multidentate Complexes of Cobalt(III)

	CD			Absolute configuration	Net chirality	Reference
	$10^3\tilde{\nu}$ cm^{-1}	$\Delta\varepsilon$	Reference			
(+)$_{589}$-[Co(linpen)]$^{3+}$	19.9 22.5	+1.68 -0.42	127	ΛΛΛΔ	Λ	128
(+)$_{589}$-[Co(penten)]$^{3+}$	19.6 22.3	+3.61 -0.49	130	ΛΔΔ	Λ	134
(−)$_{589}$-[Co(R-mepenten)]$^{3+}$	19.6 22.0	-3.31 +0.79	130	ΔΛΔ	Δ	135
(+)$_{546}$-[Co(edta)]$^{-}$	17.0 19.4	-1.7 +0.9	178	ΔΛΔ	Δ	87
(+)$_{546}$-[Co(trdta)]$^{-}$	16.7 18.9	-1.7[a] +2.5	142	ΔΛΔ	Δ	144
[Co(edda)(R-pn)]$^{-}$	20.3	-1.7	179	ΔΔΔΔ	Δ	120

[a]Taken from figure.

161

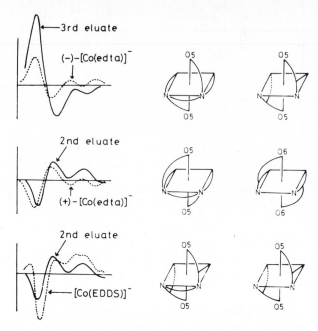

Fig. 52. CD spectra of the two *cis*(*N*) isomers of
[Co(*S*-asp)$_2$]$^-$ and of the related complexes used as reference
complexes.

two of the possible three isomers of [Co(S-asp)$_2$]$^-$. They are
cis(N) isomers, separated by ion-exchange column chromatography. In these complexes the three rings join on the face of
an octahedron at the asymmetric carbon atom in such a manner
that the rings do not define the edges of the octahedron (Fig.
21). In fact, two opposite assignments were made. On the basis
of the CD and PMR studies of (+)$_{546}$- and (-)$_{546}$-[Co(edta)]$^-$ of
known absolute configuration, the second and the third
eluates were assigned as *cis*(N)-*trans*(O$_6$) and *cis*(N)-*trans*(O$_5$),
respectively (167). On the other hand, opposite assignments
were made on the basis of CD and PMR studies of the closely
related (*S,S*)-ethylenediamine-*N,N'*-disuccinic acid complex
of cobalt(III), [Co(edds)]$^-$, in which three chelate rings
join on the face of an octahedron at noncoordinating atoms
(168). As seen in Figure 52, pretty good correlations
are recognized between CD spectra of the reference complexes
and those of the isomers in question. The X-ray study verified
the latter assignment for the two isomers: the second eluate
cis(N)-*trans*(O$_5$), the third eluate *cis*(N)-*trans*(O$_6$) (84).

1. Ion Pairing

CD spectra are generally measured in solution, but sometimes they are measured in KBr or polystyrene matrices or in
single crystals. A number of resolved tris-diamine complexes
exhibit marked changes in CD when various electrolytes are
added. This effect has been extensively studied, and the
appreciable change caused by the addition of oxo anions such
as selenate or phosphate ions is now one of the useful methods
of assigning the symmetry of CD bands in the first absorption
region (169). The interpretation has been that the selenate or
phosphate ion forms a specific ion pair (10,170,171).

On the other hand, the influence of the counterions that
comprise the complex salts has been considered to be negligibly
small. This is generally true for those complexes with five-
membered chelate rings. Δ-(lel)$_3$-[Co(R,R-ptn)$_3$]$^{3+}$, however,
affords an exception to this general observation. The solution
CD spectra of this complex ion are seriously affected by its
counterions (172).

The CD spectra in solution of some tris-chelated complexes
containing six-membered chelate rings differ from those in KBr
matrix.

2. Exciton CD

In a tris-bidentate complex containing unsaturated ligands
a major source of optical activity is the coulombic coupling
of the allowed $\pi \rightarrow \pi^*$ transitions in the individual ligands.
This coupling gives rise to component transitions that are
intrinsically optically active and that have Cotton effects
whose signs can be determined by the phase relationships of
the individual dipoles (173,174). The following are typical
unsaturated ligands:

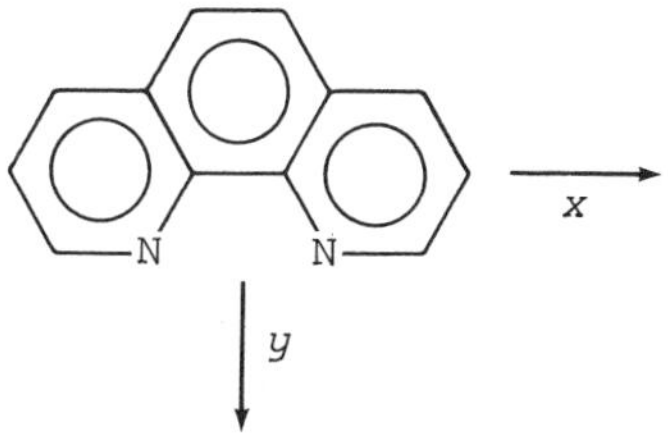

phen

12

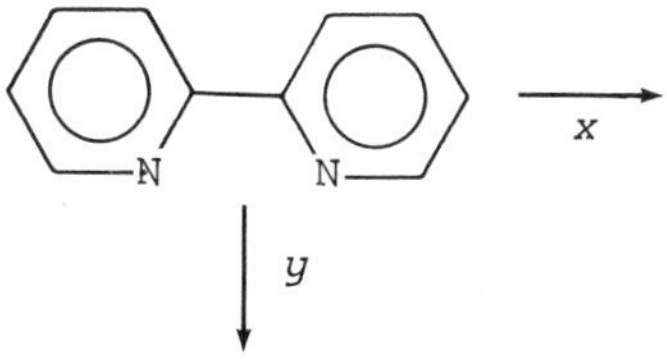

bpy

13

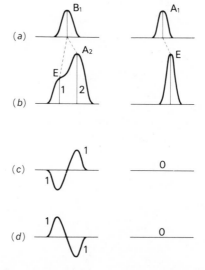

cat

14

acac

15

They possess C_{2v} symmetry, and the electric moments associated with their $\pi \to \pi^*$ transition are parallel to the long axis of the ligand (x axis) or perpendicular to it (y axis) and in the molecular plane. When three such ligands enter into combination with a metal atom to form a trigonal complex having D_3 symmetry, the x-polarized transitions of the three ligands in the

Fig. 53.   Absorption spectra and exciton CD bands of a tris-bidentate complex containing unsaturated ligands: (*a*) absorption spectrum of the ligand; (*b*) absorption spectrum of tris-chelated complex; (*c*) CD spectrum of a Δ form; (*d*) CD spectrum of a Λ form. The left side of the figure shows the case when the transition moment is parallel to the x axis; the right side, when the moment is parallel to the y axis. The wave number increases toward the right side of the abscissa. Numerals stand for relative intensity.

complex couple to each other and give rise to two electronic
transitions, $A_1 \rightarrow A_2$ and $A_1 \rightarrow E$. On the other hand the y-
polarized transition cannot produce zero-order rotatory power,
since it has no magnetic moment. It was found that the $A_1 \rightarrow E$
band appears at longer wavelength than the $A_1 \rightarrow A_2$ band (175,
176). The theory predicts that these two transitions will
give rise to typical exciton CD bands having opposite signs and
equal magnitudes. The signs of the two bands depend on the
absolute configuration of the complex, as illustrated in
Figure 53. The X-ray determinations of the absolute configurations of such complexes verified that the exciton treatment
indeed leads to the correct configuration. One example is
illustrated. 1,2-Benzenediol exhibits three electronic
absorption bands in the UV region. Figure 54 (broken line)
illustrates its absorption spectrum with the intensity scale
multiplied by a factor of 3 (curve 2). The full line indicates
the absorption spectrum of $(-)_{589}$-K[As(cat)$_3$]. It also contains

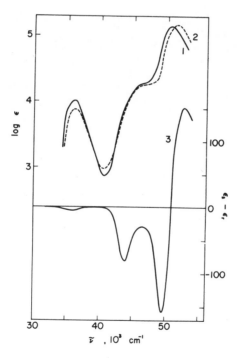

Fig. 54. Absorption spectrum of Catechol (1,2-benzenediol)
in aqueous solution (curve 2) (with the intensity scale multiplied by a factor of 3), the absorption spectrum (curve 1), and
the CD spectrum (curve 3) of $(-)_{589}$-K[As(cat)$_3$].1.5H$_2$O in
aqueous solution (177).

three bands with slightly different frequencies, indicating that
the metal ligand interaction by charge transfer is small. The
CD spectrum (curve 3) resembles typical exciton CD spectra in
the region of 52×10^3 cm^{-1}. The longer-wavelength CD band
has negative sign, and the absolute configuration can be
designated as Δ, in agreement with the result of X-ray structure determination (57). Thus the exciton treatment of optically
active coordination compounds affords reliable stereochemical
assignments of absolute configuration. This method is, however,
subject to two major qualifications: (1) It is essential that
the absorption bands of the free ligand be distinguishable
in the spectrum of the complex and appear with an intensity
appropriate to the number of ligands in the complex at frequencies close to the values for the ligand. (2) It is necessary
that the CD connected with a given ligand absorption band have
the characteristic exciton form shown in Figure 53; otherwise
it is difficult to identify it.

To sum up, it may be said that the absolute configurations of transition metal complexes can be assigned with reasonable certainty on the basis of their CD spectra. In making such
assignments reference complexes are needed for which both
crystal and molecular structure and CD data are known in detail.
Preferably these reference compounds should have high conformational and configurational stability to minimize the possibility of alteration during phase changes from solid to solution.
Reference complexes having the same chromophore should be used
to assign relative configuration.

VII. CONCLUSIONS

This chapter surveys the structure and conformation of
optically active chelate complexes whose absolute configurations
were established by means of X-rays. The choice of topics has
been largely determined by the author's own interests. As a
result of the accumulation of structural knowledge on metal
chelate complexes, it is now possible to predict with reasonable
certainty the conformation and strain energy of an unknown complex. A consistent force-field approach, as used for biopolymers
(183,184), is desirable for metal chelate complexes and, in fact,
such a program is now being elaborated (185). The absolute configuration of the complex can be determined on the basis of its
CD spectrum if a reference complex of known absolute configuration is appropriately selected. The cobalt(III) complexes containing nitrogen as ligating atoms have been most extensively
studied. It is hoped that complexes containing metals other
than cobalt and ligating atoms other than nitrogen will be
investigated in similar detail.

Recent improvements in experimental and computational
techniques in X-ray crystallography have made it possible to

estimate the atomic charge density in a transition metal complex, based on accurate intensity data (186). This information, combined with the absolute configuration, will eventually enable one to rationalize the stereospecificity of reactions of metal chelate complexes at the quantum mechanical level.

REFERENCES

1. *Nomenclature of Inorganic Chemistry*, 2nd ed., Butterworths, London, 1970, pp. 75-83.
2. *Inorg. Chem.*, 9, 1 (1970).
3. E. J. Corey and J. C. Bailar, Jr., *J. Am. Chem. Soc.*, 81, 2620 (1959).
4. Y. Saito, K. Nakatsu, M. Shiro, and H. Kuroya, *Acta Crystallogr.*, 8, 729 (1955); *Bull. Chem. Soc. Jpn.*, 30, 795 (1957).
5. K. Nakatsu, *Bull. Chem. Soc. Jpn.*, 35, 832 (1962).
6. M. Iwata, K. Nakatsu, and Y. Saito, *Acta Crystallogr.*, B25, 2562 (1969).
7. S. Baggio and L. N. Becka, *Acta Crystallogr.*, B25, 946 (1969).
8. R. J. Williams, A. C. Larson, and D. T. Cromer, *Acta Crystallogr.*, B28, 858 (1972).
9. H. Shintani, S. Sato, and Y. Saito, *Acta Crystallogr.*, B32, 1184 (1976).
10. S. F. Mason and B. J. Norman, *Proc. Chem. Soc.*, 1964, 339.
11. S. F. Mason and B. J. Norman, *Chem. Commun.*, 1965, 73.
12. A. J. McCaffery, S. F. Mason, and B. J. Norman, *Chem. Commun.*, 1965, 49.
13. J. K. Beattie, *Acc. Chem. Res.*, 4, 253 (1971).
14. K. N. Raymond, P. W. R. Corfield, J. A. Ibers, *Inorg. Chem.*, 7, 1362 (1968).
15. K. N. Raymond and J. A. Ibers, *Inorg. Chem.*, 7, 2333 (1968).
16. A. Whuler, C. Brouty, P. Spinat, and P. Herpin, *Acta Crystallogr.*, B31, 2069 (1975).
17. H. J. Haupt, F. Huber, and H. Preut, *Z. Anorg. Allg. Chem.*, 422, 255 (1976).
18. S. E. Harnung, S. Kallesoe, A. M. Sargeson, and C. E. Schäffer, *Acta Chem. Scand.*, A28, 385 (1974).
19. F. P. Dwyer and F. L. Garvan, *J. Am. Chem. Soc.*, 81, 1043 (1959).
20. H. Iwasaki and Y. Saito, *Bull. Chem. Soc. Jpn.*, 39, 92 (1966).
21. T. E. MacDermott, *Inorg. Chim. Acta*, 2, 81 (1968).
22. M. Kojima, Y. Yoshikawa, and K. Yamasaki, *Inorg. Nucl. Chem. Lett.*, 9, 689 (1973).
23. R. Kuroda and Y. Saito, *Acta Crystallogr.*, B30, 2126 (1974).
24. R. Kuroda and Y. Saito, unpublished work.

25. F. P. Dwyer, T. E. MacDermott, and A. M. Sargeson, *J. Am. Chem. Soc.*, 85, 2913 (1963).

26. F. Marumo, Y. Utsumi, and Y. Saito, *Acta Crystallogr.*, B26, 1492 (1970).

27. A. Kobayashi, F. Marumo, and Y. Saito, *Acta Crystallogr.*, B28, 2709 (1972).

28. S. Sato and Y. Saito, *Acta Crystallogr.*, B33, 860 (1977).

29. S. E. Harnung, B. Søndergaard Sørensen, I. Creaser, H. Malgaard, U. Pfenninger, and C. E. Schäffer, *Inorg. Chem.*, 15, 2123 (1976).

30. F. M. Jaeger and H. B. Blumendal, *Z. Anorg. Chem.*, 175, 161 (1928).

31. J. F. Phillips and D. F. Royer, *Inorg. Chem.*, 4, 616 (1965).

32. M. Ito, F. Marumo, and Y. Saito, *Acta Crystallogr.*, B27, 2187 (1971).

33. T. Nomura, F. Marumo, and Y. Saito, *Bull. Chem. Soc. Jpn.*, 42, 1016 (1969).

34. R. Nagao, F. Marumo, and Y. Saito, *Acta Crystallogr.*, B29, 2438 (1973).

35. J. R. Gollogly and C. J. Hawkins, *Inorg. Chem.*, 11, 156 (1972).

36. S. R. Niketić and F. Woldbye, *Acta Chem. Scand.*, 27, 621 (1973).

37. P. G. Beddoe, M. J. Harding, S. F. Mason, and B. J. Peart, *Chem. Commun.*, 1971, 1283.

38. F. A. Jurnak and K. N. Raymond, *Inorg. Chem.*, 11, 3149 (1972).

39. F. A. Jurnak and K. N. Raymond, *Inorg. Chem.*, 13, 2387 (1974).

40. F. Mizukami, H. Ito, J. Fujita, and K. Saito, *Bull. Chem. Soc. Jpn.*, 43, 3633, 3973 (1970).

41. A. Kobayashi, F. Marumo, and Y. Saito, *Acta Crystallogr.*, B29, 2443 (1973).

42. A. Kobayashi, J. Marumo, and Y. Saito, *Acta Crystallogr.*, B28, 3591 (1972).

43. I. Oonishi, S. Sato, and Y. Saito, *Acta Crystallogr.*, B30, 2256 (1974).

44. M. Kojima and J. Fujita, *Chem. Lett.*, 1976, 429.

45. S. Sato and Y. Saito, *Acta Crystallogr.*, B34, 420 (1978).

46. K. R. Butler and M. R. Snow, *Chem. Commun.*, 1971, 550.

47. K. R. Butler and M. R. Snow, *J. Chem. Soc. Dalton, Trans.*, 1976, 251.

48. K. Matsumoto and H. Kuroya, *Bull. Chem. Soc. Jpn.*, 45, 1755 (1972).

49. S. Sato and Y. Saito, *Acta Crystallogr.*, B31, 1378 (1975).

50. J. Fujita and H. Ogino, *Chem. Lett.*, 1974, 58.

51. F. Galsbøl, P. Steenbøl, and B. S. Sørensen, *Acta Chem. Scand.*, 26, 3605 (1972).

52. R. Kuroda, Y. Sasaki, and Y. Saito, *Acta Crystallogr.*, B30, 2053 (1974).

53. H. Miyamae, S. Sato, and Y. Saito, *Acta Crystallogr.*, B33, 3391 (1977).

54. A. Zalkin, D. H. Templeton, and T. Ueki, *Inorg. Chem.*, 12, 1641 (1973).

55. K. R. Butler and M. R. Snow, *J. Chem. Soc.*, A, 1971, 565.

56. A. Wada, C. Katayama, and J. Tanaka, *Acta Crystallogr.*, B32, 3194 (1976).

57. A. Kobayashi, T. Ito, F. Marumo, and Y. Saito, *Acta Crystallogr.*, B28, 3446 (1972).

58. K. R. Butler and M. R. Snow, *Inorg. Nucl. Chem. Lett.*, 8, 541 (1972).

59. M. G. B. Drew, J. H. Dunlop, R. D. Gillard, and D. Royers, *Chem. Commun.*, 1966, 42.

60. R. B. von Dreele and R. C. Hay, *J. Am. Chem. Soc.*, 93, 4936 (1971).

61. G. R. Brubaker and W. E. Webb, *J. Am. Chem. Soc.*, 91, 7199 (1969).

62. W. DeW. Horrocks, Jr., D. J. Johnston, and D. MacInnes, *J. Am. Chem. Soc.*, 92, 7620 (1970).

63. H. V. F. Schousboe-Jensen, *Acta Chem. Scand.*, 26, 3413 (1972).

64. T. Aoki, K. Matsumoto, S. Ooi, and H. Kuroya, *Bull. Chem. Soc. Jpn.*, 46, 159 (1973).

65. K. Matsumoto, H. Kawaguchi, H. Kuroya, and S. Kawaguchi, *Bull. Chem. Soc. Jpn.*, 46, 2424 (1973).

66. K. Matsumoto, S. Ooi, and H. Kuroya, *Bull. Chem. Soc. Jpn.*, 43, 3801 (1970).

67. C. F. Liu and J. A. Ibers, *Inorg. Chem.*, 9, 773 (1970).

68. G. A. Barclay, E. Goldschmied, and N. C. Stephenson, *Acta Crystallogr.*, B26, 1559 (1970).

69. R. Herak, B. Prelesnik, L. J. Manojlović-Muir, *Acta Crystallogr.*, B30, 229 (1974).

70. H. C. Freeman, *Adv. Protein Chem.*, 22, 257 (1967).

71. W. H. Watson, D. R. Johnson, M. B. Celap, and B. Kamberi, *Inorg. Chim. Acta*, 6, 591 (1972).

72. K. Matsumoto, M. Yonezawa, H. Kuroya, H. Kawaguchi, and S. Kawaguchi, *Bull. Chem. Soc. Jpn.*, 43, 1269 (1970).

73. Y. Saito and H. Iwasaki, *Bull. Chem. Soc. Jpn.*, 35, 1131 (1962).

74. W. T. Robinson, D. A. Buckingham, G. Chandler, L. G. Marzilli, and A. M. Sargeson, *Chem. Commun.*, 1969, 539.

75. R. G. Ball, N. J. Bowman, and N. C. Payne, *Inorg. Chem.*, 15, 1704 (1976).

76. K. Matsumoto, S. Ooi, M. Sakuma, and H. Kuroya, *Bull. Chem. Soc. Jpn.*, 49, 2129 (1976).

77. T. Ito, F. Marumo, and Y. Saito, *Acta Crystallogr.*, B27, 1062 (1971).

78. F. R. Keene, G. H. Searle, Y. Yoshikawa, A. Imai, and K. Yamasaki, *Chem. Commun.*, 1970, 784.

79. M. Kobayashi, F. Marumo, and Y. Saito, *Acta Crystallogr.*, B28, 470 (1972).

80. M. Konno, F. Marumo, and Y. Saito, *Acta Crystallogr.*, B29, 739 (1973).

81. K. Hosaka, H. Nishikawa, and M. Shibata, *Bull. Chem. Soc. Jpn.* 42, 277 (1969).

82. S. Yamada, J. Hidaka, and B. E. Douglas, *Inorg. Chem.*, 10, 2187 (1971).

83. I. Oonishi, M. Shibata, F. Marumo, and Y. Saito, *Acta Crystallogr.*, B29, 2448 (1973).

84. I. Oonishi, S. Sato, and Y. Saito, *Acta Crystallogr.*, B31, 1318 (1975).

85. W. A. Freeman and C. F. Liu, *Inorg. Chem.*, 7, 764 (1968).

86. C. F. Liu and J. A. Ibers, *Inorg. Chem.*, 8, 1911 (1969).

87. K. Okamoto, T. Tsukihara, J. Hidaka, and Y. Shimura, *Chem. Lett.* 1973, 145.

88. R. M. Wing and E. Eiss, *J. Am. Chem. Soc.*, 92, 1929 (1970).

89. S. F. Mason and R. D. Peacock, *Inorg. Chim. Acta*, 19, 75 (1976).

90. M. Mikami, R. Kuroda, M. Konno, and Y. Saito, *Acta Crystallogr.*, B33, 1485 (1977).

91. D. A. Buckingham, P. A. Marzilli, and A. M. Sargeson, *Inorg. Chem.*, 6, 1032 (1967).

92. H. C. Freeman and I. E. Maxwell, *Inorg. Chem.*, 8, 1293 (1969).

93. S. Yoshikawa, M. Saburi, T. Sawai, and M. Goto, *Proceedings of the Twelfth International Conference on Coordination Chemistry*, Sydney, 1969, p. 155.

94. M. Ito, F. Marumo, and Y. Saito, *Acta Crystallogr.*, B26, 1408 (1970).

95. M. Ito, F. Marumo, and Y. Saito, *Acta Crystallogr.*, B28, 457 (1972).

96. M. Ito, F. Marumo, and Y. Saito, *Acta Crystallogr.*, B28, 463 (1972).

97. R. H. Boyd, *J. Chem. Phys.*, 49, 2574 (1968).

98. D. A. Buckingham and A. M. Sargeson, *Top. Stereochem.*, 6, 219 (1971).

99. Professor S. Yoshikawa, private communication.

100. K. Toriumi and Y. Saito, *Acta Crystallogr.*, B31, 1247 (1975).

101. L. G. Marzilli and D. A. Buckingham, *Inorg. Chem.*, 6, 1042 (1967).

102. D. A. Buckingham, P. J. Cresswell, R. J. Dellaca, M. Dwyer, G. J. Gainsford, L. G. Marzilli, I. E. Maxwell, Ward T. Robinson, A. M. Sargeson, and K. R. Turnbull, *J. Am. Chem.*, 96, 1713 (1974).

103. H. C. Freeman and I. E. Maxwell, *Inorg. Chem.*, 9, 649 (1970); H. C. Freeman, L. G. Marzilli, and I. E. Maxwell, *Inorg. Chem.*, 9, 2408 (1970).

104. D. A. Buckingham, L. G. Marzilli, I. E. Maxwell, A. M. Sargeson, and H. C. Freeman, *Chem. Commun.*, 1969, 583.

105. D. A. Buckingham, J. Dekkers, A. M. Sargeson, and M. Wein, *Inorg. Chem.*, 12, 2019 (1973).

106. B. F. Anderson, D. A. Buckingham, G. J. Gainsford, G. B. Robertson, and A. M. Sargeson, *Inorg. Chem.*, 14, 1658 (1975).

107. R. D. Gillard, N. C. Payne, and G. B. Robertson, *J. Chem. Soc.*, *A*, 1970, 2579.

108. C. Tam Liu and B. E. Douglas, *Inorg. Chem.*, 3, 1356 (1964).

109. J. H. Dunlop, R. D. Gillard, N. C. Payne, and G. B. Robertson, *Chem. Commun.*, 1966, 874.

110. J. Meisenheimer, L. Angermann, and H. Holsten, *Justus Liebigs Ann. Chem.*, 438, 261 (1924).

111. D. A. Buckingham, S. F. Mason, A. M. Sargeson, and K. R. Turnbull, *Inorg. Chem.*, 5, 1649 (1966).

112. J. F. Blount, H. C. Freeman, A. M. Sargeson, and K. R. Turnbull, *Chem. Commun.*, 1967, 324.

113. M. Fujita, Y. Yoshikawa, and H. Yamatera, *Chem. Lett.* 1976, 959.

114. A. Fujioka, S. Yano, and S. Yoshikawa, *Inorg. Nucl. Chem. Lett.* 11, 341 (1975).

115. N. C. Payne, *Inorg. Chem.*, 11, 1376 (1972).

116. N. C. Payne, *Inorg. Chem.*, 12, 1151 (1973).

117. S. Yaba, S. Yano, and S. Yoshikawa, *Inorg. Nucl. Chem. Lett.* 12, 267 (1976).

118. B. Bosnich and J. MacB. Harrowfield, *Inorg. Chem.*, 14, 853 (1975).

119. S. Yaba, Y. Sano, and S. Yoshikawa, *Inorg. Nucl. Chem. Lett.*, 12, 831 (1976).

120. J. L. Halloran, R. E. Caputo, R. D. Willet, and J. I. Legg, *Inorg. Chem.*, 14, 1762 (1975).

121. H. A. Weakliem and J. L. Hoard, *J. Am. Chem. Soc.*, 81, 549 (1959).

122. D. A. House and C. S. Garner, *Inorg. Chem.*, 5, 2097 (1966).

123. P. A. Marzilli, Ph. D. Thesis, Australian National University, Canberra, 1969.

124. M. R. Snow, Proceedings of the Twelfth International Conference on Coordination Chemistry, Communication W4, 33, 92, Sydney, 1969.

125. M. R. Snow, *J. Chem. Soc.*, *Dalton Trans.*, 1972, 1627.

126. M. R. Snow, *J. Am. Chem. Soc.*, 92, 3610 (1970).

127. Y. Yoshikawa and K. Yamasaki, *Bull. Chem. Soc. Jpn.*, 46, 3448 (1973).

128. S. Sato and Y. Saito, *Acta Crystallogr.*, B31, 2456 (1975).

129. Y. Yoshikawa, *Bull. Chem. Soc. Jpn.*, 49, 159 (1976).

130. G. Schwarzenbach and P. Moser, *Helv. Chim. Acta*, 36, 581 (1953).

131. E. P. Emmenegger and G. Schwarzenbach, *Helv. Chim. Acta*, 49, 625 (1966).

132. Y. Yoshikawa, E. Fujii, and K. Yamasaki, *Bull. Chem. Soc. Jpn.*, 45, 3451 (1972).

133. J. R. Gollogly and C. J. Hawkins, *Chem. Commun.*, 1966, 873.

134. A. Muto, F. Marumo, and Y. Saito, *Acta Crystallogr.*, B26, 226 (1970).

135. A. Kobayashi, F. Marumo, and Y. Saito, *Acta Crystallogr.*, B30, 1495 (1974).

136. J. R. Gollogly and C. J. Hawkins, *Aust. J. Chem.*, 20, 2395 (1967).

137. G. S. Smith and J. L. Hoard, *J. Am. Chem. Soc.*, 81, 556 (1959).

138. J. B. Terril and C. N. Reilley, *Inorg. Chem.*, 5, 1988 (1966).

139. J. L. Sudmeier and G. Occupati, *Inorg. Chem.*, 7, 2524 (1968).

140. P. F. Coleman, J. I. Legg, and J. Steele, *Inorg. Chem.*, 9, 937 (1970).

141. J. L. Sudmeier and G. Occupati, *Inorg. Chem.*, 10, 90 (1971).

142. N. Tanaka and H. Ogino, *Bull. Chem. Soc. Jpn.*, 37, 877 (1964).

143. J. W. Weyh and R. E. Hamm, *Inorg. Chem.*, 7, 2431 (1968).

144. R. Nagao, F. Marumo, and Y. Saito, *Acta Crystallogr.*, B28, 1852 (1972).

145. J. P. Mathieu, *J. Chim. Phys., Physicochim. Biol.*, 33, 78 (1936).

146. A. J. McCaffery and S. F. Mason, *Mol. Phys.*, 6, 359 (1963).

147. R. Kuroda and Y. Saito, *Bull. Chem. Soc. Jpn.*, 49, 433 (1976).

148. A. J. McCaffery, S. F. Mason, and R. E. Ballard, *J. Chem. Soc.*, 1965, 2883.

149. B. E. Douglas, *Inorg. Chem.*, 4, 1813 (1965).

150. T. S. Piper and A. G. Karipides, *J. Am. Chem. Soc.*, 86, 5039 (1964).

151. J. R. Gollogly and C. J. Hawkins, *Chem. Commun.*, 1968, 689.

152. D. A. Buckingham, S. F. Mason, A. M. Sargeson, and K. R. Turnbull, *Inorg. Chem.*, 5, 1649 (1966).

153. R. G. Denning and T. S. Piper, *Inorg. Chem.*, 5, 1056 (1966).

154. J. Hidaka and B. E. Douglas, *Inorg. Chem.*, 3, 1724 (1964).

155. A. J. McCaffery, S. F. Mason, and B. J. Norman, *J. Chem. Soc.*, 1965, 5094.

156. K. Matsumoto, S. Ooi, and H. Kuroya, *Bull. Chem. Soc. Jpn.*, 44, 2721 (1971).

157. G. A. Barklay, E. Goldschmied, N. C. Stephenson, and
     A. M. Sargeson, *Chem. Commun.*, 1966, 540.
158. C. J. Hawkins and E. Larsen, *Acta Chem. Scand.*, 19, 185,
     1969 (1965).
159. J. I. Legg and B. E. Douglas, *J. Am. Chem. Soc.*, 88, 2697
     (1966).
160. S. F. Mason, *J. Chem. Soc.*, A, 1971, 667; *Pure Appl.
     Chem.*, 24, 335 (1970).
161. B. Bosnich and J. MacB. Harrowfield, *J. Am. Chem. Soc.*,
     94, 3425 (1972).
162. S. F. Richardson, *Inorg. Chem.*, 10, 2121 (1971).
163. L. J. DeHayes, M. Parris, and D. H. Busch, *J. Chem.
     Soc.*, D, 1971, 1398.
164. B. Bosnich and J. MacB. Harrowfield, *Inorg. Chem.*, 14,
     828 (1975).
165. B. Bosnich and J. MacB. Harrowfield, *Inorg. Chem.*, 14,
     847 (1975).
166. B. Bosnich and J. MacB. Harrowfield, *Inorg. Chem.*, 14,
     861 (1975).
167. L. R. Froebe, S. Yamada, J. Hidaka, and B. E. Douglas,
     J. Coord. Chem., 1, 183 (1971).
168. J. I. Legg and J. A. Niel, *Inorg. Chem.*, 12, 1805 (1973).
169. H. L. Smith and B. E. Douglas, *Inorg. Chem.*, 5, 784 (1966).
170. R. Larson, S. F. Mason, and B. J. Norman, *J. Chem. Soc.*,
     A, 1966, 301.
171. S. F. Mason and B. J. Norman, *J. Chem. Soc.*, A, 1966,
     307.
172. R. Kuroda, J. Fujita, and Y. Saito, *Chem. Lett.* 1975,
     225.
173. A. J. McCaffery, S. F. Mason, and B. J. Norman, *J. Chem.
     Soc.*, A, 1968, 1428.
174. B. Bosnich, *Acc. Chem. Res.*, 2, 266 (1966).
175. T. Ito, N. Tanaka, I. Hanazaki, and S. Nagakura, *Bull.
     Chem. Soc. Jpn.*, 41, 365 (1968).
176. I. Hanazaki, F. Hanazaki, and S. Nagakura, *J. Chem.
     Phys.*, 50, 265 (1969).
177. T. Ito, A. Kobayashi, F. Marumo, and Y. Saito, *Inorg.
     Nucl. Chem. Lett.*, 7, 1097 (1971).
178. R. D. Gillard, *Spectrochim. Acta*, 20, 1431 (1964).
179. L. J. Halloran and J. I. Legg, *Inorg. Chem.*, 13, 2193
     (1974).
180. K. P. Larsen and H. Toftlund, *Acta Chem. Scand.*, A31,
     182 (1976).
181. R. J. Geue and M. R. Snow, *Inorg. Chem.*, 16, 231 (1977).
182. I. I. Creaser, J. MacB. Harrowfield, A. J. Herlt, A. M.
     Sargeson, J. Springborg, R. J. Geue, and M. R. Snow, *J.
     Am. Chem. Soc.*, 99, 3181 (1977).
183. S. Lifson, *J. Chim. Phys., Physicochim. Biol.*, 65, 40
     (1968).

184. S. Lifson, "Molecular Forces," in *Protein-Protein Interactions*, R. Jaenicke and H. Helmreich, Eds., Springer-Verlag Heidelberg, 1972.

185. S. R. Niketić, K. Rasmussen, F. Woldbye, and S. Lifson, *Acta Chem. Scand.*, <u>A30</u>, 485 (1976); S. R. Niketić and K. Rasmussen, *The Consistent Force Field*, Lecture Notes in Chemistry, Vol. 3, Springer-Verlag, Heidelberg, 1977.

186. M. Iwata, *Acta Crystallogr.*, <u>B33</u>, 59 (1977).

New Approaches in Asymmetric Synthesis

H. B. Kagan and J. C. Fiaud

Université Paris Sud,
Laboratoire de Synthèse Asymétrique,
Orsay, France

I. Introduction. 176
   A. Definition 176
   B. Criteria for a Good Asymmetric Synthesis 177
   C. Factors Affecting Asymmetric Induction 178
      1. Introduction 178
      2. Ruch-Ugi Model 179
      3. Salem Model 182
      4. Steric and Conformational Effects 184
      5. Polar Effects 193
      6. MO Methods 195
   D. Strategy in Asymmetric Synthesis 198
      1. Recovery of the Chiral Agent 198
      2. Structure of the Chiral Inducer 198
      3. Importance of Experimental Conditions and
         Prochiral Substrate 200
   E. Progress in Experimental Methods 201
II. Chiral Reagents and Chiral Substrates 203
   A. Chiral Reagents 203
      1. Asymmetric Hydroboration 203
      2. Other Chiral Reducing Agents 205
      3. Asymmetric Olefin Mercuration 213
      4. Asymmetric Alkylation 213
   B. Chiral Substrates 215
      1. Total Synthesis of Steroids 215
      2. Total Synthesis of Prostaglandins 217
      3. Synthesis of Chiral Dewar Benzenes 217
      4. Asymmetric Synthesis of α-Amino Acids 217
      5. Chiral Oxazolines 220
      6. Asymmetric Alkylation of Aldehydes, Ketones,
         or Esters 225
      7. Asymmetric Conversion of a Racemic Mixture . 229

      C. Catalysis 233
          1. Asymmetric Aldolization 234
          2. Asymmetric Michael Reactions 235
          3. Asymmetric Phase-Transfer-Catalyzed Reactions 236
III. Catalysis with Transition Metal Complexes 238
      A. Asymmetric Hydrogenation of C=C Bonds 238
      B. Asymmetric Hydrogenation of C=O and C=N Bonds . . 243
      C. Asymmetric Hydrosilylation 244
      D. Asymmetric Hydroformylation 247
      E. Asymmetric C-C Bond Formation 248
      F. Asymmetric Oxidation 251
      G. Asymmetric Polymerization 252
 IV. Photochemical Asymmetric Synthesis 252
      A. Photochemistry with CPL 254
      B. Chiral Sensitizers 258
      C. Diastereoselective Photochemical Synthesis . . . 259
      D. Enantioselective Photosynthesis 264
  V. Asymmetric Electrochemistry 264
 VI. Asymmetric Synthesis with Enzymes 266
VII. Conclusions 269
      Acknowledgments 272
      References 272

I. INTRODUCTION

A. Definition

"Asymmetric synthesis" is a term first used in 1894 by
E. Fischer and defined in 1904 by Marckwald (1) as "a reaction
which produces optically active substances from symmetrically
constituted compounds with the intermediate use of optically
active materials but with the exclusion of all analytical
processes." A broader definition was proposed by Morrison and
Mosher (2): "Asymmetric synthesis is a reaction in which an
achiral unit in an ensemble of substrate molecules is converted by a reactant into a chiral unit (3) in such a manner that
the stereoisomeric products (enantiomeric or diastereomeric)
are formed in unequal amounts." With this definition we may
consider the stereoselective (4) reduction of 2-methylcyclo-
hexanone to *trans*-2-methylcyclohexanol (5) an asymmetric
synthesis, whether the ketone is used as a single enantiomer
or as the racemic mixture.

Asymmetric synthesis requires that asymmetric induction*

*A chemical process is said to involve asymmetric induc-
tion if it leads to the creation of a new chiral unit [as
defined by Cahn, Ingold, and Prelog(3)] with one configuration
present in excess over the other. According to the type of
asymmetric synthesis (see Sec-II-B-6) the result may be the
formation of two diastereomers or of two enantiomers in
unequal amounts.

occur. Conversely, however, asymmetric induction does not
necessarily result in the formation of optically active products
(Sect. I-E.). Nevertheless the understanding of factors affecting asymmetric induction is essential for an understanding of
asymmetric synthesis.

This Chapter concentrates mainly on the methods useful
in obtaining optically active compounds (Marckwald definition).
The synthesis of chiral molecules is an important problem from
a practical point of view, since many substances are needed
as the pure enantiomers. The main interest lies in the pharma-
ceutical area. An increasing number of drugs, food additives,
and flavoring agents are being prepared by total synthesis,
and asymmetric synthesis is an appropriate way to obtain
optically active intermediates. However, in this introductory
section we attempt to present all new ideas on asymmetric
induction which may be of interest to the synthetic organic
chemist, even if the chemical system under consideration cannot
give rise to practicable asymmetric synthesis.

In addition to the synthetic application of asymmetric
synthesis we must mention its usefulness for determinations of
mechanism and for configurational assignments. This latter
point was the subject of a recent article (6) and will not be
developed here. Basically, the purpose of this chapter is to
review significant progress in preparative asymmetric synthesis
since 1974 and, especially, to describe and discuss the most
recent approaches.

Generally, in a multistep total synthesis, asymmetric
synthesis is most conveniently applied at the point where the
first chiral center is created. In some cases several chiral
centers are formed in the same reaction, for example, in an
asymmetric Diels-Alder reaction (7). This chapter, however,
largely concentrates on asymmetric induction of one or two
chiral centers in a single reaction step. Asymmetric reaction
of a racemic mixture is a special case, requiring separation
of optically active epimers; this is discussed on p. 229.

The field of asymmetric synthesis has been reviewed
several times during the last 10 years (8-12). The most complete review, covering all literature data up to 1970, is the
book by Morrison and Mosher (2). The literature between 1970
and early 1974 is available in the review article of Scott and
Valentine (13). Several recent reviews on asymmetric catalysis,
a field that is expanding rapidly, are quoted in Sect. III.

B. Criteria for a Good Asymmetric Synthesis

The requirements for an efficient asymmetric synthesis have
been reviewed by Eliel (14):
1. The synthesis must lead to the desired enantiomer with high
    stereoselectivity and high chemical yield.
2. The chiral product must be readily separable from the

chiral auxiliary reagent that is needed in the synthesis.
3. Unless the chiral auxiliary reagent is very much cheaper
    than the desired product, the auxiliary reagent must be
    capable of being recovered in good yield with undiminished
    enantiomeric purity.

In addition to these conditions we must consider the
amount of optically active material produced relative to the
amount of chiral auxiliary material. The best balance is obtained if the latter is incorporated into a catalyst. In
principle a chiral catalyst is able to produce an unlimited
amount of chiral product. Catalytic asymmetric synthesis thus
embodies the most promising and spectacular progress achieved
during recent years. Sect. II-C and III are devoted to this
area. The choice of the substrate in an asymmetric synthesis
is not always obvious. Asymmetric synthesis operates most
efficiently in the first steps of a multistep total synthesis.
The advantages and disadvantages of the various strategies for
obtaining an optically active molecule are presented in ref.
15 for the total synthesis of steroids.

For a long time it was questioned whether high optical
yields* could be effectively attained by the organic chemist
without the help of enzymes. An increasing amount of recent
data (optical yields greater than 90%) demonstrates that
versatile and efficient nonenzymatic asymmetric syntheses are
indeed possible. Of course, much work must be done to find
general methods. The most serious obstacle continues to be the
lack of basic understanding of factors mediating asymmetric
induction. The optimization of optical yields remains essentially empirical, requiring many experiments. Nevertheless some
concepts are slowly emerging from the large amount of experimental data, and it is to be hoped that the element of mystery
about asymmetric induction may be gradually diminishing.

C. Factors Affecting Asymmetric Induction

1. Introduction

A typical case of asymmetric synthesis, involving in this
case addition of hydrogen or a hydride to a (C=O) double bond
is shown in Figure 1, where Z* represents a chiral moiety. In
(I) (Fig. 1), where a and b are achiral groups, the two faces
of the carbonyl groups are enantiotopic. Izumi proposed (16)
the term *enantioselective asymmetric synthesis* for a reaction
in which an enantiotopic face or ligand is selectively

*By "optical yield" we mean the enantiomeric excess (opti-
cal purity) of the product divided by that of the starting
material (or auxiliary chiral reagent or catalyst), and multiplied by 100.

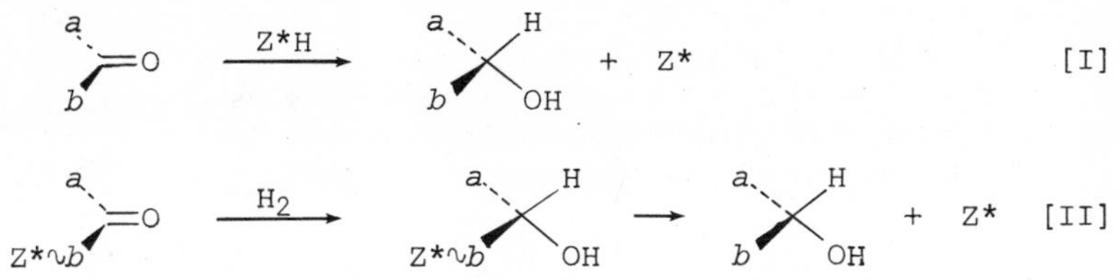

Fig. 1

attacked. For (II) the asymmetric synthesis is the result of
a selective reaction on one of two diastereotopic faces; such
a reaction may be called a *diastereoselective asymmetric synthesis*. The desired optically active product is obtained after
cleavage and removal of the chiral inducer Z*. The ratio of
enantiomers of *ab*CHOH, of course, will be identical to the
ratio of the diastereomeric precursors if the cleavage is
quantitative. If, however, one diastereomer is selectively
cleaved, the enantiomeric excess of the final product may be
changed from the diastereomeric excess in the reaction involving
asymmetric induction. The origin of the asymmetric induction is
generally hard to define, especially in those enantioselective
syntheses in which the transition state is not cyclic. In many
diastereoselective syntheses the inducing chiral centers are
more or less fixed with respect to the prochiral center, particularly in the case of cyclic structures. If information is
available concerning the reaction mechanism (which is often not
well understood), it may become possible to predict or interpret the steric course of the asymmetric synthesis. All the
different factors on which reaction rates depend can in principle affect the extent of asymmetric induction, which is the
net result of competitive reactions. Since all effects are
interrelated and overlap each other, a rigorous classification
of the various parameters is very difficult.

Before beginning to discuss these effects in the light of
recent experimental results, we shall mention an attempt to
predict optical yield and absolute configuration in an asymmetric synthesis without knowledge of any details about the
conformation of the pertinent transition states.

2. Ruch Ugi Model

The *stereochemical analogy model* of Ruch and Ugi (17)
starts with the idea that much uncertainty is introduced when
a model of asymmetric induction is proposed, since the resulting
prediction depends on the assumed conformation of the transition
state, which is often not precisely known. A mathematical treatment with the help of group theory led Ruch and Ugi to a general

expression correlating Q (the ratio of the two stereoisomers P_1 and P_2 formed in the reaction) with parameters (λ values) that are specific to the groups placed around the asymmetric centers of the system. This mathematical analysis applied well to "corresponding reactions" such as

$$P_1 \leftarrow [S_1^{\ddagger}] \xleftarrow{k_1} A \xrightarrow{k_2} [S_2^{\ddagger}] \rightarrow P_2 \qquad [1]$$

where a substrate A undergoes transformations to products P_1 and P_2 (enantiomers or diastereomers according to the type of asymmetric synthesis). In corresponding reactions the two transition states $[S_1^{\ddagger}]$ and $[S_2^{\ddagger}]$ are considered to be in thermodynamic equilibrium. In this type of reaction Q is independent of the initial concentrations of the starting materials. If a diastereoselective asymmetric synthesis is performed on a system such as that depicted in Figure 2, a very simple expression

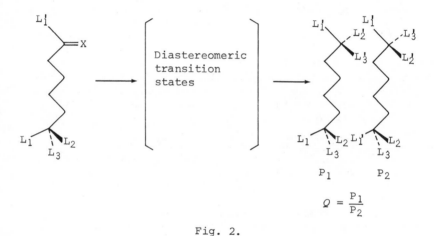

$$Q = \frac{P_1}{P_2}$$

Fig. 2.

can be used to correlate Q and the nature of the inducing and induced chiral centers:

$$\log Q = \delta\rho \, (\lambda_1 - \lambda_2) \, (\lambda_2 - \lambda_3) \, (\lambda_3 - \lambda_1) \qquad [2]$$

The λ values are numbers that characterize the substituents around the inducing asymmetric center, with the scale chosen such that $\lambda_H = 0$ and $\lambda_{CH_3} = 1$. The numbers were originally calculated from the A-values ($-\Delta GR_{a \rightarrow e}$ values) for various R groups. These values are well known from studies on axial-equatorial equilibria in cyclohexane (18). It was assumed that the apparent "size" of R as deduced from $\Delta G_{a \rightarrow e}$ is proportional to that deduced from the λ = values. Values of λ_R were calculated by means of the following equation:

$$\lambda_R = 0.82 \ \sqrt[3]{\overline{\Delta GR_{a \to e}}} \qquad\qquad [3]$$

The ρ-value, which is specific to the reaction under consideration, encompasses all experimental details (e.g., temperature, solvent, and reagent). The absolute configuration of the newly created asymmetric center is given by δ, which can only take the values +1 and -1.

The Ruch-Ugi equation works remarkably well in predicting quantitatively (19) most of the results obtained by Prelog in his atrolactic acid asymmetric synthesis. Other experimental data in agreement with eq. [2] were published by Tomoszkosi

and co-workers (20), who investigated the LAH reduction of seven 2-alkylcyclopentanones (1). The cis-trans ratio Q of diastereomers cis-2 and trans-2 obeys the eq. [4]

$$\log Q = -a_1 \lambda_R + a_2 \lambda_R^2; \qquad a_1 = 1.95, \ a_2 = 1.46 \qquad [4]$$

where the λ_R's are the original values given by Ruch and Ugi for the various R substituents (17). Equation [4] is a modified form of eq. [2] where

$$\lambda_3 = \lambda_H = 0; \qquad \lambda_1 = \lambda_R, \ \lambda_2 = \lambda_{ring}$$

Unfortunately, as stated by Ruch and Ugi (19), the λ-values generally depend on the chemical system that is examined. This removes most of the predictive power of eq. [2], since for a given system not only ρ but also a minimum set of λ-values have to be determined experimentally. For example, let us compare the phenyl and t-butyl groups in the Prelog atrolactic synthesis (17) and in the stereoselective acylation of a racemic alcohol by phenyltrifluororomethyl ketene (19). In the first reaction $\lambda_{phenyl} = 1.24$, $\lambda_{t-butyl} = 1.45$; in the second $\lambda_{phenyl} = 1.60$, $\lambda_{t-butyl} = 1.25$. There is evidently a reversal of the values which cannot be predetermined without experimental data. Recently (21) the Ruch-Ugi model was applied to the Horeau method of kinetic resolution. However, a further detailed study by Horeau and Vigneron (22) demonstrated that the Ruch-Ugi equation cannot be used in this particular case. The exact scope of the Ruch-Ugi method in asymmetric synthesis is still uncertain, but its successful use at present is clearly limited to very specific systems. The main difficulty lies in finding values that remain specific for groups when the other substituents are changed. For example, it is well known that even A-values are not invariant, thus they do not apply to geminal substituents on the cyclohexane ring (23).

3. Salem Model

Salem (24) analyzed several aspects of the asymmetric induction occurring in a diastereoselective synthesis by choosing specific cases where a chiral center lies in the prochiral plane of the molecule (Fig. 3). He compared the energy of pairs of transition states with the reagent W attacking above or below the prochiral plane. By assuming, in the transition state, an additivity of the interaction energies between W and the a, b, c groups that form the chiral centers, he demonstrated the great importance of the rotation barriers. Let us assume that the chiral center turns rapidly in the transition state (relative to the time spent by the reagent) and that there is a low asymmetric barrier component in the transition state (the three components of the potential, V_a, V_b, and V_c, which measure the contribution to the asymmetry of the barrier from substituents a, b, and c, are ca. 1 kcal/mol).

Since internal rotation in the transition state is fast, the reagent W will see all possible conformations.* The average energy difference $\overline{\Delta E}$ between the two approaches[†] can be calculated, taking into account the Boltzmann probability of each conformation. The ratio Q of the two diastereomers formed in the reaction is related to $\overline{\Delta E}$ and is given by the following equation for a small assumed onefold potential V_1:

$$\log Q = \sqrt{\frac{2}{3}} \; \frac{Z}{Ro} \; \frac{1}{(KT)^Z} \left\{ \frac{Va}{2} \; (r_bF_b - r_cF_c) + \right.$$

$$\left. \frac{Vb}{2} \; (r_cF_c - r_aF_a) \; + \; \frac{Vc}{2} \; (r_aF_a - r_bF_b) \right\} \qquad [5]$$

where Z, r_a, r_b, and r_c are defined in Figure 3; F_a, F_b, and

*This situation is included in the Curtin-Hammett principle (4), which states that if two conformers react slowly relative to their interconversion rate, the product distribution is related to the difference in energy between the two competing transition states, and is independent of the ground state conformations. For a recent discussion see (306).

[†]It is assumed that the quantity Z (in the transition state) is the same in the two approaches. This is only an approximation (see also Sect. I-C-6.).

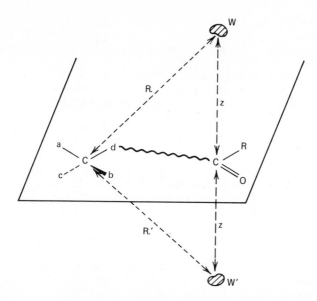

Fig. 3. Model (24) for diastereomeric transition states in asymmetric induction (W and W' represent the two competitive diastereomeric approaches of the reagent; r_a, r_b, and r_c are the lengths C-a, C-b, and C-c, respectively).

F_C represent the forces resulting from the a, b, c substituents on W. It is interesting that eq. [5] possesses a symmetry similar to the Ruch-Ugi equation, since eq. [2] can be expressed in the form

$$\log Q = \lambda_a(\lambda_b{}^2 - \lambda_c{}^2) + \lambda_b(\lambda_c{}^2 - \lambda_a{}^2) + \lambda_c(\lambda_a{}^2 - \lambda_b{}^2) \quad [6]$$

However, Salem pointed out that eqs. [5] and [6] are identical only in very special cases. In comparison with [6] the Salem equation has the advantage of relating λ to physical values. Interestingly, $\log Q$ in eq. [5] is a function that varies as T^{-2}, and not as T^{-1}, as classicaly postulated. The existence of this unusual temperature effect will be difficult to demonstrate for molecules having a geometry compatible with that shown in Figure 3. Usually the temperature range for an asymmetric synthesis is quite limited and the measure of Q is not highly accurate. In the few asymmetric syntheses for which the influence of temperature has been studied it is easy to see that $\log Q = f(1/T)$; that is, $\log Q = f(1/T^2)$ can equally well correlate the experimental data, because of the errors in Q and T.

If rotation in the transition state is s<u>low</u>, a different averaging procedure is necessary to evaluate $\overline{\Delta E}$. Calculations

were performed for the case where the onefold potential energy
is dominated by one substituent ($V_a \gg V_b, V_c$). Log Q is then
a function of V_a and an inverse function of absolute temperature T (24). Several limiting cases were discussed by Salem and
related to experimental data. The Salem calculations show how
difficult it is to quantify the asymmetric induction by a
general expression, even assuming a model as simple as that in
Figure 3.

4. *Steric and Conformational Effects*

The most popular way of predicting the direction and size
of asymmetric induction has been, and remains, based on simple
steric considerations. Only a minimal understanding of the
mechanism of the reaction is necessary. It is particularly
important to be sure that the reaction is under kinetic control.
It is then safe to estimate the relative energies of the diastereomeric transition states involved in the process. One
cannot always ascertain where the transition state lies. It
may be "reactantlike" (steric approach control according to the
Dauben nomenclature) (25) or "productlike" (product development
control). Isotopic effects sometimes give useful information on
the structure of the transition state (26).

It is generally assumed that ketone reduction proceeds
through a reactantlike transition state (27), although some
uncertainties remain (28). A very elegant study of the
reduction of some ketones by simple or complex hydrides gave
strong evidence for the importance of steric approach control
(29).

Rigid ketones (*3-5*) were selected; their geometry is such
that steric approach control or product development control can
be clearly distinguished. Relative rates of reaction on faces
(*a*) and (*b*) of ketones *3-5* were calculated. It was clearly
demonstrated that attack on face (*a*) (exo attack) is not
perturbed by the methyl group in *4* or *5*. This rules out
any contribution of product development control, since an
alkoxide on the (*b*) face strongly interferes with the methyl
group of *4* (as demonstrated by equilibration under Meerwein-
Ponndorf conditions, which gives exclusively the *anti*-alcohol).
As expected on the basis of a steric approach control, attack
on face (*b*) of *4* is inhibited by the *syn*-methyl group but the
anti-methyl group of *5* has no effect on the relative rate of
reaction at the (*b*) face.

The situation in cyclohexanones is more complex because
other factors can influence the overall stereochemistry (29).
Nevertheless good predictions concerning the prevalent stereo-
isomer can be obtained if it is assumed that a sterically
hindered reagent cannot easily react on the axial side of the
carbonyl group.

$(a) \rightarrow \underset{}{} \leftarrow (b)$

3   4   5

LAH | *a* attack

Thus, for example, in the reduction of substituted cyclohexanones remarkable stereochemical control of the reduction was obtained by Brown (5) through use of boro-hydrides substituted with bulky groups. Equatorial attack is highly preferred, and the axial alcohols are formed in ex-cellent yields.

Interaction between gauche vicinal hydrogens has received much attention from Wertz and Allinger (30). They concluded from molecular mechanics calculations that the *gauche*-butane interaction is the result of a minimization of the number of gauche vicinal H/H interactions. This effect is found again in the cyclohexane, with the consequence that a hydrogen in an *equatorial* position should be more sterically hindered than one in the *axial* position. The well-known equatorial preference of a substituent is then interpreted as the result of mostly of the axial preference of the hydrogen on the carbon bearing the sub-stituent. With this concept Wertz and Allinger (30) explained the axial attack of unhindered hydrides which give the equatori-al alcohol. The driving force here results from the hydrogen being less hindered in the axial than in the equatorial posi-tion. This explanation is not contradictory to one based on torsional strain (see p. 191).

Simple steric considerations often lead to straightforward predictions of the direction of an asymmetric synthesis if the inducing asymmetric centers are in a cyclic structure or in a cyclic transition state. For example, optically pure aspartic acid can be prepared in excellent chemical yield (31) by the three-step synthesis outlined in Figure 4. *erythro*-1,2-Diphenyl-aminoethanol (*6*), was deliberately chosen to place two phenyl groups cis to each other in the cyclic compounds 7. Without knowing the actual conformation (*7a* or *7b*), it may be expected

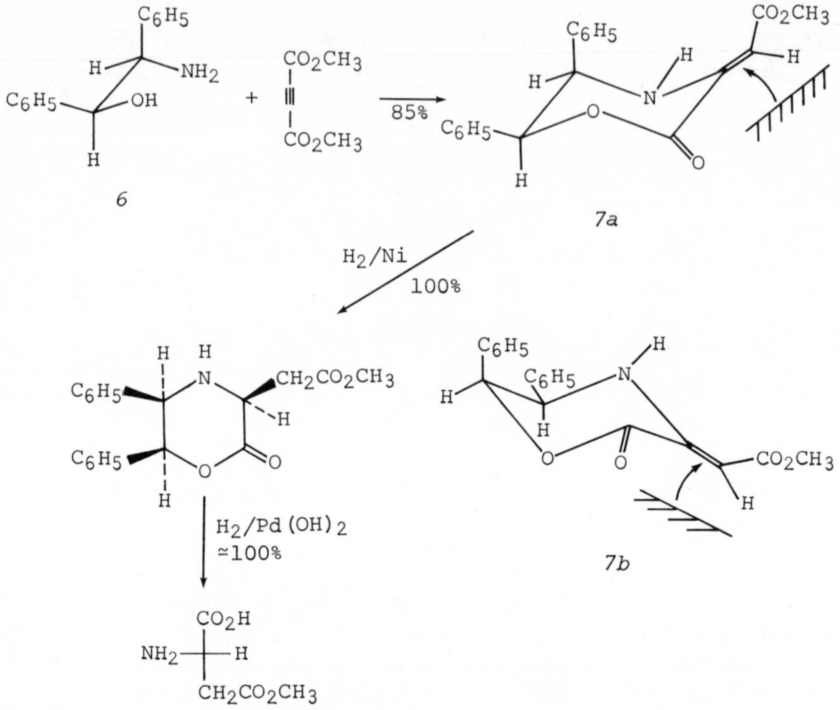

Fig. 4. Asymmetric synthesis of optically pure aspartic acid (31).

that a phenyl group will remain pseudoaxial, strongly interfering with the catalyst in the adsorption step. This argument
leads to the prediction of a reduction trans to the axial phenyl
group. If 7a is the conformation involved, then equatorial
attack is operating. If 7b is the actual conformation, the
stereochemistry is controlled by a remote substituent (here the
axial phenyl group in position 4). Such long-range effects are
not unusual for heterogeneous reduction: for some examples see
ref. 32.

Predictions based on simple steric grounds also operate in
Meyers' asymmetric syntheses with chiral oxazolines, which are
presented in Sect. II-B-5.

Uncertainties can arise when predictions are based on simple
steric considerations. This is well illustrated by reduction
of the bicyclic ketone 8. When R is phenyl, the attack is trans
to R, giving the cis-alcohol (attack a) as expected. But when
R is methyl or isopropyl, the trans-alcohol is predominantly
formed by attack from the b side (33). An explanation was re-

cently given by Lefour (34), based on small deformations of the bicyclic skeleton and on calculations analogous to those developed in Sect. I-C-6. (For an alternative explanation, see p. 195).

A conformation-dependent measure of steric environment was proposed by Wipke (35) to predict the preferred addition on diastereotopic faces of a ketone. A new concept was defined, the steric congestion $C_{(x)}$ at a reaction center x. Steric congestion is a property of the substrate molecule in its ground state, and is independent of reaction partners and transition state structure. It represents only a part of the total effect called steric hindrance. $C_{(x)}$ is the inverse of the accessibility $A_{xa}(i)$ of x, on side a, for example, with respect to an atom i that hinders the approach of the reagent W. If r_i is the van der Waals radius surrounding group i, a cone of preferred approach can be defined (Fig. 5) which intersects a

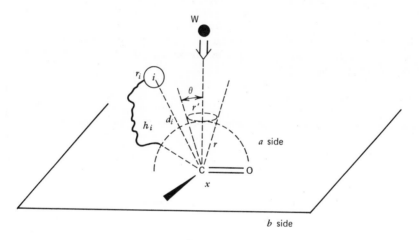

Fig.   5.

sphere of unit radius centered on x. The spherical cross section of the preferred approach is obtained in terms of a solid angle which is then equated with A_{xa}. Equation [7] expresses $A_{xa}(i)$ using the conventions of Figure 5.

For each hindering atom i, there is defined a cone of preferred approach, centered on the perpendicular and tangent to the sphere of van der Waals radius r_i of surrounding atom i. Intersection of this cone with a sphere of unit radius centered on x defines a spherical cross section of preferred approach; $A_{xa}(i)$ is the accessibility of x on one side a, with respect to i. The angle θ is easily derived from r_i, d_i (distance from x to i) and h_i (height of i about the plane), and is expressed as

$$A_{xa}(i) = 2r^2 (1 - \cos \theta) \qquad [7]$$

The congestion for side *a* is the sum of the contributions from each atom located on this side:

$$C_{xa} = \sum_i C_{xa}(i) = \sum_i \frac{1}{A_{xa}(i)} \qquad [8]$$

It is assumed in this calculation that the attack of the nucleophilic reagent is perpendicular to the carbonyl, as indicated in Figure 5. A nonperpendicular attack on the double bond was considered the preferred approach only in highly congested ketones. If atom *i* overlaps the line of perpendicular approach, a corrective displacement term is added to C_{xa}. Computations of congestion have been made for several sterically hindered ketones (35). The calculated congestions allow correct prediction of the stereoselectivity but do not correlate quantitatively with the absolute rates of reaction, which depend on the transition state. When the difference between the congestion on the two sides is small, and when one is not very congested ($C < 30$), predictions are not always correct, since steric control as defined here is no longer strictly operative. The size of the reagent can, for example, influence the side of attack on an unhindered ketone such as 3-cholestanone, and a more refined approach is needed.

Steric congestion is a useful concept that allows prediction of stereochemistry in additions to congested prochiral ketones or olefins. It works if the steric congestion is high and if the transition state is more reactantlike than productlike (25,27). Perhaps better correlations between rates of reaction and congestion will be obtained by using corrective terms characterizing other effects such as the torsional effect (27).

Wipke and Gund (35) did take into consideration a nonperpendicular attack of the double bond as the preferred approach only for highly congested ketones. It has recently been found, however, that *nonperpendicular* attack is quite general for any carbonyl group, the angle of approach being 110° (Fig. 6).

This finding, which rests on both experimental data (36) and calculations (37), should have important consequences in stereochemistry. Until now, with rare exceptions (38) perpendicular attack has always been considered to be preferred (39). Recently Baldwin (40) pointed out that the postulate of nonperpendicular attack is of great help in predicting the stereochemistry of addition to the carbonyl function of cyclohexenones (in both mono- and polycyclic systems). He developed a vector-analysis method for estimating the direction of

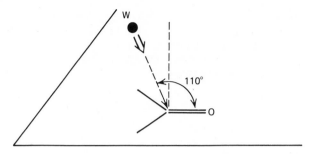

Fig. 6. Preferred attack of a nucleophilic reagent on a ketone (36,37). According to the distance at which the transition state is established the carbonyl will be more or less pyramidal.

approach for the nucleophile. If the carbonyl is considered to be essentially nonconjugated to the double bond, the approach will be as described in Figure 6. If complete conjugation occurs, the enolate structure must be considered, and the 110° attack with respect to the C=C double bond will force the nucleophile to arrive from a different direction. The vector-analysis method is based on the mixing of the two preceding extreme cases, by assuming a mesomeric structure for a cyclo-hexenone. This allows one to estimate the preferred approach of the reagent, and it is then easy to see which substituents in the molecule mediate the stereoselectivity of the addition by steric effects.

In the preceding examples, as well as in most of the cases given in the literature, emphasis is placed mainly on *repulsive* steric effects. It is, however, well known that *attractive* van der Waals forces can operate at long range, though they are seldom taken into account when asymmetric induction is concerned. Attractive steric effects should play a role in chemical reactions, but it is difficult to devise the proper experiment to demonstrate them. Evidence for such effects may be found in some simple molecules. For example, it was shown by NMR studies (41) that 1,3,5-trineopentyl-benzene prefers the syn conformation 9, presumably because of

8   9

the attraction of the alkyl groups for each other. Discussions
of attractive steric effects may be found elsewhere (42,43).

It may be expected that more attention will be given to
steric attractions in asymmetric synthesis.

An example of attractive forces increasing asymmetric
induction was described in the total synthesis of prostaglandin.
Highly stereoselective reduction of *10* to *11* (15-*S* configuration) is very difficult. It would require a highly preferred
attack (*a* attack) on a single enone conformation. With various
small R groups *10* shows approximatively 50:50 distribution
between the *s-cis*- and *s-trans*-enone conformers. It was
found (44) that if R is a *p*-phenylbenzoyl moiety, the reduction
by *12* of *10a* is very stereoselective, leading to *11a*. In the
case of *11b* the stereoselectivity is even higher (84% e.e. of
the *S* configuration). Molecular model analysis of the conformation of precursors *10* indicates excellent contact between
the long chain R and the enone chain in the *s-cis* conformation. There is thus a "freezing" of one reactive conformation
of a chain by attractive interactions with another chain. The
reagent *12* is allowed to approach only from the *a* (*pro-R*) side,
hence produces the desired 15-*S* alcohol *11*.

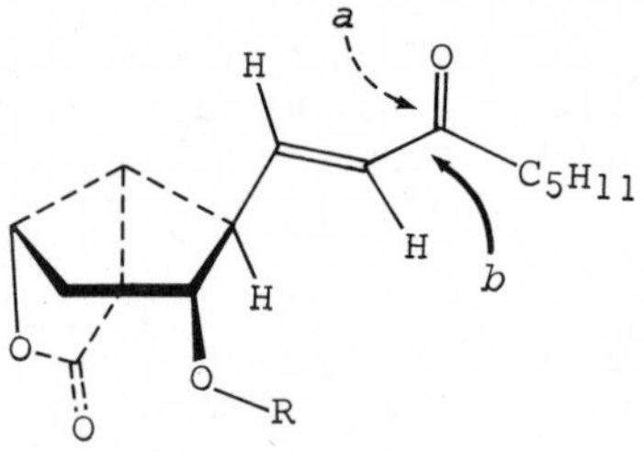

10   (s-cis) 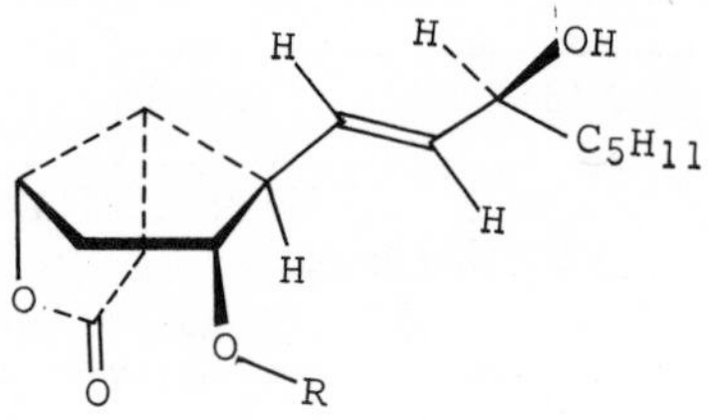   11

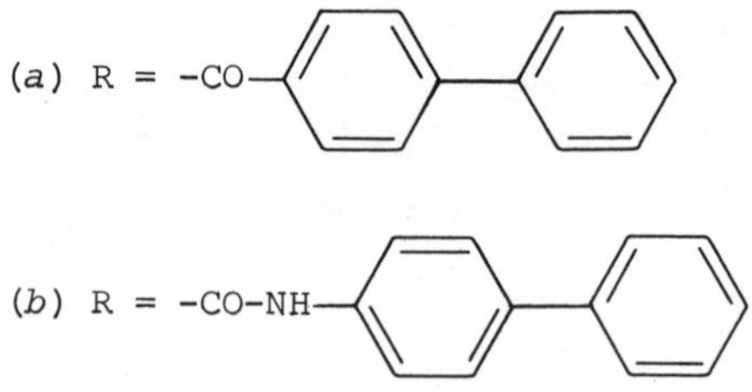

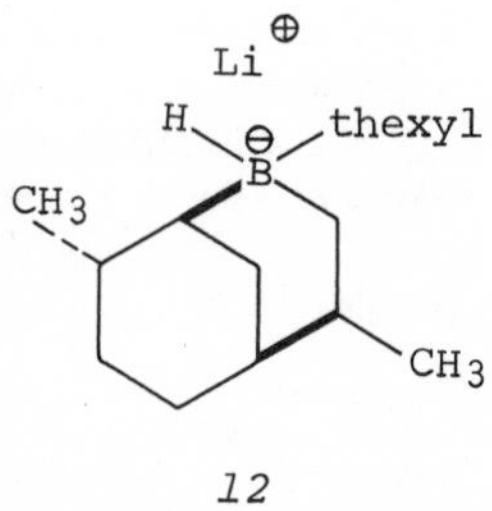

(a) R = -CO-⟨biphenyl⟩

(b) R = -CO-NH-⟨biphenyl⟩

12

Conformational factors are important in control of the
steric course of reactions (18). Conformational, steric, and
polar effects are inextricably mixed. Preferred as well as
disfavored conformations are well known for certain standard
structural units of organic compounds, and it may be useful to

take them to into account when models of asymmetric induction
are elaborated. For example, a C_{sp_3}-C_{sp_3} bond generally avoids
eclipsed conformations.

The eclipsing effect, hypothesized to be of importance
in the transition state for ketone reduction, led to the
Felkin model (27) for acyclic ketones. Karabatsos (45) proposed a different model of asymmetric induction for the same
compounds, taking into account the experimental fact that very
often a carbonyl group likes to eclipse a vicinal C-H or C-C
bond. These two modifications of the classical Cram's rule
are discussed on p. 199. Prelog's rule (46) was established in
1953 to accommodate the known conformational data on α-
dicarbonyl compounds and esters. The Prelog conformations of
phenylglyoxylate esters have even been used in a semiquantita-
tive analysis of asymmetric induction in atrolactic ester
synthesis (47).

In fact, however, it appears that the preferred conformation of α-keto esters such as (-)-menthyl phenylglyoxylate is
different from that assumed by Prelog (48). According to their
authors (see note 9 in ref. 24) the Cram and Prelog rules
remain empirical models that successfully correlate experimental
data with configuration of starting material. However, the conformations chosen are not necessarily those actually found in
the transition state. In all models of asymmetric induction,
especially in acyclic molecules, one of the more difficult
tasks is deciding which are the actual conformations involved
in the transition state.

Many six-membered ring transition states are known or
postulated in organic chemistry, as in some Grignard reductions
of ketones in Diels-Alder reactions, in [3,3]-sigmatropic
rearrangements, and in Meerwein-Ponndorf reactions. A chair-
like transition state is generally considered to be preferred
to a boatlike conformation, as in cyclohexane itself. This
hypothesis is routinely employed to correlate absolute
configuration of reactants and products when an asymmetric
reduction passes through a six-membered cyclic transition
state. Many examples may be found in the Morrison and Mosher
book (2) (see also citations in ref. 49). The detailed geometry of transition states for hydrogen transfer in the
reduction of ketones by Grignard reagent was discussed more
recently (50).

The extension of conformational analysis from cyclohexane
to six-membered transition states (in which some sp^2 character may be found on several atoms of the ring) has until now
been surprisingly successful. It has even been possible to
predict the cis/trans ratio of olefins produced by a Claisen
rearrangement of substituted vinyl allyl ethers, as well as
by other [3,3]-sigmatropic rearrangements. Through examination of a chair ("cyclohexanelike") model for the transition
state, Faulkner (51) was able to correlate the axial/equatorial

free-energy differences of the substituents in this model
with the cis/trans ratio of the olefinic products.

Nevertheless it would be dangerous to envisage chair
conformations for six-membered transition states in all
instances. For example, the pyrolysis of β-hydroxy olefins
goes through a concerted cyclic process, and may be described
as a suprafacial 1,5-hydrogen migration.

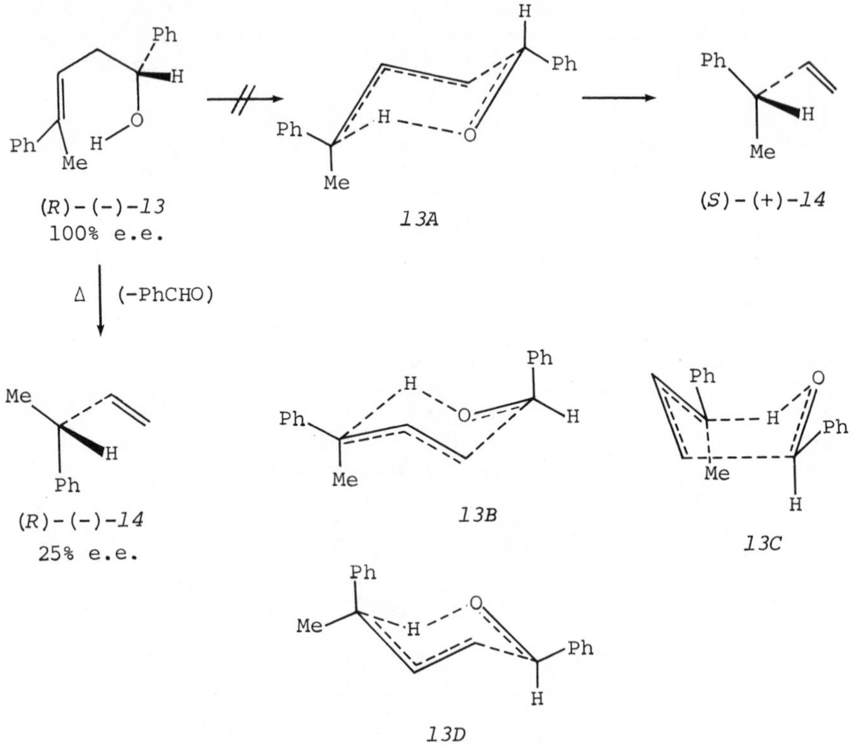

$(R)-(-)-13$
100% e.e.

$13A$

$(S)-(+)-14$

Δ | (-PhCHO)

$(R)-(-)-14$
25% e.e.

$13B$

$13C$

$13D$

Fig. 7.   Asymmetric synthesis in the pyrolysis of β-hydroxy
olefin. Conformation of the transition state (49).

The rearrangement of acyclic β-hydroxyolefin 13, like a
[3,3]-sigmatropic rearrangement, may take place through a
transition state with a cyclohexanelike (chair or boat) geometry. Examination of the asymmetric induction in the "self
immolative" asymmetric synthesis of $(R)-(-)-14$ from $(R)-(-)-13$
(Fig. 7) clearly demonstrates that transition state $13A$ is not
preferred (49). The experiment does not, however, tell us that
the less stable transition states $13B$ or $13C$ are to be considered as leading to $(R)-(-)-14$. In fact, the more plausible

explanation, (52) involves a well-documented effect in cyclohexane rings the "gem effect," which forces phenyl groups with
geminal substituents to be axial. It is known that in 1-phenyl-
1-methylcyclohexane, there is a reversal of the apparent
"size" of groups as compared to their A-values (53). Thus *13D*
will be the preferred transition state, leading to (R)-*14*.

5. *Polar Effects*

Polar effects, which may be defined quite broadly, cover
dipole-dipole and ion-dipole interactions. These interactions
may influence the conformation of reactants and the structures
of transition states, especially in ionic reactions. Some
aspects of polar effects are discussed here.

One of the first classical examples of polar effects in
stereochemical control of reaction comes from the work of
Henbest and co-workers (54), who demonstrated that a CN group
far away from a double bond can determine the direction of
epoxidation at that double bond. Thus in compounds *15* and *16*

the attack is mainly trans to the –CN group, the proposed
explanation being a dipole-dipole interaction between the
electrophilic reagent and the nitrile group. This interaction decreases as the distance between the double bond and
the –CN function increases; it is also decreased by polar
solvent molecules. Similar polar effects have sometimes been
invoked to explain steric control by a function far away
from a double bond (for one example see ref. 55), and could be
in part responsible for long-range effects in steroids. A
special situation is the formation of charge transfer complexes.
The first case of asymmetric induction directly connected with
charge transfer complexes was recently demonstrated (56a). It
concerns (Fig. 8) the asymmetric reduction of *p*-substituted
benzophenones by chiral Grignard reagents (which are known to
react through a cyclic transition state). The para substituents
X (on the ketone) and Y (on the Grignard reagent) cannot give
rise to steric effects. No asymmetric induction could be
expected from such experiments. However, when X = CF_3 and Y =
OCH_3, one aromatic ring becomes the acceptor and the other the

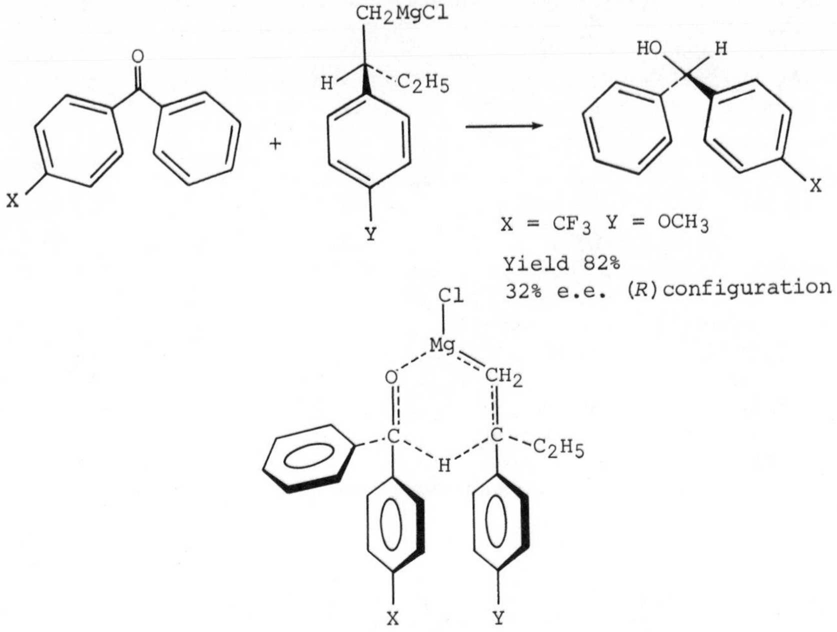

$X = CF_3 \quad Y = OCH_3$

Yield 82%

32% e.e. (R)configuration

$A^{\dagger}$

Fig. 8. Asymmetric induction by charge transfer (56).

donor for charge transfer interactions.* The preferred transition state is predicted to be $A^{\dagger}$ (Fig. 8), which is suitable for charge transfer. Experiments show an appreciable asymmetric induction (32% optical yield), the absolute configuration of the formed chiral benzhydrol being in agreement with the transition state $A^{\dagger}$. It is interesting to note that asymmetric induction here is brought about by an attractive force. In this case steric repulsions lead to the wrong prediction of the direction of asymmetric induction. When $X = CH_3$, $Y = OCH_3$, the optical yield drops to 2%, presumably because charge transfer is no longer operating. When $X = Y = CF_3$, the optical yield is 16% with a transition state $A^{\dagger}$ prevailing; this interesting result (56b) demonstrates an attractive effect between two CF_3-C_6H_4 rings in the transition state. Interpretation of hydride reductions of bridged ketones such as 8 is difficult (see p. 186). A detailed study (57a) of various

*Chiral sensitizers are able to induce asymmetric reactions via exciplex formation (see Sect. IV-B), which probably involves a charge transfer.

models related to *8* led to a new hypothesis: an "anisotropic
induction effect." The electron-releasing induction effect of
alkyl groups would be anisotropic, decreasing, in *8*, the
reactivity of the (a) face which is trans to the alkyl groups.
This factor could be of importance in the addition of various
nucleophiles to carbonyl compounds.

Asymmetric induction was observed in some 1,3-dipolar additions of *N*-alkoxyalkylnitrones derived from sugars to activated
or isolated olefins. The isoxazolidine ribosides that are
formed can be cleaved to chiral *N*-unsubstituted isoxazolidines
of high optical purity (57b). The stereoselectivity of the
reaction (in the cycloaddition step) was ascribed to a new
stereoelectronic effect, a "kinetic anomeric effect."

6.   *MO Methods*

MO methods are developing rapidly, and calculations often
give qualitative or quantitative information in good agreement
with experimental data. It is only very recently that MO methods
have been used in an attempt to determine the origin of asymmetric induction in specific systems.

A prochiral carbonyl group connected to a chiral center will
lose its symmetry plane, whatever its molecular conformation.
The electron density on each face is no longer the same on the
basis of this symmetry argument. It seems reasonable to assume
that a driving force in the steric course of the addition of a
nucleophilic reagent will be its preference to react on the
face bearing the least electron density. *Ab initio* calculations
(58) have been performed on one or two fixed conformations of
simple molecules such as 2-chloropropanal. Calculations for a
given conformation show that the π-electron cloud indeed
becomes dissymmetric and that the difference between the
electron densities of the two faces can be estimated.

A similar treatment was later published (59) for norbornene, a rigid olefin for which exo stereoselectivity in electrophilic additions is well recognized. Results of the calculation
show that electron density is higher at the exo face.

Electronic effects on the stereochemistry of attack on
trigonal atoms in six-membered rings have also been considered
(60). The steric course of attack was rationalized on the
basis of an interaction of the π-orbital of the trigonal atom
with one σ-orbital of the ring, leading to different electron
densities on the two faces of the trigonal atom. However,
calculations were not performed to support these hypotheses.

The preceding approaches do not take into account the
nature of the reagent or all the interactions occurring in the
two competing transition states. Recently Nguyen Trong Anh and
his co-workers (61) published a complete *ab initio* study of
1,2-asymmetric induction. The calculations were performed

on the "super molecule" formed from a nucleophile and a prochiral aldehyde such as $CH_3CH(C_2H_5)-CHO$. All conformations of
the system were considered. The result is expressed in terms
of two curves for the conformational energy of the diastereomeric
supermolecules. The curves for 2-methylbutanal are shown in

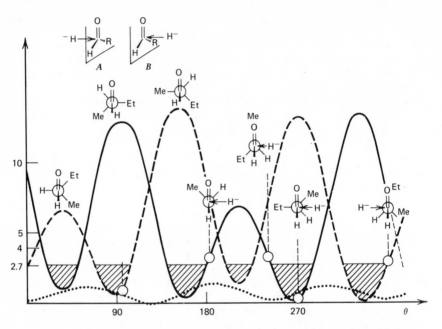

Fig. 9. Analysis of the various transition states for the
nucleophilic addition of H^- to Me(Et)CHCHO (61). Solid line,
conformational energy of A; dashed line, conformational energy
of B. F, Felkin-like transition state; C, Cram-like transition
state; K, Karabatsos-like transition state. Part of this figure
is taken from ref. 61, and is reproduced with the permission of
the copyright holder, *Nouveau Journal de Chimie*.

Figure 9. These curves allow one to predict the stereochemistry
of the predominant product if it is assumed that the Curtin-
Hammett principle applies. One needs only to compare the
relative energies of the minima of the two curves. Calculations were refined by changing the distance between H^- (the
nucleophilic reagent) and the carbonyl group, or by optimizing
the angle of attack of H^- [taking into account the possibility
of a nonperpendicular approach (36,37)]. A further optimization
was obtained by introducing electrophilic assistance by Li^+,
or solvation of H^-. The general shapes of the curves are only
slightly changed. The main conclusion is that the minima of the

curves always have a conformation similar to that previously postulated by Felkin (27) for the transition states of the reaction. The conformations proposed by Cram (62) or Karabatsos (45) also lead to correct predictions of the stereochemistry, but do not seem to have a physical meaning. The origin of the stability of staggered "Felkin-type" conformations has been discussed (61). It is thought that this geometry allows a stabilization of the π^*_{CO} MO by a good overlap with $\sigma^*_{C\alpha-L}$ (C_α is an asymmetric carbon atom; L is the largest group on C_α). The interaction of the C=O LUMO with the H^- HOMO will be enhanced, especially if the nucleophilic reagent is placed antiperiplanar with respect to $C\alpha$-L. This antiperiplanar disposition is assumed to be of importance also in nucleophilic additions of substituted cyclohexanones (63). Without entering into the details of this paper (63), we point to another of its conclusions, namely that, other things being equal, the proportion of axial attack will increase with increase in charge control (or hardness of the reagent) whereas the proportion of equatorial addition will increase with increase in orbital frontier control (or the softness of the reagent). Of course the usual steric factors such as sizes of substituents also need to be considered. In conclusion one may hope that MO methods will be able, in the near future, to give insights into the complex phenomena of stereoselectivity if more emphasis is given to solvation and the nature of the nucleophilic reagent. These new approaches should permit the prediction of the configuration of the preponderant product and the ratio of the the stereoisomeric products if the nature of the transition states is reasonably well known. It is sometimes difficult by these methods to decide when steric factors play a greater role than orbital factors. Steric factors are implicitely considered when the molecular geometry of the transition state is taken into account for detailed calculations.

An interesting idea pointed out in ref. 63 (and in some references cited there) involves consideration of the dichotomy between charge control and frontier orbital control. In orbital control (soft reactants), since the reagent is closer to the substrate in the transition state, steric effects should be more pronounced and stereoselectivity should be enhanced. This provides a guideline for modifying in a predictable way the structure of reactants for improving the optical yield in an asymmetric synthesis. It is difficult to gain experimental evidence, in a given reaction, that the length or structure of the transition state changes with the nature of the reactants. It should be possible, however, to investigate this field, as was done recently for the S_N2 reaction (64) and for the $NaBH_4$ reduction of some (alkyl-substituted) indanone-tricarbonylchromium complexes (65) where isotope effects were examined.

D. Strategy in Asymmetric Synthesis

1. *Recovery of the Chiral Agent*

Asymmetric synthesis is becoming a preparative method in organic chemistry. New approaches are employed that fulfill several of the criteria given in Sect. I-A.

The recovery of the chiral inducing moiety often presents a difficult task, especially in a diastereoselective synthesis, since it is necessary to cleave the bond connecting the chiral inducing moiety to the remaining part of the molecule. Of course the reaction must be mild enough to avoid any racemization of the newly created asymmetric center.

Cleavage reactions have been designed to regenerate, in the chiral product, certain functional groups: OH, NH_2, CO_2H, and CHO. More work is necessary to improve the efficiency in these processes.

In an enantioselective synthesis the optically active product is usually not connected to the chiral inducer. For example, in asymmetric reductions of ketones by LAH modified by quinine the chiral alcohol that is formed is not bonded to the alkaloid and can be easily recovered. However, in some enantioselective syntheses, such as the Diels-Alder condensation of a prochiral diene with dimenthyl acetylenedicarboxylate, the primary product has to be transformed with removal of the inducing alcohol.

Asymmetric catalysis combines two advantages: the amount of chiral catalyst is small with respect to the number of molecules synthesized, and by its essence the catalyst does not remain bound to the optically active product.

2. *Structure of the Chiral Inducer*

The main problem in an asymmetric synthesis is achieving a high optical yield. As mentioned previously, in recent years optical yields of 90% or higher have been reported, but unfortunately there are no general rules to allow one to predict such yields in a specific case.

The choice of the chiral inducer is still largely empirical, the custom being to use available optically active materials such as terpenes, alkaloids, and aminoacids. If a modest optical yield is initially obtained with a given system, it is sometimes possible to optimize the stereoselectivity by making systematic changes in the structures of the reactants. Spectacular results have been obtained in several cases by this method. The necessary modification may be based on an evaluation of the steric effects involved in the synthesis. If it is suspected that a given group behaves as a "large" group, it may be useful to prepare a molecule in which this group is replaced by an even bigger group. For example, the efficacy of menthol

in many asymmetric syntheses is well known, and can be tentatively related to the steric relation between the oxygen atom
and the isopropyl group. The replacement of the isopropyl
group by the $-C(CH_3)_2C_6H_5$ group gives a phenylmenthol in which
a very bulky group is vicinal to the hydroxyl. This alcohol
gives rise to exceptional stereoselectivity in catalyzed
asymmetric Diels-Alder addition (6) when used as a acrylate
ester (in place of menthyl acrylate).

Evaluation of steric effects is very difficult, since the
familiar "L, M, S" approach of Cram's and Prelog's asymmetric
syntheses can be applied only to very simple molecules having
one asymmetric center. Even here the effect of change on L, for
example, cannot be safely predicted, since conformational properties may be changed. As previously stated, cyclic structures
or cyclic transition states permit an easier prediction of the
direction of the asymmetric induction because fewer degrees of
freedom are left in the system. Perhaps as a result, high
optical yields are very often observed. It seems that the more
organized and the tighter the transition state is, the easier
it is to attain good asymmetric induction. Secondary interactions (attractive or repulsive) do not necessarily by
themselves introduce rigidity in the competing transition
states, but they help increase the free energy difference
between them. It should be recalled here that a difference in
activation energies of $\Delta\Delta G_{25}^{\ddagger} = 1.5$ kcal/mol is large enough
to lead to an optical yield of 92% at 25°C. It may thus be
very useful to devise systems in which the chiral inducer is
modified in such a way that the chiral unit is involved in a
ring. This approach was used (31) in the synthesis described
in Figure 4. An efficient general asymmetric synthesis of
α-amino acids is based on the same principle (66). Some
asymmetric hydrogenations described in Sect. III-A are
relevant to this concept. Hydrogen bonding or chelation can
replace covalent bonding in the introduction of one or
several rings in the system. A very good example is the use
of metalated chiral oxazolines in asymmetric synthesis. As
described in Sect. II-B, alkylation of these compounds is
highly stereospecific only when the oxazoline bears a methoxyl
group on a side chain. This group coordinates the metal
(lithium bound to a nitrogen atom), forming a chelating ring.
The oxazoline ring is now transformed temporarily into a
bicyclic system whose shape induces a substantial difference
between the faces of the prochiral center. Whenever the transition state involves secondary or temporary bonding, it is
difficult to rationalize the asymmetric induction by direct
interactions between the inducing asymmetric center and the
reagent. An analysis of the preferred conformations (necessarily
involving molecular chirality) in the transition states would
be helpful in deciding how to modify the chiral inducer. Lack
of understanding of the details of reaction mechanisms is one

of the major obstacles to the improvement of optical yields.
For example, in a given case it may be more useful to check
the effect of solvent or temperature changes, salt effects,
and so forth, rather than to perform extensive structural
modifications of the chiral inducer.

3.   Importance of Experimental Conditions and Prochiral Substrate

Very often asymmetric induction is strongly solvent
dependent. Sometimes opposite configurations are obtained in
two different solvents (see the example given in Sect. III-A).
The E,Z configuration of a prochiral center (olefin, imine, etc.)
should influence the configuration of the chiral center formed
in a diastereoselective synthesis if the same side remains

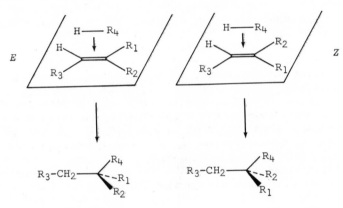

Fig. 10.   Influence of E, Z configuration on asymmetric
induction in a diastereoselective synthesis. If the isomeric
chiral olefins are assumed to be strongly hindered on the
underside, the opposite configurations for the products should
be obtained.

shielded from the reagent (Fig. 10). The extreme specificity
depicted in Figure 10 may not necessarily be in evidence;
nevertheless some examples of obtaining opposite configurations
in related asymmetric synthesis have been described in recent
years (67). Also the first examples have been observed in
asymmetric reduction of E,Z olefins through use of a chiral
catalyst (see Sect. III-A). If there is a slow E,Z interconversion during the reaction, the optical yield will of
course decrease, as the result of an averaging effect. Such
an effect has not been clearly demonstrated yet, however.

A decrease in temperature is classically related to an
increase in stereoselectivity (this is true below the isokinetic

point, where k_1 and k_2 are equal). Unfortunately the temperature
effect is often more complex than that. Salem (24), as indicated previously, predicted in some circumstances a T^{-2}
dependency of log k_1/k_2. A serious difficulty lies in the fact
that actual reactions are much more complicated than a simple
set of two corresponding reactions (see eq. [1]). The following type of kinetic scheme may frequently apply:

$$P_1 \xleftarrow{\quad k_1 \quad} B \xrightleftharpoons[k_3]{k_4} A \xrightarrow{\quad k_2 \quad} P_2 \qquad\qquad [9]$$

where A and B are two isomeric chiral species involving the
prochiral center that are interconverted. In this simple
scheme it is assumed that diastereomeric products P_1 and P_2
are stereoselectively formed from B and A, respectively. In
a 1,2 asymmetric induction (such as the reaction of Grignard
reagents with an α-aminoketone) (68); the reaction follows a kinetic
scheme such as that given in eq. [9]. The situation is similar
in the alkylation of chiral oxazolines (see Sect. II-B-5). The
asymmetric induction in eq. [9] will depend on four rate constants, and the temperature effect becomes very difficult to
predict. Because of the temperature dependence of the four
rate constants, the optical yield must be optimal in a given
range of temperature.

It is difficult to give general guidance to the organic
chemist planning an asymmetric synthesis. The method must fulfill the conditions expressed by Eliel (ref. 14, Sect. I-B). The
maximum efficiency of the synthesis will be attained after
optimization of the structure of the reactants (one of which is
chiral) as well as of the experimental procedure. Diminishing the flexibility of the transition state or the number of
competitive pathways often improves the optical yield. Finally
a greater insight into the reaction mechanism involved in a
synthesis can be of great help in making a rational change in
the makeup of the system leading to asymmetric synthesis.

E. Progress in Experimental Methods

A correct evaluation of the optical yield is absolutely
necessary for both preparative and mechanistic considerations.
Many earlier reports of asymmetric syntheses were erroneous
because optical rotations were measured on compounds purified
by crystallization, which may change the enantiomeric composition. When a mixture of enantiomers needs to be purified,
distillation, solvent extraction, or chromatographic methods
can be safely used. Contamination by impurities of high specific rotation remains a serious problem when optical yields are
measured with a polarimeter (see ref. 69 for a recent case).
Great care must be observed in diastereoselective asymmetric

synthesis, since the primary products are diastereomeric. Subsequent destruction or purification of the diastereomers may lead to alteration of the enantiomeric purity of the product. In an enantioselective synthesis the safest procedure is to analyze the enantiomeric purity by nonpolarimetric methods.* NMR analysis with chiral solvents (71) or chiral shift reagents (72) is very useful but not general. Sometimes the spectroscopic properties of a partially racemic compound differ from those of the pure enantiomer (73). GLC on chiral phases is the most sensitive way of measuring enantiomeric excess (74,75), and can be applied to α-amino acids and to some acids or alcohols, In "direct methods" a derivatization with a chiral reagent converts a mixture of enantiomers into a mixture of diastereomers. To obtain good results it is necessary to use an excess of the reagent (to avoid partial reaction which would give rise to a kinetic resolution). It is also important to have an optically pure reagent. The most popular reagent is $Ph(CF_3)C(OMe)COCl$ (MTPACl), developed by Mosher (76) for derivatization of chiral amines or alcohols. NMR, GLC, or high pressure liquid chromatography can often be used to determine the diastereomeric composition. A recent illustration of the combined use of these methods may be found in ref. 77. In diastereoselective synthesis the asymmetric induction is related to the ratio of the two diastereomers that are formed. A simplification, as far as the extent of asymmetric induction is concerned, is to start from a racemic mixture. If reaction [II] of Figure 1 is combined with its mirror image, the system becomes racemic. The diastereomers formed in the reduction are racemic, but the ratio of racemic diastereomers is the same as in [II]. The asymmetric induction is not changed.[†]

---

*Morrison and Mosher (2) have proposed that the expression "optical purity" be replaced by "enantiomeric excess." The percentage of enantiomeric excess (% e.e.) in the product is now widely used to express the result of an asymmetric synthesis, since it avoids any reference to the method involved for analyzing the enantiomeric composition. Optical purity (calculated from specific rotations) does not always coincide with enantiomeric excess. Horeau (70) found such a discrepancy with 2-methyl-2-ethylsuccinic acid, whose enantiomeric composition can be safely deduced only by nonpolarimetric methods.

[†]As pointed out by Wynberg (78), this is only a first approximation because with the optically active system (eq. [II] in Fig. 1) the medium is chiral as a result of the presence of the chiral reagent. This is equivalent to a chiral solvent effect, if we assume some weak interactions (R/R for example) between the chiral molecules of substrate (which are all of the same configuration, R, for example). In a racemic system the

This approach was taken (79) to reinvestigate an asymmetric synthesis of α-amino acids where optical yields of 95% were claimed (80), the chemical yield being in the range of 50%. The reaction involved HCN addition to $C_6H_5CH(CH_3)-N=CHR$. The resulting α-aminonitrile was hydrolyzed and hydrogenolyzed to $NH_2-CH(R)CO_2H$. Optical purity of the latter does not, however, reflect the optical yield because of the experimental procedure used (purification by crystallization of a diastereomeric mixture of amino acids). Indeed when the extent of asymmetric induction was appropriately measured (79) on the initial α-aminonitrile mixture by NMR, it proved to be ca. 60%.

II. CHIRAL REAGENTS AND CHIRAL SUBSTRATES

A. Chiral Reagents

The classical, and still the most common, method of generating optical activity involves the use of chiral reagents; many examples can be found in ref. 2. We do not cover all the new results obtained since 1972, but confine ourselves to new trends or new approaches in this area. The following arbitrary classification (II-A-1 through II-A-4) may be found useful.

1. Asymmetric Hydroboration

Asymmetric hydroboration remains an efficient method for the conversion of prochiral olefins to chiral products, especially chiral alcohols (81). An additional attractive feature of this reaction lies in the accessibility of both enantiomers of di-3-pinanylborane, the classical chiral hydroboration reagent. This feature is examplified in the asymmetric hydroboration (82) of carbomethoxymethycyclopentadienes *17* by (-)- or (+)-di-3-pinanylborane, which leads to the enantiomeric (R,R) or (S,S) product *18* in high yield. (Comparison with the optical rotation of fully resolved *18* indicates 96% e.e. for the product synthetised.) Compound *18* is easily transformed into *19* and *20*, which possess the four chiral centers required to prepare natural prostaglandin F_2. In an analogous way and with the same reagents (R,R)-*21* and (S,S)-*22* were prepared from 5-methyl-1,3-cyclopentadiene in 30% yield. The alcohols produced were essentially pure trans, and showed at least 95% e.e. (by Mosher's method of enantiomeric purity measurement

---

interactions are now different (R/R, S/S, and R/S), and some changes may be expected in the asymmetric induction. Wynberg states, "When a chiral substance undergoes a reaction, the reaction rate and the product ratio will depend-inter alia-upon the enantiomeric excess present in the starting material."

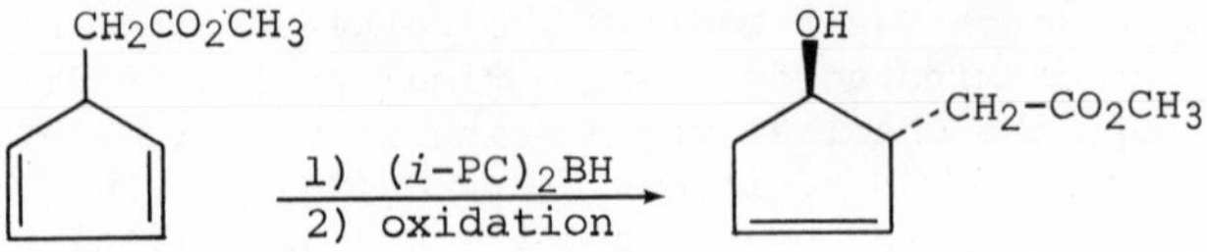

$$17 \xrightarrow[\text{2) oxidation}]{\text{1) } (i\text{-PC})_2\text{BH}} 18$$

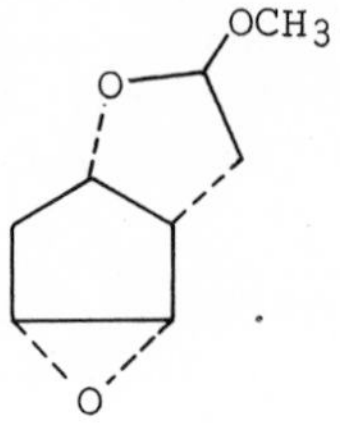

19 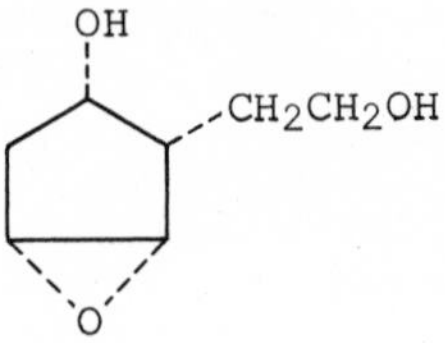   20

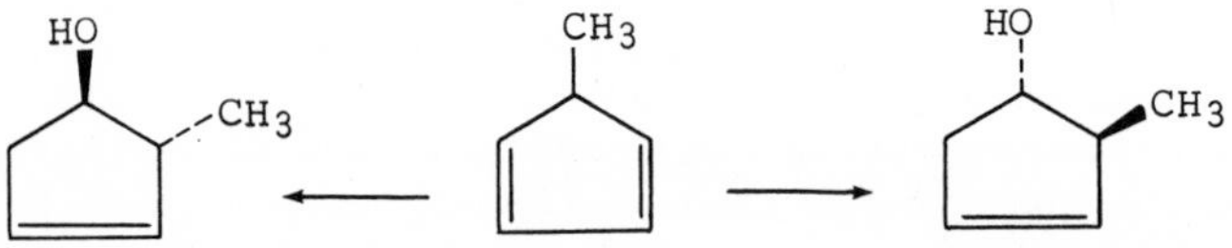

(R,R)   (S,S)
21   22

with MTPACl*) or 98% e.e. (by comparison with optical rotation of resolved material). This preparation allowed the asymmetric synthesis (83) of loganin pentaacetate.

A promising new route for the asymmetric synthesis of iodoalkanes from olefins was recently described by Brown (84). (-)-Diisopinocampheylborane (23) (91% e.e.), when treated with

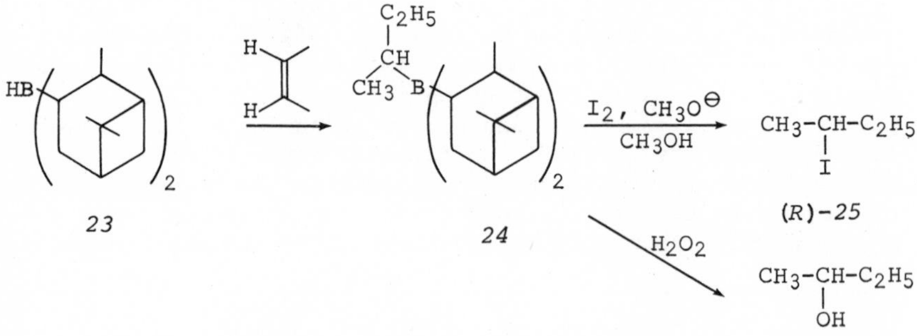

23   24

$$\xrightarrow[\text{CH}_3\text{OH}]{I_2,\ \text{CH}_3\text{O}^\ominus} \begin{array}{c} \text{CH}_3\text{-CH-C}_2\text{H}_5 \\ | \\ \text{I} \end{array}$$

(R)-25

$$\xrightarrow{H_2O_2} \begin{array}{c} \text{CH}_3\text{-CH-C}_2\text{H}_5 \\ | \\ \text{OH} \end{array}$$

(S)-26

* (R)-(+)-α-Methoxy-α-trifluoromethylphenylacetyl chloride.

cis-2-butene, affords the borane *24*. Reaction of iodine with *24* in the presence of sodium methoxide-methanol yields (*R*)-2-iodobutane (*25*) in 84% e.e. From the (*S*) stereochemistry of the alcohol *26* produced via treatment of *24* with H_2O_2 it was concluded that the sodium methoxide-induced reaction of organoboranes with iodine proceeds generally with inversion of configuration at the carbon bearing the boron atom.

2. *Other Chiral Reducing Agents*

New modifications of lithium aluminium hydride (LAH) have been introduced, and studies have been continued on the already known complexes between LAH and alcohols, aminoalcohols, or sugars (13).

(-)-Menthol was successfully used for the reduction of β-aminoketones to (*R*)-(+)-β-aminoalcohols (*27*). A considerable increase in optical yield, from a few percent to 70%, was achieved by changing from a 1:1 to a 1:3 LAH/(-)-menthol ratio (85).

A 1:1 mixture of *cis*-2,3-pinanediol (*28*) and $C_6H_5CH_2OH$ was effective in modifying LAH so that it would reduce diphenylmethyl alkyl ketones (*29*) asymmetrically (86).

Darvon alcohol* (*30*) is among the more efficient complexing amino alcohols, yielding, with ethereal LAH solution, quite a puzzling reagent (87). Depending on the length of time that the reagent has been allowed to stand before it is used in reduction of a carbonyl substrate, one or the other enantiomeric alcohol may be obtained. With respect to the parameters of the reaction, Mosher and co-workers made the following observations: when acetophenone is added to a freshly prepared and heterogeneous solution of the reagent (LAH:*30* in 1:2.3 ratio) in ether, one obtains (*R*)-methylphenylcarbinol of 68% enantiomeric purity, in a nearly quantitative yield; if the reagent is allowed to stand for a few hours or is refluxed for a few minutes, the now homogeneous solution reduces the same substrate to the *S* enantiomer of the alcohol in 43% yield and 66% enantiomeric purity. Additional experiments showed that the reversal of stereoselectivity was not merely originating from the heterogeneous or homogeneous nature of the reducing mixture. The nature and reactivity of the reducing moieties are not well defined, but the variation of the stereoselectivity in the reduction of ketones probably reflects some change in the reducing species with the age of the reducing mixture (88).

*Darvon alcohol is (+)-(*2S*,*3R*)-4-dimethylamino-3-methyl-1,2-diphenyl-2-butanol (see Table 3). It reacts with LAH according to the equation

$$LiAlH_4 + nR\text{-}OH \rightarrow LiAl(OR)_nH_{4-n} + \frac{n}{2} H_2 .$$

The use of LAD with *30* led to the synthesis of
$C_6H_5CD(CH_3)OH$ and $C_6H_5D(H)OH$ with 80% and 40% e.e., respectively (89).

LAH-*30* has also been applied to the preparation of the
optically active cis,cis and trans,trans diols *32* and *33*
through reduction of *dl-cis*-α-acetoxy ketone (*31*) with 64% and
67% optical purity, respectively (90). It is pertinent to this
case that the reaction of chiral reagents with racemic substrates to give two sets of diastereomers, each one being
composed of enantiomers, has been discussed by Horeau and
Guetté (91). These authors established that the optical
purities of the diastereomers produced in such cases are in
inverse ratio to their respective amounts.

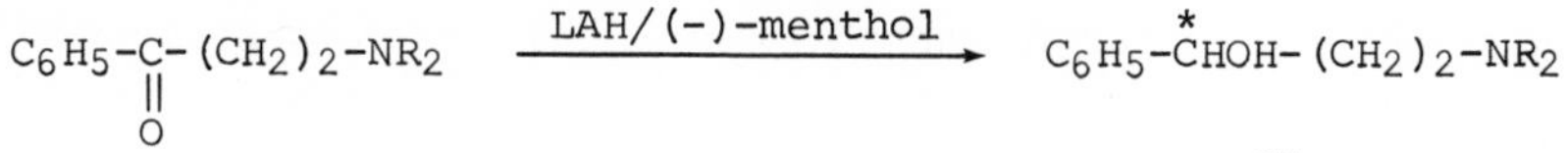

$$C_6H_5-\underset{\underset{O}{\|}}{C}-(CH_2)_2-NR_2 \xrightarrow{\text{LAH/(-)-menthol}} C_6H_5-\overset{*}{C}HOH-(CH_2)_2-NR_2$$

27

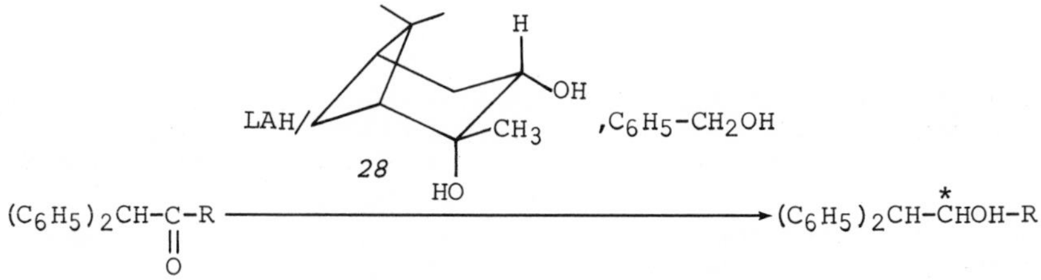

$$(C_6H_5)_2CH-\underset{\underset{O}{\|}}{C}-R \xrightarrow{\hspace{5cm}} (C_6H_5)_2CH-\overset{*}{C}HOH-R$$

29

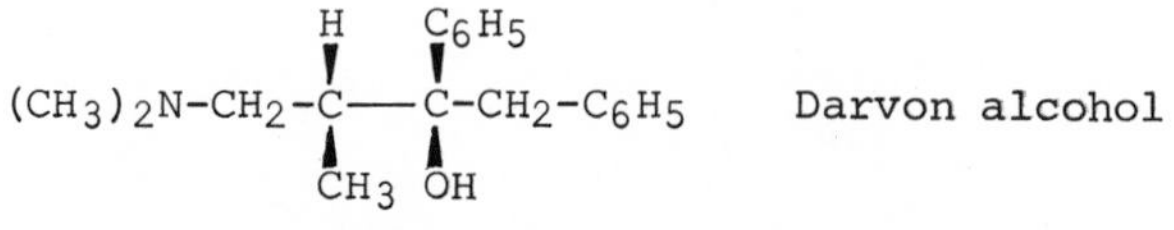

$$(CH_3)_2N-CH_2-\overset{\overset{H}{|}}{\underset{\underset{CH_3}{|}}{C}}-\overset{\overset{C_6H_5}{|}}{\underset{\underset{OH}{|}}{C}}-CH_2-C_6H_5 \qquad \text{Darvon alcohol}$$

30

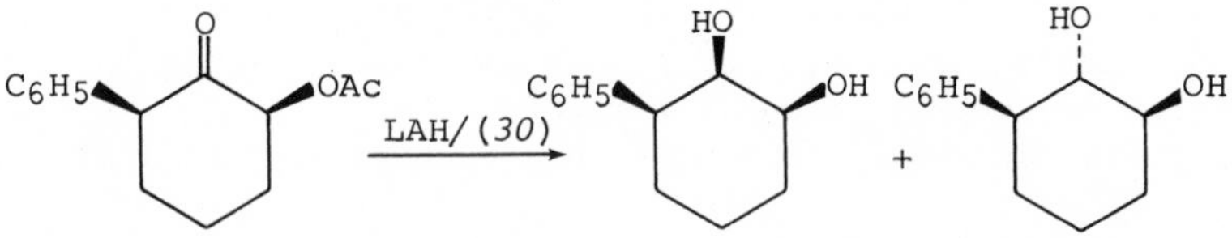

By reduction of various substituted aromatic alkyl
ketones (*34*) with a 1:1:2 mixture of LAH, *N*-methylephedrine,

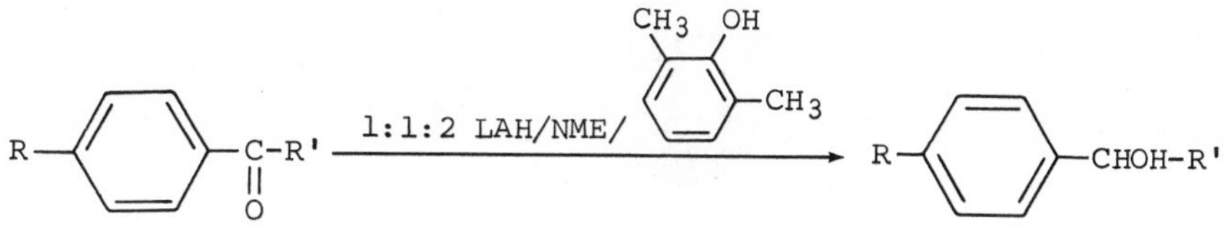

R—⬡—C(=O)—R' $\xrightarrow{\text{1:1:2 LAH/NME/}}$ R—⬡—CHOH—R'

(reagent shown: 2,6-dimethylphenol, CH₃, OH, CH₃)

34 → 35

NME = (−)-N-Methylephedrine

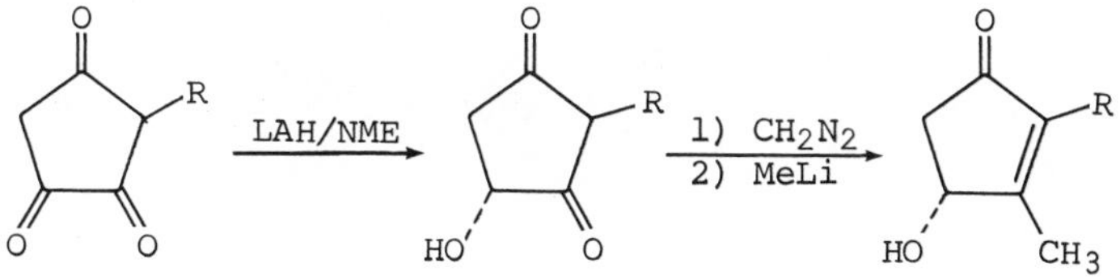

36 $\xrightarrow{\text{LAH/NME}}$ 37 $\xrightarrow[\text{2) MeLi}]{\text{1) CH}_2\text{N}_2}$ 38

R = n-Pr, allyl

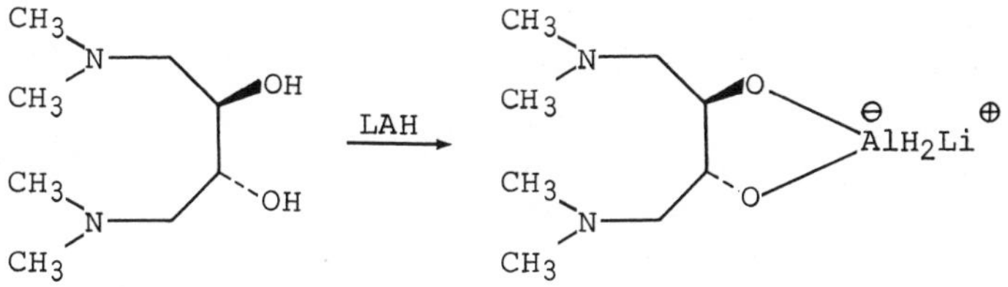

39 $\xrightarrow{\text{LAH}}$ 40

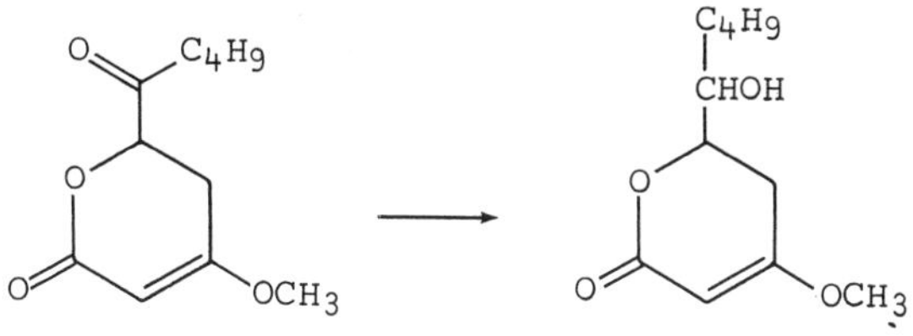

41 → 42

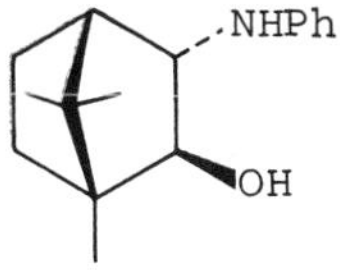

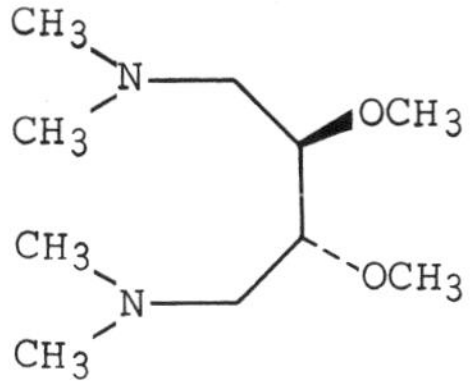

44 43

207

and 2,6-dimethylphenol, Vigneron (92) obtained the alcohols *35* in enantiomeric purities as high as 88%. For these reductions he also investigated the influence of temperature, a critical factor in asymmetric induction.

The (-)-N-methylephedrine-LAH reagent also reduces 2-allyl-1,3,4-cyclopentanetricne (*36*) regiospecifically and stereoselectively to the alcohol *37* with 55 to 58% e.e. This alcohol can be further transformed to *38* (93).

Seebach (94) used stoichiometric amounts of aminoalcohol *39* (1,4-dimethylamino-2,3-butanediol, DDB) to modify LAH, giving reagent *40*, which reduces aldehydes, ketones, and ozonides to optically active alcohols. Enantiomer *40* yielded (*S*)-carbinols from dialkyl and alkyl ketones. Since DDB is readily available from diethyl tartrate, both enantiomers of DDB are accessible, leading to levo- or dextrorotatory products; DDB is easily separated from the products and is recovered without any loss of enantiomeric purity. The hydridoaluminate *40* was shown (95) to be useful for the reduction of pyrone (*41*) to the (-) isomer of pestalotin (*42*). (1*R*)-3-*endo*-phenyl-amino-2-*exo*-norbornanol (*44*) was also used to complex in the asymmetric reduction of (-)- and (+)-dehydropestalotin to pestalotin (*42*) (95).

From the reductions of alkyl aryl, dialkyl, or diaryl ketones with LAH complexed with chiral amino alcohols, Cervinka (96) deduced an empirical rule for establishing the configuration of the preponderant enantiomer of the alcohol obtained. Cervinka's rule states that reduction of a ketone R-CO-Ar with the LAH-(-)-quinine reagent produces alcohols of (*R*) configuration.

The use of LAD with an equivalent amount of (-)-quinine makes possible the preparation of (+)-1-deuterioalcohols R-CHDOH in 10 to 40% e.e. from the aldehydes *R*CHO (97).

Welvart (67) showed that Cervinka's correlation is valid for the reduction of alkylidenemalononitrile with LAH complexes of amino alcohols. As previously mentioned (Sect. I-D-3), he observed opposite asymmetric induction in the reduction of *E* and *Z* isomers, respectively, of alkylidenecyanoacetic esters with the same reagents.

Landor (98) used an LAH-3-*O*-benzyl-1,2-*O*-cyclohexylidene-α-D-glucofuranose complex (*45*) for the asymmetric transformation of the enyne *46* to allenol *47*. The complex *45* also yields optically active amines *49* from oximes or their derivatives (*48*, R^2 = *O*-tetrahydropyranyl or *O*-methyl). Aluminohydride *45* tranfers two active hydride atoms. The first, which is the more accessible, goes to the carbon atom of the oxime group in *48*; examination of models for the two diastereomeric transition states shows that the re face (99) of *48* is preferentially attacked by the most reactive hydride to give optically active amine *49* of the (*S*) configuration. In the ethanol-modified reagent an ethoxy group replaces the more reactive hydride, leaving the other one for the transfer to the carbon of the imine. Examination of the model indicates that attack on the side of *48* is now preferred, leading to the amine of (*R*) configuration. Only slight differences in the stereoselectivity were found as a function of the nature of the R^2 substituent in the oxime *48*. For alkyl aryl ketoximes steric factors do not determine the stereoselectivity. Desoxybenzoin oxime shows higher stereoselectivity (25%) than do acetophenone oxime (11%) and α-naphthyl methyl ketone oxime (10%) in contrast to what would be expected from steric considerations only. It is thought (100) that interactions between the benzyl residue of the furanose ring and the aromatic group in the oxime counteract the effect of difference in the bulkiness of substituents.

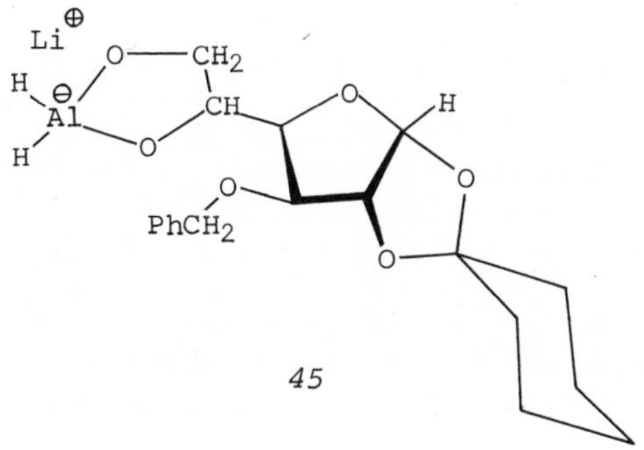

45

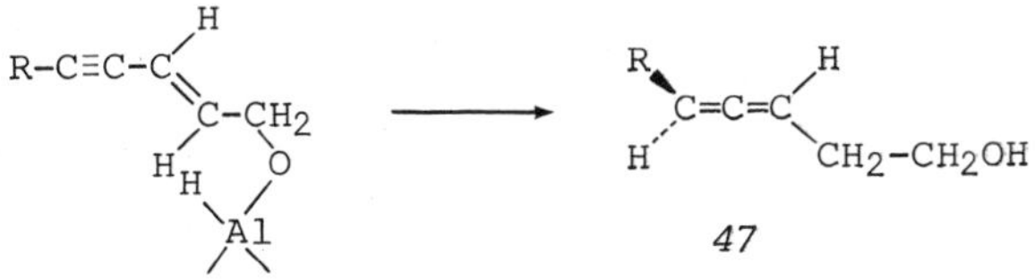

46   *47*

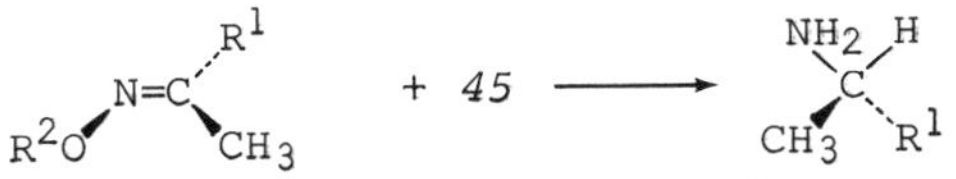

48   *49*

$$C_6H_5-CO-(CH_2)_n-N\big< \xrightarrow[\text{or } (-)-\text{isorbornylMgCl}]{(+)-2-\text{Methylbutyl MgBr}} C_6H_5-CHOH-(CH_2)_n-N\big<$$

50

$n = 1, 2$

51   52   53

Studies continue on reagents which reduce ketones or activated olefins via a β-hydrogen transfer. To explain the observed stereochemistry of the reaction, both steric interactions and polar or electronic factors must be taken into account.

β-Hydrogen transfer of asymmetric Grignard reagents has been used by Klabunovskii (101) for the reduction of α- or β-amino ketones 50 in satisfactory enantiomeric purities, although the yields were low (3-38%).

Guetté (56) has rationalized the results of the reduction of p-substituted benzophenones by chiral Grignard reagents in terms of charge transfer interactions between the aromatic rings of the substrate and reagent (see Sect. I-A-3).

Nasipuri (102) carried out an investigation of the reduction of phenyl alkyl ketones with chiral alkoxyaluminium dichlorides (51), producing chiral alcohols with the configuration shown in 53, via the preferred transition state depicted in 52.

It was shown once more that Prelog's rule, which predicts the stereoselectivity of nucleophilic addition to the α-keto esters of chiral alcohols, is also valid for the reduction of α,β-ethylenic esters by LAH (103). However, the rule is not applicable to these substrates when Grignard reagents are used as reducing agents. Reduction, by chiral Grignard reagents, of alkylidenecyanoacetic esters (ethyl Z- and E-2-cyano-3-phenyl-butenoates) and acetophenones gives opposite stereochemical results (67). This finding can be rationalized on the basis of a cyclic mechanism for Grignard additions to ketones previously proposed by Whitmore (Fig. 11), whereas a noncyclic mechanism (Fig. 12) (which may involve polar interactions between the reagent and the substrate) is probably involved in Grignard additions to ethylenic compounds.

N-(α-Methylbenzyl)-N-methylaminoalane has proved to be highly enantioselective (84% e.e.) in the reduction of acetophenone (104).

Optically active (+)-tris-(S)-2-methylbutylaluminum was used (105) to reduce ketones R-CO-Me or R-CO-Ph to the

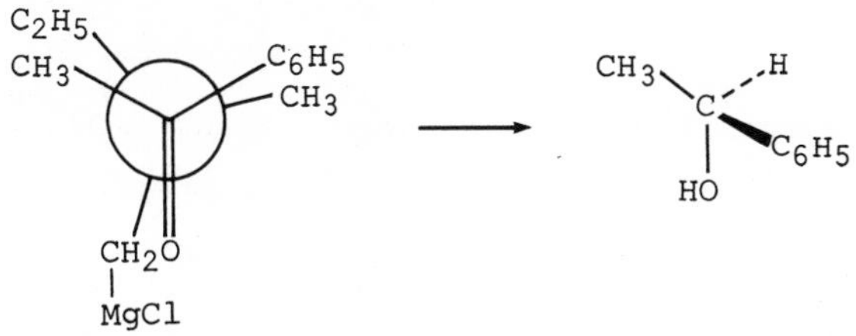

Preferred transition state
for the reduction of acetophenone with (S)-(+)-methyl-
2-butylmagnesium chloride

Predominant product

Fig. 11.

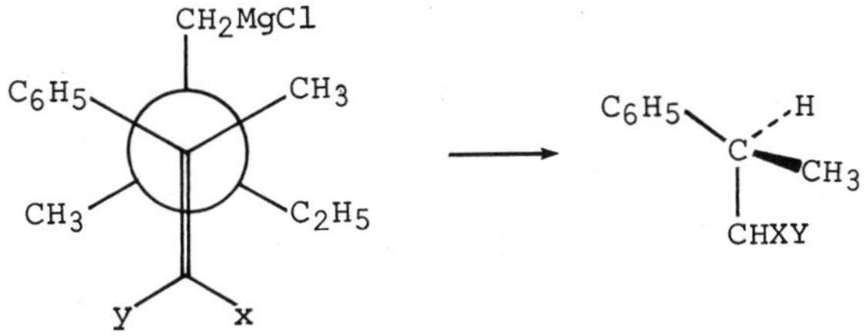

x and y = CN or $CO_2C_2H_5$

Preferred transition state
for the reduction of butenoates
with (S)-(+)-methyl-2-butyl-
magnesium chloride

Predominant product

Fig. 12.

corresponding alcohols with up to 46% e.e. The steric and
electronic interactions in the competing cyclic transition
states for the β-hydrogen-transfer step were rationalized. It
was also shown (106) that chiral dialkyl zinc reagents such as
$\{[(S)\text{-EtMeCH}(CH_2)_n]\}_2Zn$ (n = 1, 2, 3) reduced isopropylphenyl
ketone to the corresponding (S)-carbinol with less than 15%
optical purity.

   Totally stereoselective reduction of pyruvate to D- or
L-lactate is known to take place with NADH, using a D- or
L-lactate dehydrogenase catalyst (107).

   A model of nonenzymatic reduction (108) is a nicotinamide
derivative, where R is a chiral moiety (R^1 = α-phenylethyl).

Though *54*, of course, does not induce asymmetry in the
reduction of achiral alkyl- or phenylglyoxylates, it reduces
(-)-menthyl benzoylformate to (-)-menthyl (R)-(-)-mandelate
(6% e.e.); the sense of this induction can be predicted by
Prelog's rule (109).

Inouye and co-workers (110) have developed a similar
reaction, also involving the nonenzymatic direct transfer of
hydrogen from 3,5-dicarboalkoxy-2,6-dimethyl-1,4-dihydro-

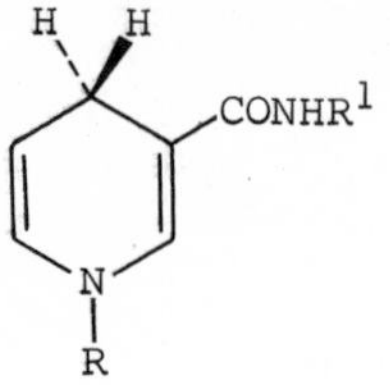

54

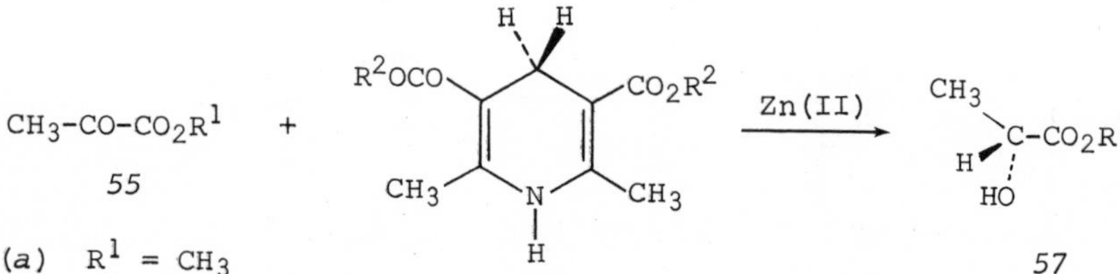

$CH_3-CO-CO_2R^1$ +

55

(a) $R^1 = CH_3$
(b) $R^1 = (-)$-menthyl

56

(a) $R^2 = CH_3$
(b) $R^2 = (-)$-menthyl

57

pyridines (Hantzsch esters *56*) to α-keto esters. They obtained
single and double asymmetric induction reactions, combining
chiral or achiral *56* with chiral or achiral *55*. Thus chiral *56b*
reacted with achiral *55a* to yield (R)-(-)-*57* in 47% e.e., while
the reaction of the chiral reagent *56b* with the chiral substrate *55b* enhanced the asymmetric induction to 78% e.e. The
reaction works in the presence of Zn (II) species prepared by
a Reformatsky reaction. These biomimetic reactions are of interest both from a mechanistic and a preparative point of view,
and improvements can be expected in the stereoselectivity
through use of more organized and sterically defined systems.

The stereochemical outcome of some asymmetric Reformatsky
reactions, involving reaction of α-bromoesters with (-)-menthyl
and (+)-bornyl pyruvates in THF or benzene, does not follow the
prediction of Prelog's generalization (111). Nevertheless, and
for the same reaction performed in the polar solvent DMF or
DMSO, the configurations of the products are in agreement with

Prelog's rule. This would imply that in the former case a
chelate-like transition state, with the two carbonyl groups of
the keto ester and the metal, is likely to be involved, and
that a competition for zinc between the polar species (solvent
or substrate) occurs in the second case.

3. Asymmetric Olefin Mercuration

Asymmetric inductions of up to 32% have been reported in
the oxymercuration of styrene by optically active mercury salts.
Thus mercuric tartrate afforded optically active 1-phenyl-
ethanol via the S_N2 attack of water on a molecular complex
whose stereochemistry was depicted as *58* (112).

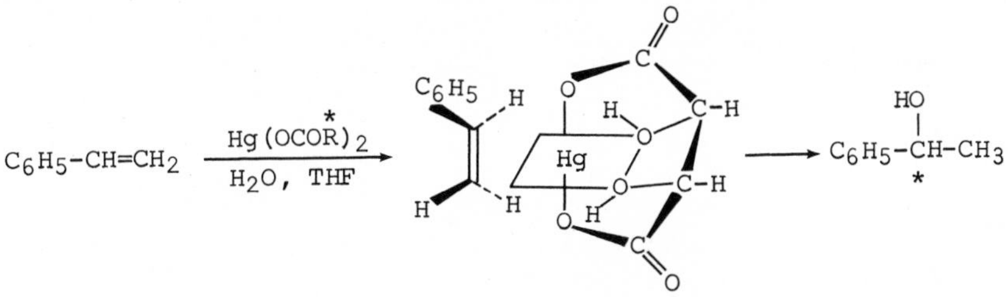

58

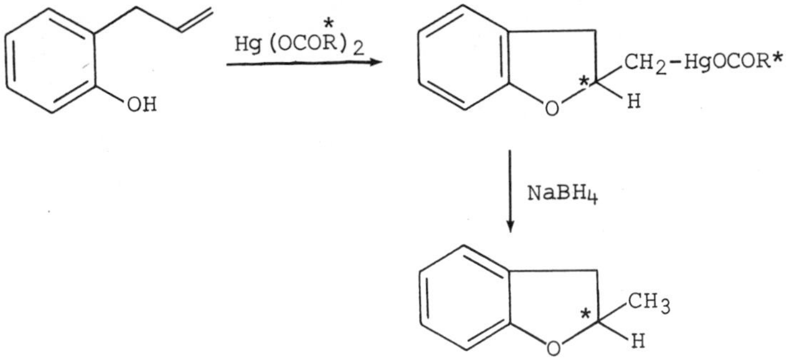

59

Oxymercuration of 2-allylphenol with mercury(II) salts of
chiral acids, followed by cyclization and demercuration, gives
optically active 2,3-dihydro-2-methylbenzofuran (*59*) (113).
Optical yields in this asymmetric cyclization are low.

4. Asymmetric Alkylation

Various chiral reagents (ylid reagents and cuprates) have

been prepared recently for the creation of C-C bonds in asymmetric syntheses.

Optically active dialkylamino alkylaryloxosulfonium fluo-
borates (60) give chiral ylides (61) when treated with base.
Reaction of 61 with benzaldehyde produces optically active
styrene oxide in 20% e.e. Reaction of 61 with an electrophilic
olefin leads to optically active cyclopropane derivatives.
Methyl *trans*-cinnamate affords methyl *trans*-2-phenylcyclopro-
panecarboxylate in 30% e.e. (114).

Reaction of chiral phosphonate carbanions 63 with ketene
62 leads to optically active allenic carboxylic acid esters
64 with 10 to 20% e.e. (115).

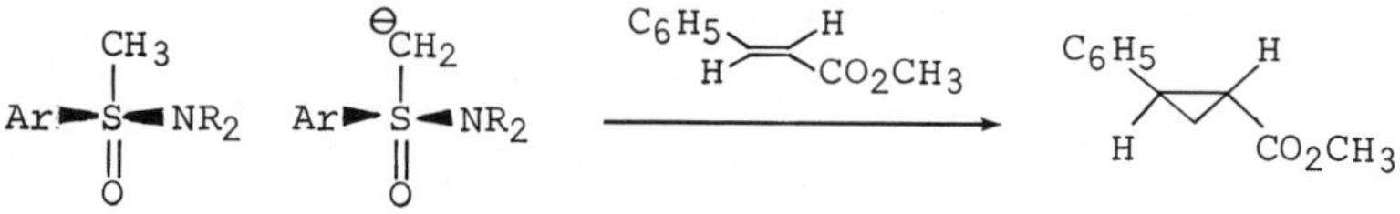

60   61

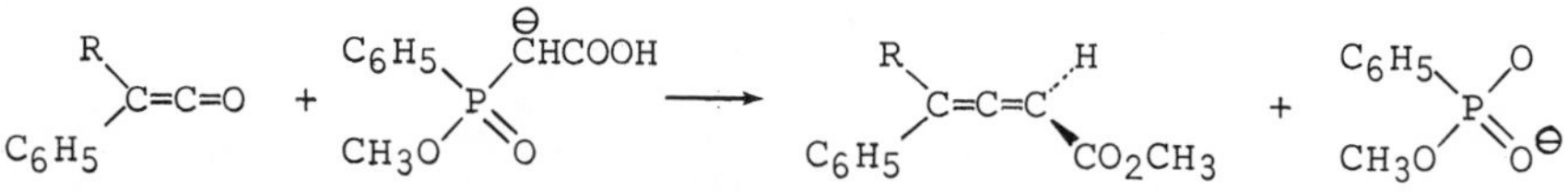

62   63   64

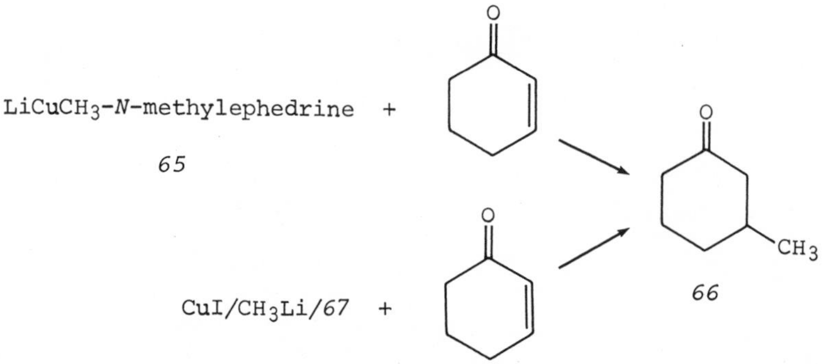

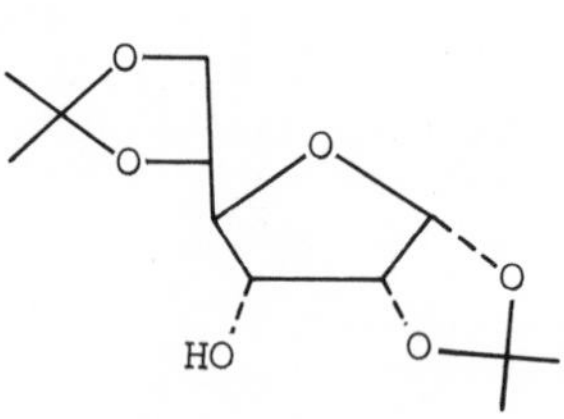

LiCuCH₃-*N*-methylephedrine +

65

CuI/CH₃Li/67 +

66

67

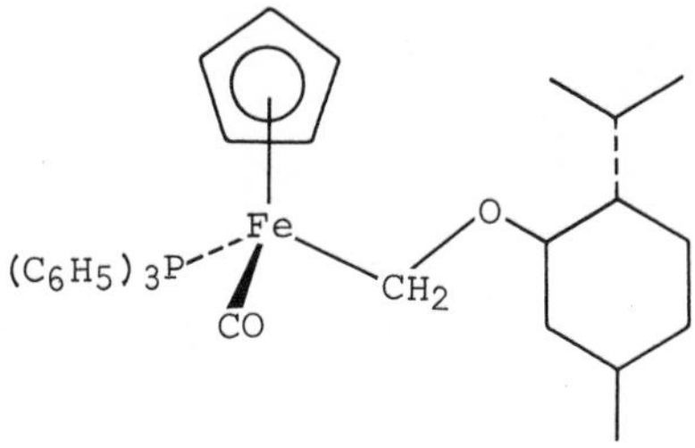

68

Use of (-)-*N*-methylephedrine with cuprate reagents induces asymmetric induction in the reaction of *65* with α,β-unsaturated ketones; as an illustration reaction of *65* with 2-cyclohexenone affords 3-methylcyclohexanone (*66*) (2% e.e.). The cuprate reaction in presence of furanose (*67*) leads to *66* with 7% e.e. (116). In view of the large variety of cuprate reactions in organic synthesis it may be expected that more research will be done on asymmetric reactions in this field.

Chiral polyaminoethers related to (*39*) (such as DAB *43*), as well as chiral diamines such as *N,N,N',N'*-tetramethyl-1,2-cyclohexanediamine (117), are useful complexing agents in a large variety of reactions involving condensation of organolithium derivatives with aldehydes and ketones (118).

Diastereomeric (+)- and (-)-(*pentahapto*-cyclopentadienyl)-(*monohapto*-methylmenthyl ether)-carbonyltriphenylphosphine-iron complexes (*68*), when cleaved by HBF_4 in solution of *trans*-1-phenylpropene, gave the (1*R*,2*R*)-*trans*-1-methyl-2-phenyl-cyclopropane (26% e.e.) and (1*S*,2*S*)-*trans* enantiomer (38% e.e.), respectively (119).

B. Chiral Substrates

This section deals with reactions in which an achiral reagent reacts with a chiral substrate containing a moiety inducing chirality which, in a following reaction, is removed and, if possible, recovered. A few useful examples of this very general process are presented.

1. *Total Synthesis of Steroids*

Saucy (119) has developed various asymmetric reactions leading to precursors of steroids. As an example the asymmetric condensation of the Mannich base *69* with 2-methyl-1,3-cyclopentanedione yielded *70* in up to 80% e.e. Further stereospecific reactions created new asymmetric centers with destruction of the inducing center, giving *71*, a precursor of estrone.

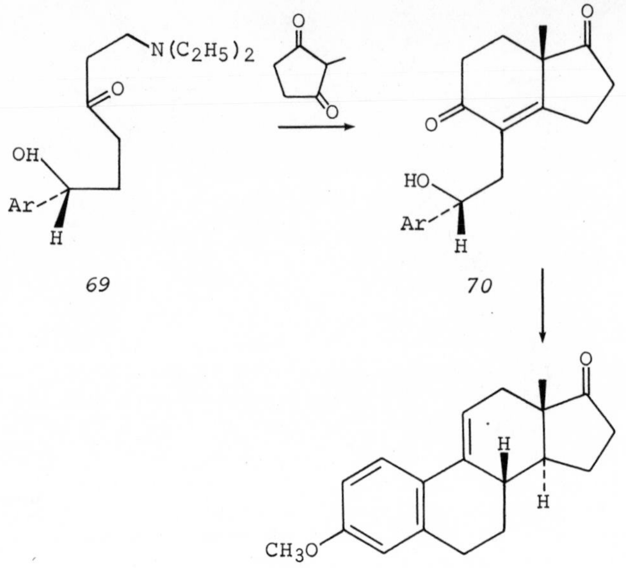

69 → 70

71

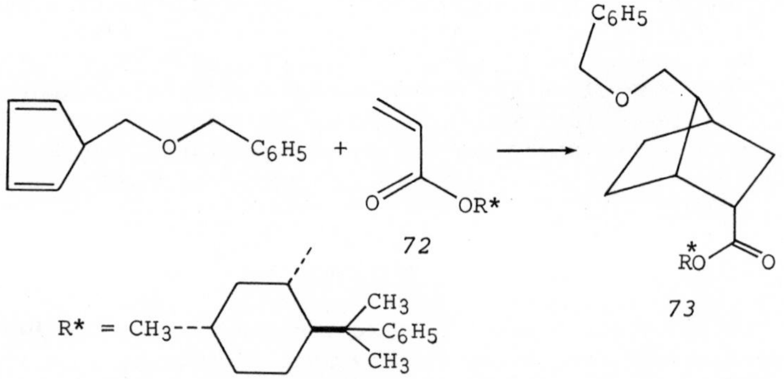

72

R* = CH₃

73

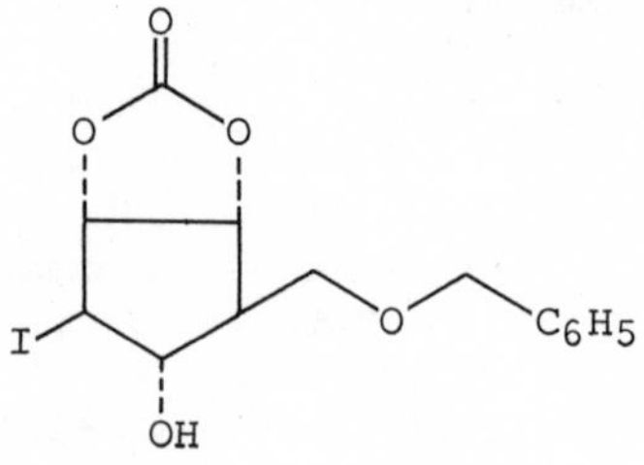

74

2. *Total Synthesis of Prostaglandins*

Corey (7) has reported a highly stereoselective Diels-Alder reaction involving addition of 5-benzyloxymethylcyclopentadiene to the chiral acrylate *72* to give *73* (89% yield); further transformations afforded the key prostaglandin intermediate *74* in optically pure form. Optically pure acrylate *72* was prepared from (-)-pulegone in 71% yield.

3. *Synthesis of Chiral Dewar Benzenes*

The first optically active "Dewar benzene" was produced by the asymmetric cycloaddition reaction of (-)-menthyl phenylpropynoate *75* to *76* in 60% yield. The enantiomeric purity of *77* was shown to be 21% (121).

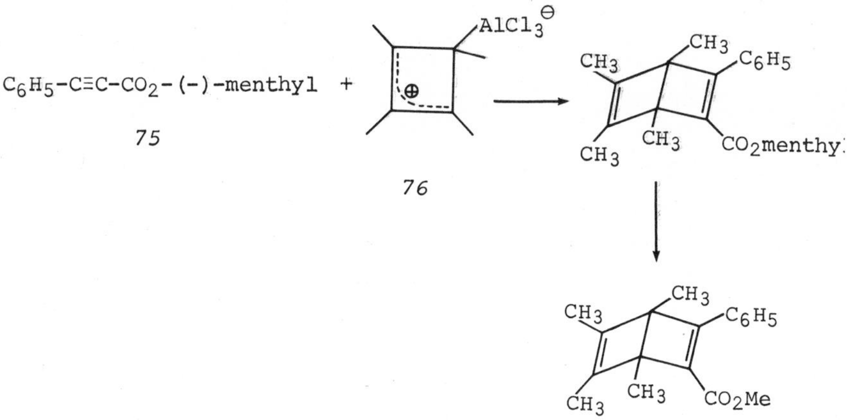

$$C_6H_5-C\equiv C-CO_2-(-)-menthyl \quad + \qquad \xrightarrow{\quad AlCl_3^{\ominus}\quad}$$

75

76

77

4. *Asymmetric Synthesis of α-Amino Acids*

Asymmetric synthesis of α-amino acids still commands considerable attention. Improvements of many previously described syntheses have been reported, involving new optically active chiral aids.

The Strecker synthesis has been reinvestigated by Harada (122), using the reaction of hydrogen cyanide with a mixture of a chiral amine and aldehyde. Optically active α-amino acids of 22 to 55% e.e. were isolated. A related synthesis consists of addition of Me_3SiCN to chiral imines (*78*) to give α-amino acids (*79*) in up to 70% e.e. (123).

Ugi (124), in studies of four-component condensations, obtained an exceptionally high degree of stereoselectivity,

using optically active ferrocenylamine (*80*) as chiral inducing
agent. He also pointed out a great difference between the rate
of cleavage of *81* to *N*-benzoyl-(*S*)-valine-*t*-butylamide (*82*),
the (*R,R*) diastereomer being cleaved about 50 times more rapidly than the (*R,S*) isomer. Thus successive cleavage of a mixture
containing a high (*R,S*)/(*R,R*) ratio would produce, after half-
reaction, a residue still more enriched in the (*R,S*) diastereomer.
This method allows the synthesis of α-amino acids of very high
optical purity and was, also, used in polypeptide synthesis
(125).

$$\begin{array}{c}\text{R}\\[-2pt]\backslash\\[-4pt]\text{C=N-R'}\\[-4pt]/\\[-2pt]\text{H}\end{array}\quad\xrightarrow[\text{ZnI}_2]{(\text{CH}_3)_3\text{SiCN}}\quad\begin{array}{c}\text{R-CH-N-R'}\\[-2pt]|\quad|\\[-2pt]\text{CN Si(CH}_3)_3\end{array}\quad\xrightarrow[\text{2) H}_2/\text{Pd(OH)}_2]{\text{1) H}^+}\quad\begin{array}{c}\text{R-CH-NH}_2\\[-2pt]|\\[-2pt]\text{COOH}\end{array}$$

78

79

$80\ (\text{R}^1\text{-NH}_2)$

$$(S)\text{-valine} \leftarrow \begin{array}{c}\text{NH}t\text{-Bu}\\|\\\text{CO}\\|\\\text{C}_6\text{H}_5\text{-CO-NH-CH}\\\backslash\text{CH}\!-\!\text{CH}_3\\\backslash\text{CH}_3\end{array} \leftarrow \begin{array}{c}\text{R}^1\quad\text{NH}t\text{-Bu}\\|\quad|\\\text{CO}\\|\\\text{C}_6\text{H}_5\text{-CO-N-CH}\\\backslash\text{CH}\!-\!\text{CH}_3\\\backslash\text{CH}_3\end{array}$$

82   *81*

　　　Optically active α-phenylglycine has been used as a chiral
auxiliary reagent for the synthesis of other α-amino acids by
hydrogenolytic asymmetric transamination (126). A linear relationship between the temperature and the inverse of log Q was
established (Q is the ratio of the diastereomeric products, as
defined in Sect. I-C-2.). An improvement of the transamination
reaction was made by Yamada (127), who used as chiral auxiliary
reagents *t*-butyl esters (*83a*) of chiral α-amino acids, thus
avoiding the hydrogenolysis step and replacing it by an oxidative decarboxylation (Fig. 13). By this process L-alanine was
prepared with up to 71% e.e. A similar procedure for transamination from an optically active acid to a ketone was used
to prepare optically active amines. Thus 2-amino-3-phenyl-

propane could be obtained in 56% yield and 87% e.e., starting
from *83b* and benzyl methyl ketone (128).

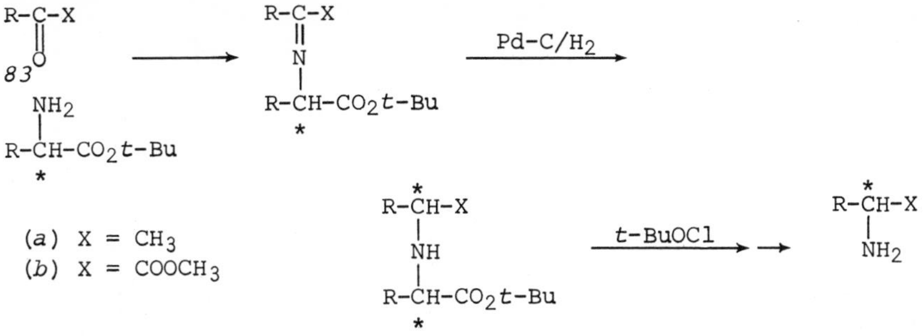

R-C-X   R-C-X
   ‖   ‖
   O   N $\xrightarrow{\text{Pd-C/H}_2}$
83 |
               R-CH-CO$_2$t-Bu
   NH$_2$ *
   |
R-CH-CO$_2$t-Bu
   *

                           R-CH-X   R-CH-X

(*a*) X = CH$_3$ NH $\xrightarrow{t\text{-BuOCl}}$ NH$_2$
(*b*) X = COOCH$_3$
                 R-CH-CO$_2$t-Bu
                   *

     Fig. 13. Asymmetric syntheses of amino acids and amines
by chemical transamination from optically active amino acids
(127,128).

     Bycroft (129) reported an asymmetric synthesis of α-amino
acids from α-keto acids and ammonia, involving the stereoselective hydrogenation of *85*, which substrate was synthesized
by coupling (*S*)-proline and an α-keto acid to give *84*, which
was dehydrated to *85*. The asymmetric induction in the hydrogenation of *85* is better than 90%. The synthesis may be modified

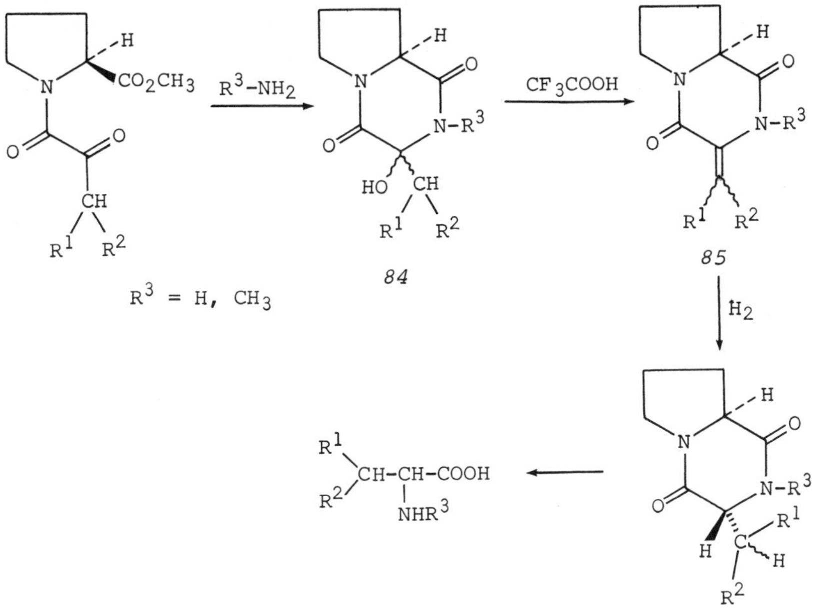

to lead to optically active N-methylamino acids. (S)-proline proved to be the best inducing amino acid, and was also recoverable.

An asymmetric synthesis of dipeptides was performed, using L-proline as chiral reagent to yield L-prolyl-α-amino acids (86). The enantiomeric purities of the resulting peptides ranged from 22 to 35% (130).

5. Chiral Oxazolines

Meyers has described the asymmetric synthesis of a great variety of optically active compounds via chiral oxazolines readily prepared from (+)-aminodiol (87). A review on asymmetric synthesis with chiral oxazolines is available (131).

A typical example is the asymmetric synthesis of α-alkanoic acids (132) (Fig. 14), which involves two steps, abstraction of a proton in 88 by a base, then alkylation of the resulting N-lithiated enamines $89A$ and $89B$. It has been shown that the temperature of the base abstraction reaction has no influence on the enantiomeric excess of acid 90. This rules out any enantioselective base abstraction of the proton in 88, which would result in the formation of different amounts of isomeric 89, and indicates that it is more likely that the second step, alkylation, is the one responsible for asymmetric induction. In fact, lowering the temperature in the alkylation step increases the degree of asymmetric induction.

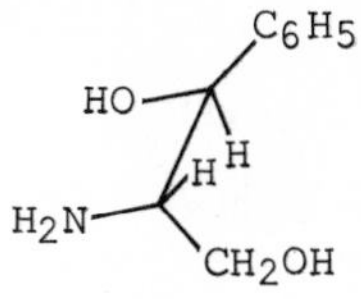

Fig. 14. Asymmetric synthesis of 2-alkylcarboxylic acids via 2-oxazolines (132).

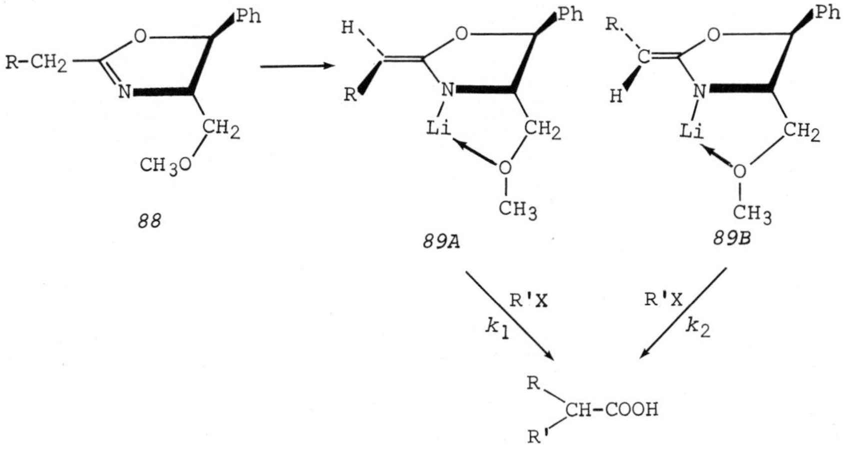

This is an example of a study in which asymmetric induction has helped in the elucidation of reaction mechanism and stereochemistry. The postulated mechanism is consistent with the fact that the attacking group R' approaches from the side opposite to the R group already present, and the stereoselectivity reflects differences in the rates k_1 and k_2 of alkylation, provided that k_1 and k_2 are small compared to the rate of equilibration between 89A and 89B. Both enantiomers of the acid 90 may be synthesized by reversing the order of introduction of R and R'. The asymmetric reactions that can be performed through the use of chiral oxazolines are listed in Table 1, which includes information on yields and e.e. The main advantages of this method are its versatility, satisfactory yields, high enantiomeric excess in products, and the easy recovery of optically active inducing material. Since 87 will presumably no longer be commercially available in the near future, chiral oxazolines derived from phenylalaninol were devised as alternate chiral reagents (133).

TABLE 1

Oxazoline	Reagent[a]	Product (% yield)	(% e.e.)
	1) LDA 2) R'X	R‑CH‑COOH (50–86) R' 2-alkylcarboxylic acid	(51–86)
	1) LDA 2) I(CH₂)₂OSiMe₃	(58–75) 2-alkylbutyrolactone	(~70)
	1) R'Li 2) MeOH	R‑CH‑CH₂‑COOH (30–76) R' 3-alkylcarboxylic acid	(90–99)
	1) LAH 2) R‑CO‑R'	R‑CHOH (78–89) R' alcohol	(up to 65)

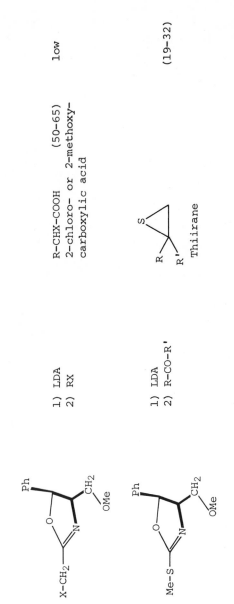

1) LDA
2) RX

R–CHX–COOH   (50–65)
2-chloro– or 2-methoxy–
carboxylic acid

low

1) LDA
2) R–CO–R'

Thiirane

(19–32)

[a]LDA = lithium diisopropylamide.

223

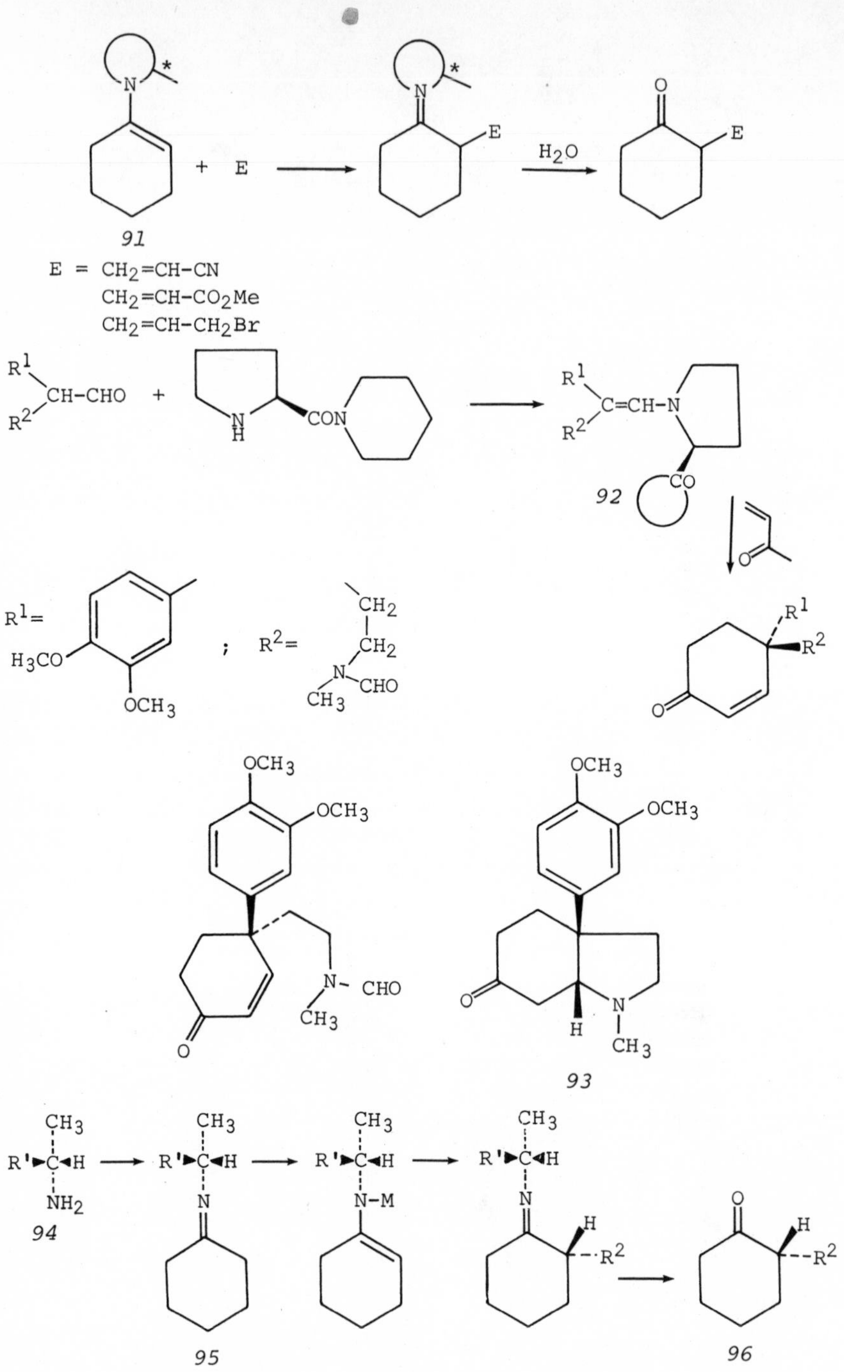

$E = CH_2=CH-CN$
$\quad\quad CH_2=CH-CO_2Me$
$\quad\quad CH_2=CH-CH_2Br$

Fig. 15. Asymmetric synthesis of mesembrine (135).

6. Asymmetric Alkylation of Aldehydes, Ketones, or Esters

Various reactions involving chiral enamines of aldehydes and ketones, or their lithio salts have been described.

Yamada performed 2-substitution of various 2- or 4-substituted proline esters (Fig. 15). The reactions with acrylonitrile, methyl acrylate, allyl bromide, and bromine (134) were investigated. This method was applied successfully to the synthesis of (+)-mesembrine (93), which involved asymmetric alkylation of chiral enamine 92 by methyl vinyl ketone (135). Lithiation of 95 improved both the chemical and optical yields of alkylation, leading to 96 in 40-50% yield and up to 37% e.e. A stereochemical model was constructed to explain the induction. Amine 94 of (S) configuration gave the ketone with the configurational arrangement shown in 96 (136).

97a   97b

A related reaction is the synthesis of optically active α-cyclocitral through asymmetric cyclization of a chiral enamine; unfortunately both chemical and optical yields were low (137).

Low-temperature alkylation of a chiral lithiated enamine 97 followed by careful hydrolysis leads to optically active α-alkylcyclohexanone 96 (138). As in other asymmetric syntheses with chiral oxazolines, the methoxyl group in 97 plays a key role. Coordination of lithium ion to the methoxy gives rise mainly to conformers 97a and 97b, which have a trans-1,2-disubstituted cyclopentyl structure in 97a and a cis-1,2-disubstituted one in 97b so that 97a would be favored. In addition it is assumed that both coordination of the halogen atom of RX with a lithium atom, and interaction between the π-bond of the ring and the alkyl group of RX would favor transition state 97a over 97b, leading to ketone 96. This is indeed the case with a large number of alkylating agents (see Table 2). This constitutes a very simple method of preparing quite a variety of 2-alkylcyclohexanones in high optical purity and of predictable configuration.

TABLE 2

$97 \rightarrow 96$

RX	Chemical yield[a]	% e.e. (configuration)
Me$_2$SO$_4$	72	82 (R)
EtI	56	(R)
nPrI	50	>95 (R)
CH$_2$=CH-CH$_2$Br	80	>90 (S)
Ph-CH$_2$Br	56	(S)

[a]Based on chiral imine.

Lithio derivatives of the chiral oxazolines 98 may react at different rates with the two enantiomers of a racemic alkyl halide 99; in kinetic resolution of 99 by 98 enantiomer 99a is recovered in excess while the 3-alkylcarboxylic acid 100a is produced predominantly (139).

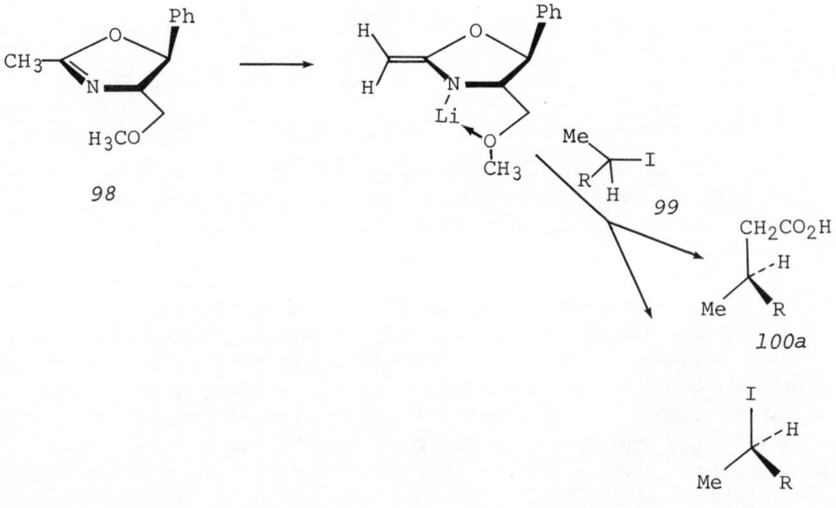

Asymmetric alkylation of ketones has also been performed via alkylation of lithiated chiral hydrazones (140). The inducing hydrazine 101 (Fig. 16) is readily available from (S)-proline, and recoverable from the nitroso compound obtained by ozonolysis or hydrolysis of 102. Alkylated ketones show enantiomeric purities reaching 87%.

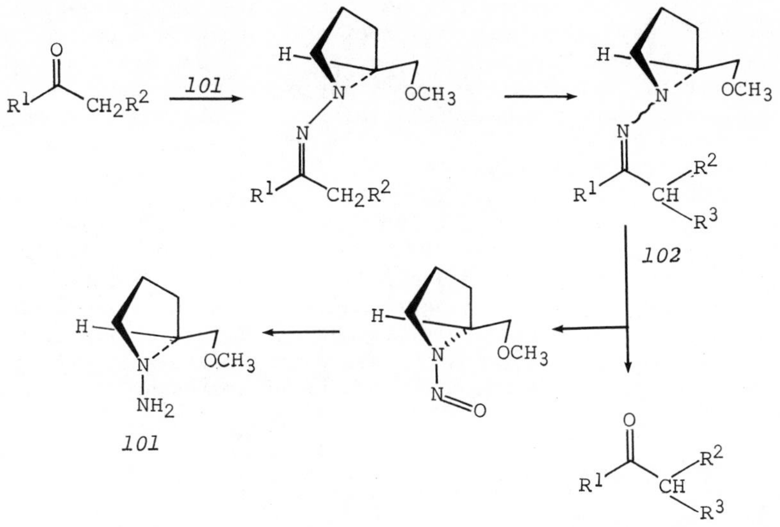

Fig. 16. Asymmetric alkylation of ketones via chiral hydrazones (130).

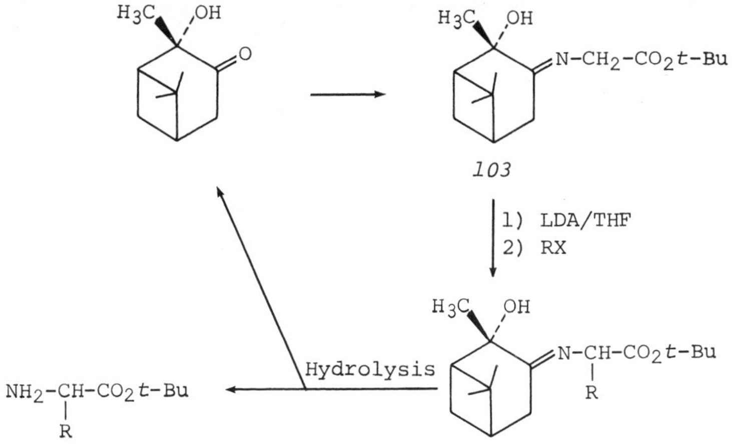

R = Me, n-Bu, benzyl

Fig. 17. Asymmetric synthesis of α-amino acids via chiral imines (141).

227

Yamada has reported an asymmetric synthesis of α-amino acids by asymmetric alkylation of the carbanion produced from the chiral Schiff base *103* (Fig. 17). This synthesis satisfies most of the criteria for a good asymmetric synthesis, that is, high asymmetric and material yields, plus the possibility of recycling the chiral inducer which is available in both enantiomeric forms (141).

Trost (142) has alkylated, in a stoichiometric reaction, diethyl sodiomalonate with *syn,syn*-1,3-dimethyl-π-allyl palladium chloride dimer *104*, in the presence of chiral phosphines, to give chiral diethyl (*E*-hex-2-en-4-yl)malonate *105*, which, upon subsequent transformations, was converted into chiral ethyl (*E*)-3-methyl-4-hexenoate. Similar alkylation of *106* with methylmalonate anion leads to a mixture of optically active *107* and *108*. The distribution of compounds *107* and *108* as well as the enantiomeric purities were dependent on the chiral inducing phosphine (143). Enantiomeric purities as high as 79% were attained.

Mikolajczyk developed interesting methods to prepare optically active sulfinates (144) and phosphinites (145), by reaction of alcohols to, respectively, sulfinyl chlorides and phosphinyl chlorides in presence of optically active tertiary amines; enantiomeric excesses may reach 43%.

7.   Asymmetric Conversion of a Racemic Mixture

Efficient transformation of racemic compounds into a single optically active form, which was considered in the total synthesis of steroids (146), remains a major challenge. The problem can be solved in certain cases: if the racemic compound is converted into a labile racemic mixture that racemizes faster than it is involved in a stereoselective process (crystallization, enantioselective reaction, etc.), then a complete asymmetric transformation is possible. The available processes for a complete or almost complete conversion of a racemic mixture into chiral products are summarized in Figure 18.

Optical activation of racemic aldehydes or ketones can be performed via hydrolysis of their chiral enamines with an achiral acid, or of an achiral enamine with a chiral acid.

Hydrolysis of *109* (one isomer) gave atropaldehyde in optically active form (37% e.e.). Hydrolysis of the enamine of optically active α-pipecoline and 2-methylcyclohexanone led to about 20% e.e. in the ketone. Analysis of the situation is more complicated in the protonolysis of enamine *110*, which exists as two isomers *110a* and *110b* (ratio 60:40), and leads to optically active *111* (147).

109	*110a*	*110b*	*111*

The formation of the enamine of a racemic α-substituted carbonyl compound with an achiral amine removes the α-hydrogen atom, leaving a compound that possesses two enantiotopic faces instead of a chiral center. That such an enamine is a labile compound is shown by the rapidly reached equilibrium of the

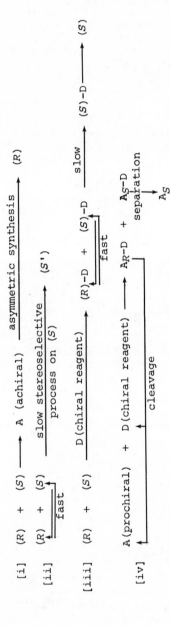

Fig. 18. Main paths for the asymmetric conversion of a racemic mixture.

cis-trans isomeric enamines *112a* and *112b* (148) (Fig. 19).
Camphosulfonate iminium salts *113a* and *113b* are said to be in
equilibrium, and the enantiomeric purity of *114* thus reflects
the position of this equilibrium (148).

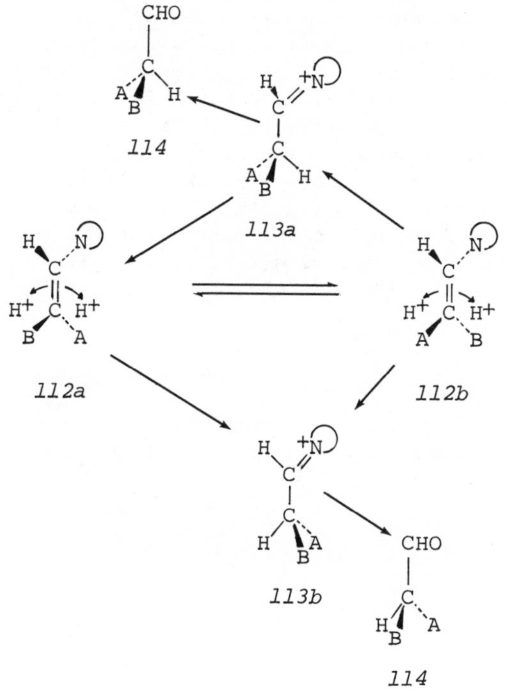

Fig. 19. Enantioselective protonation of achiral enamines
by chiral acids (148).

Hydrolysis of a *Z:E* mixture of the morpholino enamine of
atropaldehyde with a concentrated aqueous solution of tartaric
acid gives enantiomerically enriched atropaldehyde (up to 10%
e.e.); when mandelic acid is used as the proton acid, an e.e.
yield of 14% is attained (149). The enantiomeric purities
obtained vary according to the *E:Z* ratio of the enamine starting mixture, indicating two asymmetric induction values, one
for each cis-trans isomer. Such an asymmetric synthesis requires, for success, an enantioselective protonation of each
isomeric enamine, and a rate of hydrolysis of the iminium
cation greater than its rate of isomerization.
    Several examples of processes [ii] (Fig. 18) are known
(see Sect. IV-A). The spontaneous resolution of a labile racemic

mixture, such as described for 1,1'-binaphthyl is another case (150).

Cases [iii] and [i] (Fig. 18) involve asymmetric transformations. The principles involved are discussed in (4). Total syntheses of optically active steroids (146) and prostaglandins (151) have been performed based on [i].

A case related to process [iii] was developed (152) involving the conversion of DL-phenylglycine methyl ester into the D- or L-enantiomer, by using tartaric acid in presence of a small amount of cyclohexanone. Slow selective crystallization of one diastereomeric salt of the amino ester occurs from the methanolic solution (Fig. 20). The soluble phenylglycine is simultaneously racemized through Schiff base formation with cyclohexanone. Up to 95% of one enantiomer was obtained in 100% e.e. In contrast to a previous report (148) it was shown that

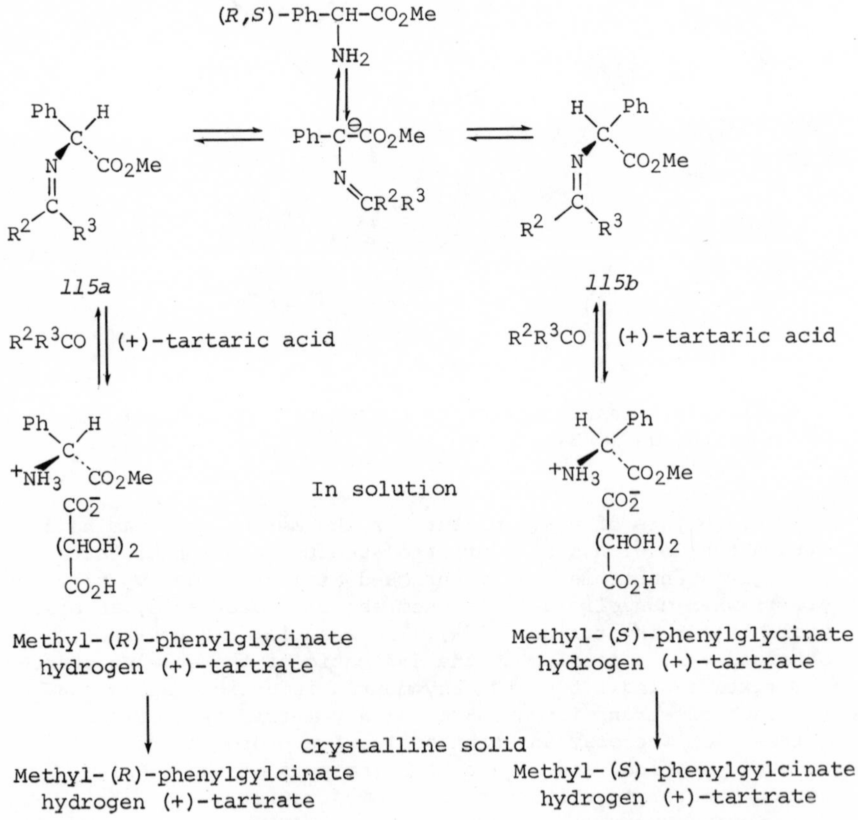

Fig. 20. Asymmetric transformation of phenylglycine esters with (+)-tartaric acid (152).

there is little, if any, induced asymmetry in solution and that
the ratio of enantiomers produced does not reflect the position of the equilibrium between *115a* and *115b*.

Investigators at the Allied Chemical Co. have described
a similar process through which L-lysine is obtained in 100%
enantiomeric purity from racemic α-aminocaprolactam (ACL)
through a process involving simultaneous resolution and racemization of the optical isomers of an ACL-nickel chloride
complex (153) (Fig. 21). A supersaturated ethanolic solution of
the complex DL-(ACL)$_3$NiCl$_2$ is formed by reaction of DL-(ACL)
with nickel chloride. Adding seed crystals of L-(ACL)$_3$NiCl$_2$ in
EtOH causes the L-complex to deposit, while the remaining D-
complex is simultaneously racemized (rapidly!) by base. One
can thus achieve full conversion of the original racemic
material into L-(ACL) complex of 97% enantiomer purity. Decomposition of complex *116* and subsequent treatment leads to
nickel-free L-amino caprolactam, which is further hydrolyzed to
L-lysine. There are two requirements for success in this process: (1) uncomplexed ACL must exchange rapidly with complexed material for efficient resolution and racemization; (2)
the racemization and crystallization rates must be of the same
magnitude (in the same solvent, at the same temperature), and
these two processes should occur in highly supersaturated
solution, while spontaneous crystallization of any other species
is avoided.

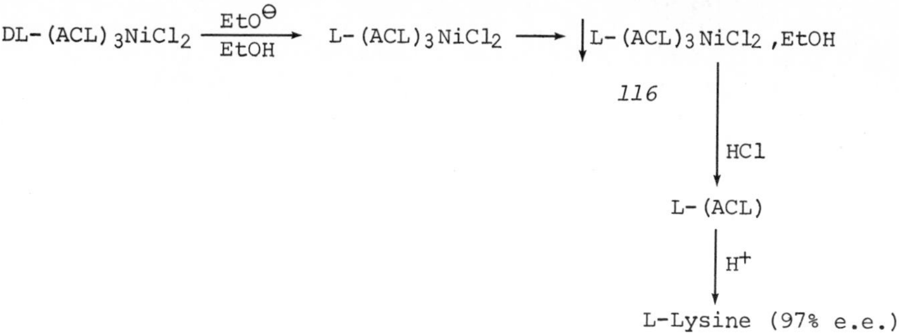

$$DL\text{-}(ACL)_3NiCl_2 \xrightarrow[\text{EtOH}]{EtO^{\ominus}} L\text{-}(ACL)_3NiCl_2 \longrightarrow \downarrow L\text{-}(ACL)_3NiCl_2\text{,EtOH}$$

116

HCl

L-(ACL)

H$^+$

L-Lysine (97% e.e.)

Fig. 21. Preparation of L-lysine via asymmetric conversion of DL-aminocaprolactam (153).

C. Catalysis

A certain number of base-catalyzed reaction (aldolization,
Michael reaction, etc.) have been performed asymmetrically by
the use of chiral amines. Some of these processes involve new
approaches to asymmetric syntheses of important compounds such
as steroids.

1. Asymmetric Aldolization

The Schering (Berlin) (154) and Hoffmann-LaRoche (New York) (155) companies both reported the annelation of triketone *118* formed through Michael reaction of 2-methyl-1,3-cyclopentanedione *117* with methyl vinyl ketone.

Cyclization using either primary or secondary amines or amino acids as chiral catalysts differentiates the enantiotopic carbonyl groups of *118*, thus converting the prochiral central carbon atom into the asymmetric carbon in *120*.

The amount of catalyst used in the first report (154) was high (1/3 molar equivalent to substrate), yielding *120* in up to 84% e.e. It was later reported that almost complete stereoselectivity can be achieved in this reaction by the use of as little as 3% (molar equivalent) (S)-(-)-proline as catalyst. Furthermore the catalyst can be recovered, and a cis relation between the rings in intermediate ketol *119* was demonstrated. Proline was shown to be the best catalyst for cyclizing *118*. Because of its insolubility the choice of the solvent was important. The poor efficiency of amino esters relative to the corresponding amino acids in asymmetric induction points to the importance of the acid function. A scheme has been proposed (156) for the differentiation of the two diastereomeric transition states, but it is not convincing. More work is needed to understand all the details of this exceptionally efficient asymmetric synthesis. The same type of asymmetric aldolization reaction has been applied by Danishefsky (157) to the synthesis of optically active *122* using an aromatic amino acid as catalyst. In this case L-proline was found to give disappointing results in terms of optical selectivity, but L-phenylalanine produced *122* in 86% e.e., the yield being in excess of 82%.

117   *118*

120   *119*

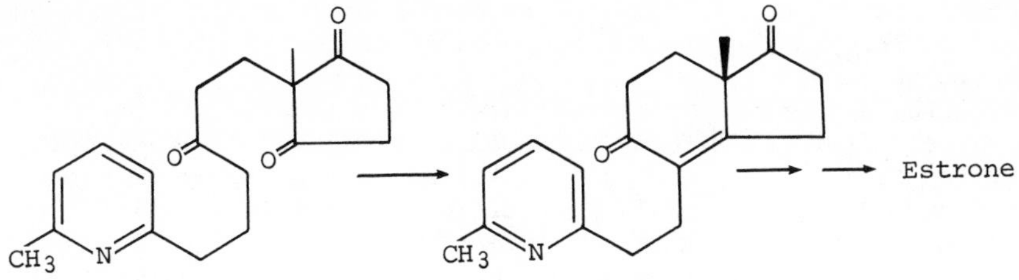

121   122   Estrone

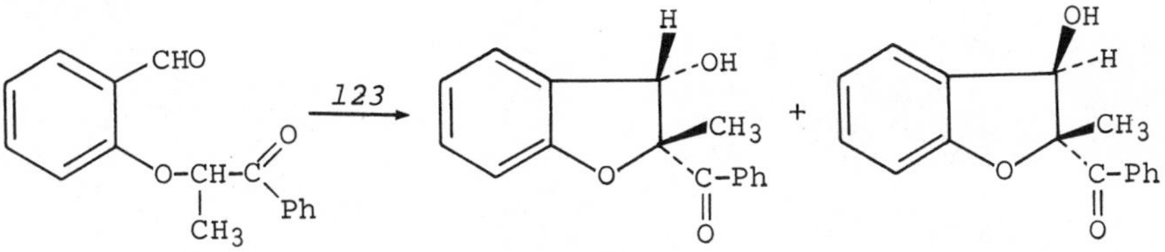

124a   124b

Both asymmetric aldol-type condensations and Michael-type reactions catalyzed by quinine and 2-(hydroxymethyl)quinuclidine (*123*) have been reported (158). Diastereomers *124a* and *124b* could be separated, and both showed optical activity; unfortunately their enantiomeric purities were not evaluated (159).

2. Asymmetric Michael Reactions

Michael addition of 2-carboxymethyl-1-indanone (*125*) to acrolein in presence of catalytic amounts of partially resolved (57% e.e.) *123* gave optically active *126*, whose

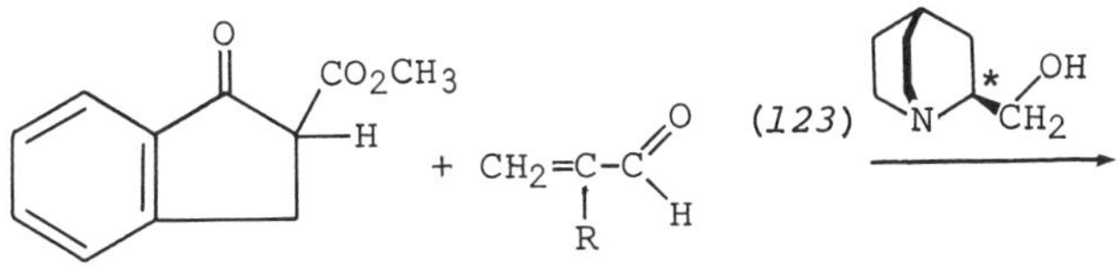

125

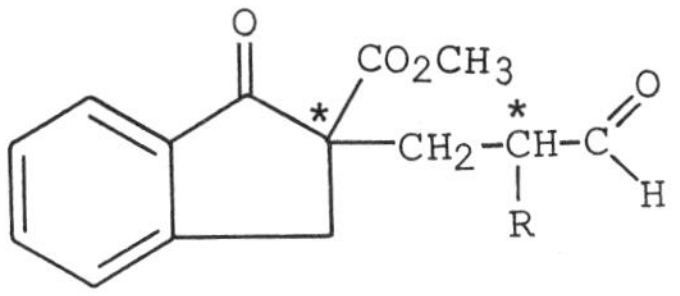

126

enantiomeric purity or diastereomeric purity was not determined. This is the first example of an asymmetric Michael
addition (159).

The appearance of optical activity in *124* and *126* means
that the protonated form of the base (after abstraction of a
proton from the substrate) is involved in the nucleophilic
addition to the carbonyl group or activated double bond,
probably via anion-pair interaction.

Wynberg (160) has studied the reaction of various Michael
donors toward methyl vinyl ketone in the presence of catalytic
amounts of quinine. Starting from 2-carbomethoxy-indanone
(*127*), he obtained *128* in 68% e.e.

$$\text{127} \quad + \quad CH_2=CH-CO-CH_3 \quad \longrightarrow$$

127

128

3. *Asymmetric Phase-Transfer-Catalyzed Reactions*

Phase-transfer-catalysis processes require quaternary
ammonium compounds as counterions for the extracted anions.
Optically active quaternary ammonium compounds have been used
in various reactions in attempts to perform asymmetric syntheses
under phase transfer conditions. The catalysts employed, *N*-
alkyl-*N*-methyl-ephedrinium salts (*129*), were usually derived
from *N*-methyl ephedrine.

The first described phase transfer catalyzed asymmetric
reaction was that of trimethylsulfonium iodide with benzaldehyde to give phenyloxirane (*130*). The authors claimed 97% e.e.;
this result was found to be erroneous (161b) because of
contamination by the chiral epoxide *131* produced through
base decomposition of *129*.

By a phase transfer technique using *129a* as catalyst (161a),
addition of dichlorocarbene or dibromocarbene to olefins led
to the corresponding chiral dichloro- or dibromocyclopropane
adducts in low optical yields. In both cases hypothetical

schemes of preferred transition states were presented for the reaction.

An asymmetric alkylation reaction of cyclic β-diketones or β ketoesters in the presence of *129b* as a catalyst led to *132* in relatively modest enantiomeric purity (162). In this reaction the initial concentration of hydroxyl ion in the aqueous phase was far less than 50% (not over 1*M* or 2*M*) so as to avoid decomposition of the catalyst and production of optically active side products.

129

(a) R = -Me
(b) R = -CH$_2$-Ph
(c) R = -(CH$_2$)$_{11}$-CH$_3$

Z = -OC$_2$H$_5$, -CH$_3$.   *132*

Initially Colonna (163) found no asymmetric induction in the reduction of dialkyl ketones by borohydride ion in a two-phase system in which *129c* was used as catalyst. More recently he showed (164) that the same catalyst could promote the asymmetric reduction of alkyl aryl ketones with a small e.e.

On the other hand, the reduction of acetophenone by NaBH$_4$ in the presence of *129c* to give methylphenyl carbinol in 39% e.e. has been described (165) by Massé.

133   *134*

Stereoselective epoxidation of chalcone *133* by H_2O_2 has been achieved in the presence of a quaternary ammonium salt of quinine, *N*-benzylquininium bromide (166), yielding epoxide *134* in 25% e.e. Curiously, epoxidation with *t*-butyl hydroperoxide as oxidant reversed the sense of the asymmetric induction.

Asymmetric phase transfer catalysis is still in its infancy and, in view of these results, further developments concerning the scope and the stereoselectivity of the process may be expected in the near future.

III. CATALYSIS WITH TRANSITION METAL COMPLEXES

Many enzymatic systems involve metalloenzymes that act as chiral catalysts and are able to perform stereospecific chemical transformations such as hydroxylations. It is surprising that for a long time biochemical processes failed to stimulate organic chemists to use chiral transition complexes as asymmetric catalysts.

The reason is that homogeneous catalysis has been developed only since 1966. This has made possible the development and subsequent improvement of chiral catalysts. Asymmetric hydrogenation was most intensively studied at first, but asymmetric induction in C-C bond formation is now gaining more attention. The main results in asymmetric catalysis through 1974 are included in refs. 167 through 180. Several reviews are devoted to reactions catalyzed by chiral transition metal complexes. Asymmetric hydrogenation has been reviewed several times (167), and a very complete review through 1975 has recently appeared (168). A chapter devoted to asymmetric hydrogenation can be found in several volumes dealing with hydrogenation reactions (169,170). We present here only the main results and new developments. Asymmetric hydrosilylation through 1976 is covered by a review article (171). Important results in asymmetric synthesis via C-C bond formation, through the end of 1973, may be found in (172).

A. Asymmetric Hydrogenation of C=C Bonds

The impetus for the large amount of work published since 1968 was the discovery by Wilkinson in 1966, of an efficient homogeneous catalytic system. The catalyst precursor is $RhCl(PPh_3)_3$. It is known that, in the mechanism of this catalytic system, two phosphines remain coordinated to the rhodium atom when olefin reduction occurs. It was therefore attractive to try to reduce prochiral olefins after replacing triphenylphosphine in the Wilkinson catalyst by chiral phosphines. Since phosphines are good ligands for a large variety of catalytic systems (hydrosilylation, hydroformylation, olefin dimerization, etc.), synthesis of chiral phosphines is taking

on an increasing importance. Several approaches can be
envisaged in the preparation of chiral phosphines:

1.   The phosphorus atom can itself be the inducing asymmetric center, and is then very close to the metal and the
coordinated prochiral olefin. This approach was investigated
independently by Horner (173) and by Knowles (174).

2.   A much simpler approach is to introduce the diphenyl-
phosphino group into optically active molecules such as natural
products. There is no resolution step in the preparation of
phosphine of general formation R^*-PPh_2 (R^* = chiral group). For
example, Morrison (175) has prepared menthyl and neomenthyl
diphenylphosphine (MDPP and NDPP) from two readily available
chiral alcohols.

3.   To decrease conformational mobility in phosphines
R^*-PPh_2 and to have sterically better defined coordinated
ligands, Kagan and Dang (176) synthesized chelating diphosphines
$Ph_2P-R^*-PPh_2$ in which the two phosphorus atoms are separated by
several carbon atoms.

4.   Since approach 3 was very successful, chiral phosphines $PhP^*(R)CH_3$ obtained in 1 were coupled, and diphosphines
$PhP^*(R)-(CH_2)_2-P^*Ph(R)$ used as chiral ligands (177).

One may expect to see, in the near future, the synthesis
of chiral monophosphines or diphosphines by a combination of
approaches 1 and 2, or 3 and 4, with asymmetric centers located
both on phosphorus atoms and on the organic chain.

The most spectacular results were observed with prochiral
olefins bearing one or two polar functions such as CO_2H or
$NHCOCH_3$. Some representative chiral phosphines are listed in
Table 3, with the most interesting results in asymmetric
catalysis compiled in Table 4.

The asymmetric synthesis of α-amino acids was achieved in
optical yields as high as 95%. L-Dopa, a useful drug in the
treatment of Parkinson disease, is now manufactured by such a
process (177). This represents the first industrial asymmetric
synthesis.

Chiral ligands show different specificities toward prochiral substrates. The efficiency of a given ligand can be
considerably increased by small structural modification. For
example, DIOP (Fig. 3) has been the subject of a considerable
amount of work (for reviews on asymmetric catalysis with DIOP
see refs. 182 and 183). Introduction of one methyl group on
each phenyl ring can increase, decrease, or maintain the optical
yield in a given reaction. Thus extra methyl groups in meta
positions change the optical yield from 82 to 88% in the asymmetric synthesis of N-acetylphenylalanine (184). Carbocyclic
analogue of DIOP behave similarly. The acetonide ring can be
replaced by a cyclopentane ring(184), for example, or by a
cyclobutane ring (185) without great changes in optical yield
in α-amino acid synthesis.

TABLE 3

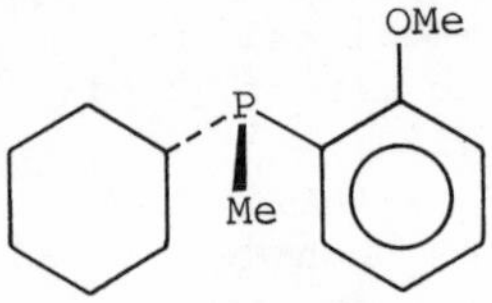

(+)-ACMP *(174)*

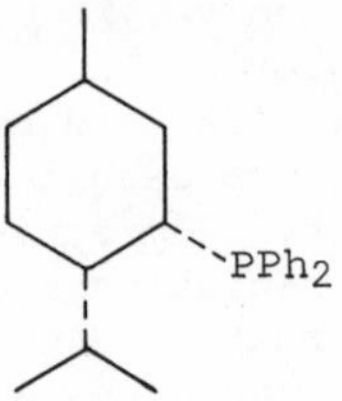

(+)-NMDPP *(175)*

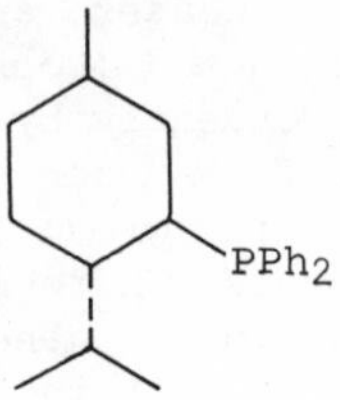

(-)-MDPP *(175)*

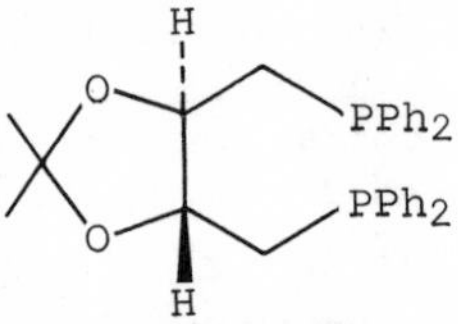

(+)-DIOP *(176)*

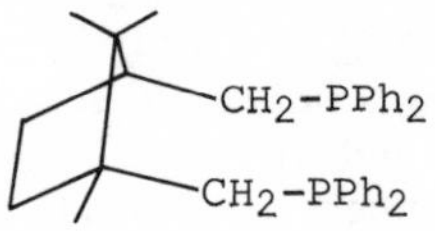

(+)-Camphos *(168)*

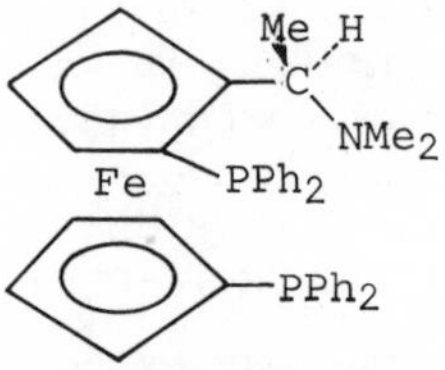

(+)-BPPFA *(181)*

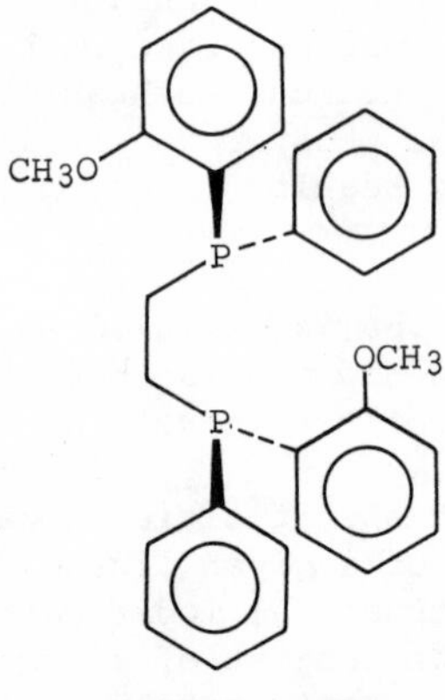

DIPAMP *(177)*

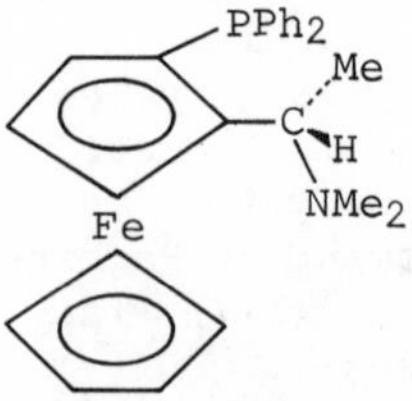

(-)-MPFA *(208)*

The mechanism of asymmetric reduction, especially when the substrates are olefins with polar groups, is not fully understood. It was hypothesized that the *N*-acetyl group could be coordinated to the rhodium atom (176). Some support for this hypothesis can be found in the good optical yields obtained with RhCl(DIOP) in the reduction of CH_2=C(Ph)NHAc (179,187).

TABLE 4

Ligand[a]	Reaction[b]	% e.e.	Product	Reference
(+)-ACMP	(Z)-ArCH=C(NHAc)CO$_2$H $\longrightarrow$ ArCH$_2$CH(NHAc)CO$_2$H Ar = 3.OMe, 4.OH-phenyl	90	(S)	174
(+)-NMDPP	(E)-PhC(Me)=CHCO$_2$H $\longrightarrow$ PhCH(Me)CH$_2$CO$_2$H	61.8	(S)	168
(-)-MDPP	(Z)-PhC(Me)=CHCO$_2$H $\longrightarrow$ PhCH(Me)CH$_2$CO$_2$H	30.6	(S)	175
(+)-DIOP	(Z)-PhCH=C(NHAc)CO$_2$H $\longrightarrow$ PhCH$_2$CH(NHAc)CO$_2$H	82	(S)	176, 178
(+)-DIOP	MeCH=C(NHAc)Ph $\longrightarrow$ EtCH(NHAc)Ph	92	(R)	179
(+)-DIOP		88	(+)	180
(+)-BPPFA	PhCH=C(NHAc)CO$_2$H $\longrightarrow$ PhCH$_2$CH(NHAc)CO$_2$H	93	(S)	181
DIPAMP	PhCH=C(NHAc)CO$_2$H $\longrightarrow$ PhCH$_2$CH(NHAc)CO$_2$H	95.7	(S)	177
(+)-Camphos	(E)-PhCH=C(Me)CO$_2$H $\longrightarrow$ PhCH$_2$CH(Me)CO$_2$H	15	(R)	17

[a]Formulas are given in Table 3.
[b]See references for experimental conditions. In all the examples reduction is almost quantitative.

In fact, however, the situation is very complicated, as opposite asymmetric inductions were observed in benzene and ethanol (187). In the latter solvent the catalyst behaves like Rh(DIOP)$^+$Cl$^-$, as demonstrated when (COD)Rh(DIOP)$^+$ClO$_4^-$ is taken as catalyst precursor.

A model of asymmetric induction was proposed (188) to explain asymmetric reduction with RhCl(DIOP). Recently the structure of IrCl(COD)DIOP has been determined by X-ray analysis. The conformation of chelated DIOP is clearly seen, and some correlations could be tentatively made with the course of asymmetric induction (186). It was hypothetized that asymmetric centers in DIOP induce a specific conformation of the chelate ring, which in turn promotes the formation of a chiral rhodium atom.

Of course the phenyl rings play a key role in the chiral recognition of the enantiotopic faces of the substrate. Asymmetric synthesis with DIPAMP should lead to less ambiguity in prediction of the steric course of reduction because of the nature of the ligand. X-Ray analysis of Rh·COD·DIPAMP$^+$PF$_6^-$ (see Table 3) was recently performed, allowing attempts toward rationalization of the origin of asymmetric induction in α-amino acid synthesis. The model involves simple steric considerations and assumes bonding between the *N*-acetyl group and the rhodium atom (189). Interestingly it was found that the (*E*) or (*Z*) configuration of the double bond can influence both the rate of reaction and the optical yield. This was observed in reduction of *N*-acetyl-α-aminoacrylic derivatives (178,189). A detailed study of the reduction of α,β-unsaturated carboxylic acids showed (168) that very often, but not invariably, the opposite configuration in the product results from cis and trans isomers around the double bond. This trend, which was found for several chiral phosphines, should have mechanistic implications (178).

A new family of simple chiral diphosphines that has recently been generated appears very promising. 2,3-Bisdiphenyl-phosphinobutane (*135b*) gives a catalyst that induces a very efficient asymmetric synthesis of *N*-acetylphenylalanine from the usual precursor (190). Almost quantitative optical yields were observed in the synthesis of leucine and phenylalanine. In this context the diphosphine *135a* as well as other diphosphines (*135*) (R = H, R' = various radicals) have been synthesized (190,191).

Some amide-rhodium complexes can catalyze homogeneous hydrogenation of olefins. A chiral catalyst was obtained by Abley and McQuillin (192) by using *N*-formyl α-phenylethylamine as the amide in Py$_2$(amide)RhCl$_2$(BH$_4$)$^+$Cl$^-$. Optical yields up to 60% were attained in the reduction of methyl(*E*)-β-methyl-cinnamate. Recently this catalyst was used by Klabunovskii (193) (75% e.e.). Phosphinites R-OPPh$_2$ are ligands that are easy to prepare, but do not give very active catalytic systems. However, chiral diphosphinites derived from *trans*-1,2-cyclo-

hexanediol (194a), *trans*-1,2-cyclopentanediol (194b), and
1,1'-binaphthol (194c) were useful in asymmetric reductions.

A chiral bisaminophosphine derived from (-)-α-methyl-
benzylamine was demonstrated to be a good ligand in the
rhodium-catalyzed reduction of several amino acid precursors.
Optical induction is generally fairly good, optical yields
being similar to those obtained with DIOP (195).

Few examples of homogeneous catalytic chiral complexes
are found in which the transition metal is not rhodium.
There is only one report of a ruthenium-DIOP system (191).
A cobaloxime-quinine catalyst, Co(dmg)$_2$-quinine, has been
developed (197) which catalyzes asymmetric reduction of α,β-
unsaturated carboxylates or ketones such as methyl atropate or
α-phenylacrylophenone to the corresponding saturated compounds
with 10 and 49% e.e., respectively.

B. Asymmetric Hydrogenation of C=O and C=N Bonds

Many chiral alcohols or amines bearing one asymmetric
center are important compounds (flavoring agents, drugs, etc.)
or starting materials for the synthesis of more complex molecules. Unfortunately there are few reports of the direct asymmetric reduction of C=O and C=N double bonds, because of a
lack of suitable homogeneous catalysts. The Wilkinson catalyst
RhCl(PPh$_3$)$_3$ is not active in ketone reduction, but Schrock and
Osborn demonstrated that certain cationic rhodium complexes
function as catalysts for ketone reduction, though catalytic
activity is low. The first publications on asymmetric reductions of ketones have recently appeared, describing (RhL$_2$diene)$^+$
as chiral catalyst (L$_2$ = chiral phosphines). When benzylmethylphenylphosphine or DIOP is the chiral ligand, acetophenone is
reduced to phenylmethylcarbinol with 8% e.e. (198,199). Better
results were observed by changing the experimental conditions,
optical yields up to 51% being obtained in acetophenone reduction with a rhodium-DIOP catalyst (200). In a detailed study
of a rhodium-ACMP catalyst (201) a strong solvent effect was
observed in the reduction of 2-octanone or methyl benzyl ketone. Carboxylic acids are the best solvents for this reduction, enantiomeric excesses of 12% and 20%, respectively,
being achieved. Methyl acetoacetate is a good substrate in
homogeneous asymmetric hydrogenation. When the chiral ligand
is cyclohexyl(2-isopropoxyphenyl)methylphosphine, the chiral
β-hydroxy ester is produced with 71% e.e. In the course of a
total synthesis of prostaglandin a selective asymmetric reduction of *136b* to *137b* was performed with ACMP as the chiral
phosphine, with about 30% optical yield (202). There is only
one example of an asymmetric reduction of a Schiff base,
namely PhMeC=NCH$_2$Ph. An optical yield of 22% was attained with
a rhodium-DIOP catalyst (198).

An interesting bis(dimethylglyoximato) cobalt(II) system

was investigated by Ohgo and co-workers (203). The complex
Co(DMG)$_2$-quinine catalyzes the hydrogenation of α-oxocarbonyl
compounds at or below room temperature at a hydrogen pressure
of 1 atm. Benzoin of 78% e.e. can be obtained from benzil. The
catalyst system resembles oxidoreductases in its behavior, and
a mechanism for the hydrogenation was proposed (203).

135

(a)   R = Me;   R' = H (191)
(b)   R = R' = Me   (190)

136   *137*

(a)   R = -CH$_2$-CH=CH-(CH$_2$)$_3$-CO$_2$Me
(b)   R = -(CH$_2$)$_6$-CO$_2$Me

C. Asymmetric Hydrosilylation

Asymmetric hydrosilylation is mechanistically related to
hydrogenation. An Si-H bond is cleaved like a H-H bond by
oxidative addition onto the transition metal. Then the two
fragments are successively transferred to the coordinated
double bond. The synthetic utility of hydrosilylation has appeared recently, especially in asymmetric catalysis. It com-
plements hydrogenation, since ketones or imines are easily
reduced to alcohols or amines by silanes. It also permits the
reduction of conjugated double bonds. The first asymmetric
hydrosilylations involved chiral palladium, nickel, or
platinum complexes. Details may be found in a review on asymmetric hydrosilylation (171). Here we shall only present
recent results obtained with rhodium complexes, which are the
most efficient asymmetric catalysts.

Some interesting asymmetric syntheses of various alcohols
are reported in Table 5. It is important to note that a great
many variations are possible in the structure of the catalyst,
allowing adaptation for successful asymmetric reduction of a

TABLE 5[a]

Substrate	Ligands (L_2) and Silane[b]	Product, configuration and % e.e.	References
PhCOMe	(+)-DIOP, H_2SiPh_2	PhCHOHMe (S) 28	204
	(+)-DIOP, $H_2SiPh(\alpha Np)$	PhCHOHMe (S) 58	205
PhCO(t-Bu)	(−)-DIOP, H_2SiPh_2	PhCHOH(t-Bu) (R) 41	206
	(R)-BMPP, $HSi(Et)Me_2$	PhCHOH(t-Bu) (R) 56	207
EtCOMe	(−)-DIOP, $H_2SiPh(\alpha Np)$	EtCHOHMe (R) 42	205
PhCOMe	(−)-(MPFA), H_2SiPh_2	PhCHOHMe (R) 49	208
![2-methyl-2-cyclohexenone]	(R)-BMPP,[c] $H_2SiPh(\alpha Np)$	![product with OH] (+) 43	209a
$MeCOCO_2n\text{-}Pr$	(+)-DIOP, $H_2SiPh(\alpha Np)$	$MeCHOHCO_2n\text{-}Pr$ (R) 81.5	209b
$C_5H_{11}COCO_2n\text{-}Pr$	(+)-DIOP, $H_2SiPh(\alpha Np)$	$C_5H_{11}CH(OH)CO_2n\text{-}Pr$ (R) 65	210
$PhCOCH_2Cl$	(+)-DIOP, H_2SiPh_2	$PhCHOHCH_2Cl$ (S) 63	210
$MeCO(CH_2)_2CO_2t\text{-}Bu$	(+)-DIOP, $H_2SiPh(\alpha Np)$	$MeCH(OH)(CH_2)_2CO_2t\text{-}Bu$ (S) 84.4	171

[a]See references for experimental conditions; yields are generally excellent.
[b]αNp: α-Naphthyl.
[c]BMPP: Benzylmethylphenylphosphine.

TABLE 6[a]

Substrate	Silane and temperature		Product	% e.e.,
Ph(Me)C=NCH$_2$Ph	H$_2$SiPh$_2$	24°C	Ph(Me)CH-NHCH$_2$Ph	50 (S)
	H$_2$SiPh$_2$	2°C	Ph(Me)CH-NHCH$_2$Ph	65 (S)
Ph(Me)C=N-Ph	H$_2$SiPh$_2$	5°C	Ph(Me)CH-NHPh	47 (S)
PhCH$_2$(Me)C=NCH$_2$Ph	(MeSiHO)$_n$	24°C	PhCH$_2$(Me)CHNHCH$_2$Ph	14 (S)
	H$_2$SiPh$_2$	24°		38.7 (R)

[a]Reaction performed in benzene solution. The product is recovered after hydrolysis with an excellent yield.

given substrate. Both the chiral ligand and the silane may be
modified. Enantiomeric excesses up to 85% were observed. Asymmetric hydrosilylation of imines to chiral amines is summarized
in Table 6.

If a prochiral silane such as $H_2SiPh(\alpha Np)$ is allowed to
react with an achiral ketone such as diethyl ketone in the
presence of catalytic amounts of $RhCl(-)$-DIOP, the primary
product is a chiral silane. In this silane, of formula
$Et_2CHO-SiHPh(\alpha NP)$, the only asymmetric center is the silicon
atom [(S) configuration]. The optical yield in asymmetric
synthesis of a typical silane is 46% (212). Additional
examples may be found in ref. 212. If a chiral ketone such as
menthone is used, the asymmetric induction at silicon reaches
85% (213). Since the alkoxy group in an alkoxysilane can be
replaced with inversion of configuration by an alkyl group
via a Grignard reaction, this process also constitutes an
asymmetric synthesis. A related asymmetric synthesis of chiral
silanes is the asymmetric alcoholysis of dihydrosilanes. Thus
benzylic alcohol and $H_2Si-Ph(\alpha Np)$ in the presence of $RhCl(+)$-
DIOP yield $HSi(OCH_2Ph)Ph(\alpha Np)$ of 19% e.e. (214).

D. Asymmetric Hydroformylation

Hydroformylation by cobalt catalysts is an industrial
reaction of long-standing importance for the large-scale
preparation of aldehydes, and has given rise to many mechanistic
studies. Recently rhodium catalysts have been introduced,
allowing hydroformylation to proceed under relatively mild
conditions. The first attempts using chiral catalysts were not
encouraging. A survey of the early positive results, obtained
in 1972-1973, may be found in a review by Pino and co-workers
(215) on asymmetric hydroformylation, which covers research
through 1974. With a catalyst composed of $Co_2(CO)_8$ and (S)-N-α-
methylbenzylsalicylaldimine optical yields up to 15% have
been obtained in the synthesis of $(+)$-(R)-hydratropaldehyde
from styrene. The same reaction was investigated by several
research groups using a rhodium-DIOP catalyst, the optical
yield being in the range of 20 to 25% (216). With phenylmethylbenzylphosphine as ligand an optical yield of 17.5% can be
obtained (217).

Interesting results were obtained (215,218) in the hydroformylation of aliphatic olefins, with $HRh(CO)(PPh_3)_3$ as the
catalyst precursor and $(-)$-DIOP as the chiral ligand. In almost all cases the isolated aldehydes are significantly optically active. A detailed investigation of the influence of experimental conditions on regiospecificity and optical yield led
to a discussion of the mechanism of the catalytic cycle and the
proposal of a simple model of asymmetric induction. The
highest enantiomeric excess (27%) was obtained in the asym-

metric hydroformylation of *cis*-butene to 2-methylbutanal. A discussion of the results (215) is outside the scope of the present article; however, it should be mentioned that the origin of the asymmetric induction was ascertained. It arises through equilibration between the diastereomeric π-olefin complexes (or less likely in the formation of these π-olefin complexes) and/or during the next step, which is olefin insertion into an Rh-H bond. A considerable improvement was obtained by a Japanese group (218), using a diphosphine structurally related to (-)-DIOP. Hydratropaldehyde in 44% e.e. is formed from styrene.Interestingly the absolute configuration of the aldehyde is opposite to that obtained with (-)-DIOP. Asymmetric induction is also observed in the hydroformylation of 1-butene or *cis*-2-butene. It is to be hoped that still more effective chiral ligands will render this method suitable for synthetic uses.

The asymmetric hydroformylation of alkenes over platinum catalysts in the presence of $SnCl_2$ has recently been described (219). Several styrene or butene derivatives were hydroformylated, the catalyst being $PtCl_2$(-)-DIOP + $SnCl_2$. The transformation of 2,3-dimethyl-1-butene into 3,4-dimethylpentanal was achieved with 15% e.e. A careful analysis of many results (219) indicated that the asymmetric induction with this system takes place after the intermediate metal-alkyl complex formation, in contrast to the rhodium (-)-DIOP system, where asymmetric induction is already achieved at this stage (215).

Alkene hydroesterification is a reaction in which alkenes are treated by a CO + H_2 mixture in a solvent such as methanol. The net result of the reaction is the addition of H and CO_2Me to the double bond. Asymmetric hydroesterification of styrene derivatives was recently achieved with $PdCl_2$(-)-DIOP as the catalyst, though the asymmetric induction was not very high (220).

E. Asymmetric C-C Bond Formation

The codimerization of 1,3-cyclooctadiene and ethylene to give optically active 3-vinyl-1-cyclooctene (VCO) (221) is the first example of a catalytic asymmetric synthesis in which the chiral center is produced in high optical purity via C-C bond formation. The reaction, carried out with an allyl-nickel catalyst and chiral phosphines, leads to VCO in up to 70% e.e.; this result was achieved through a careful study of the influence of the nature of the chiral phosphine, the temperature, and the phosphine-nickel ratio in the catalyst.

Catalytic codimerization of norbornene, norbornadiene, and (±)-2-bornene with ethylene involved an extensive study of asymmetric synthesis using π-allyl nickel halide catalysts. One finding from this study is the linear dependence of the reaction temperature on the optical yield of *138* formed by

codimerization of ethylene and norbornene using (-)-isopro-
pyldimenthylphosphine as chiral ligand.

X-Ray structure study, combined with knowledge of the
absolute configuration of the complex formed from methyl (π-1-
methyl-2-butenyl)nickel and (-)-dimenthyl(methyl)phosphine, has
led to a model from which a correlation between the absolute
configurations of the phosphine and of the major product was
made (1972).

Hidai (222) has reported an asymmetric synthesis of
citronellol (*140*) involving, as the asymmetric reaction, the
telomerization of isoprene and methanol through use of a
palladium complex. The use of neomenthyldiphenylphosphine
as a chiral ligand gave dimer *139*, which was converted into
(+)-citronellol (*140*) (8.6% e.e) , while (-)-citronellol (17.6%
e.e.) was produced with a menthyldiphenylphosphine ligand.

Catalytic alkylation of π-allylic systems by malonate is
performed over a Pd(0)-DIOP catalyst (223). A quantitative
yield of *141* is obtained with up to 38% e.e.

Asymmetric allyl transfer from allyl phenyl ether to
2-carboethoxycyclohexanone occurs in the presence of a
Pd(0)-DIOP catalyst. 2-Allyl-2-carboethoxycyclohexanone is
thus produced quantitatively in about 7% e.e. (224).

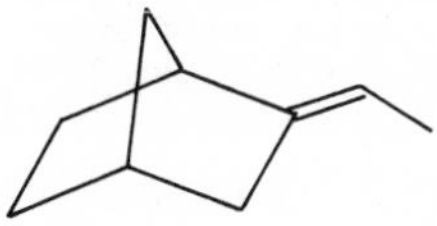

138

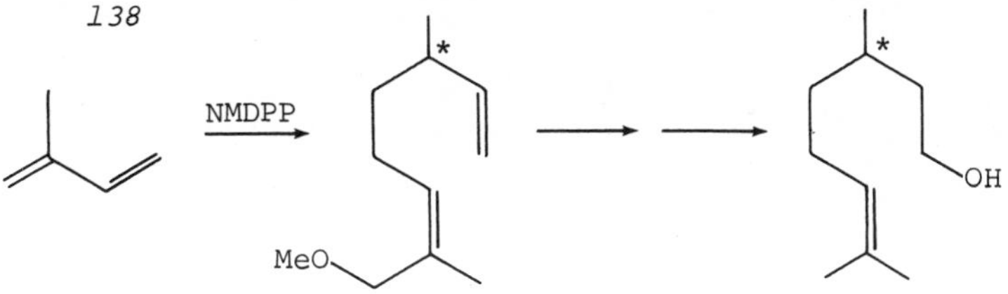

139   *140*

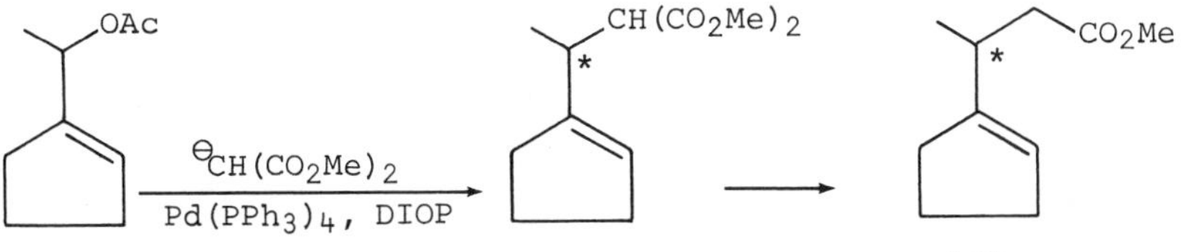

141

Various papers and patents report the preparation of
optically active chrysanthemates (mixture of cis and trans
isomers) by the catalytic decomposition of ethyl diazoacetate in the presence of 2,5-dimethyl-2,4-hexadiene (225).

Aratani (226) achieved a 60 to 70% e.e. of both *cis-142* and *trans-142*. The catalyst used was rather sophisticated: the amino alcohol *144,* prepared from reaction of the Grignard reagent of *143* with L-alanine ethyl ester, was the chiral ligand, effectively making *145* the catalyst for the reaction. An approximately 50:50 cis-trans mixture was obtained with 68% e.e. for the trans isomer and 62% e.e. for the cis isomer. The enantiomeric purities of the products were shown to increase with the bulkiness of the substituent R in *145*. A carbene complex is thought to be responsible for the reaction.

Otsuka (227) reported an asymmetric cyclopropanation of styrene and 1,1-diphenylethylene with ethyl diazoacetate. The cobalt catalyst was prepared by reaction of (+)-camphorquin-one-α-dioxime (*146*) with cobalt chloride. The reaction probably proceeds via a cobalt-carbene complex. Spectroscopic studies suggest that the bidentate dioxime ligand coordinates to Co(II) through the N and O atoms to form a six-membered

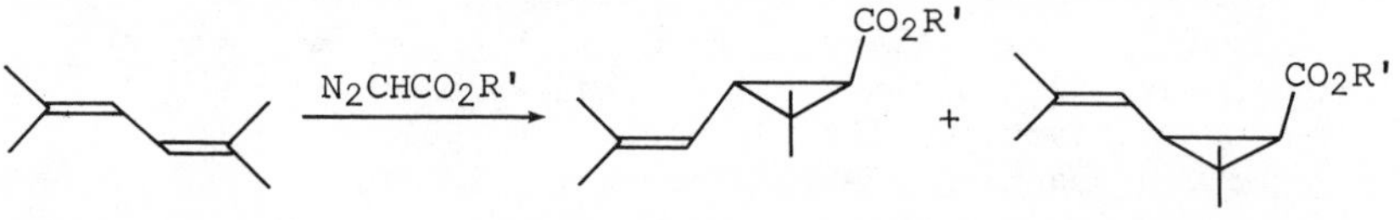

trans-142   +   cis-142

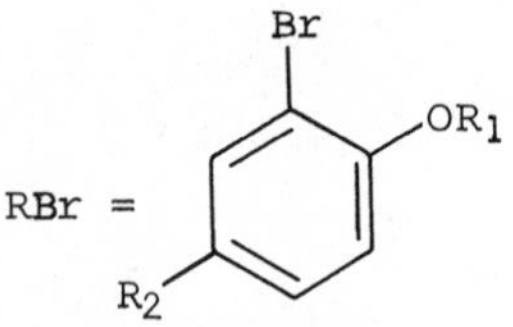

RBr =

143 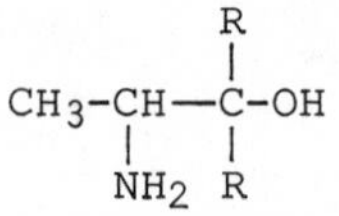   144 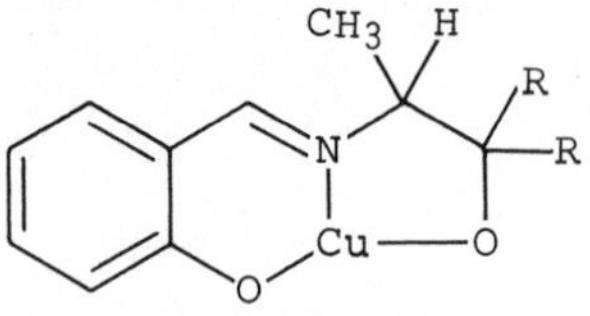   145

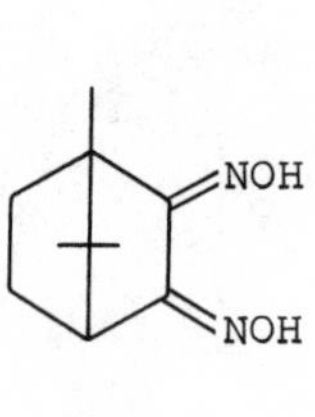

146

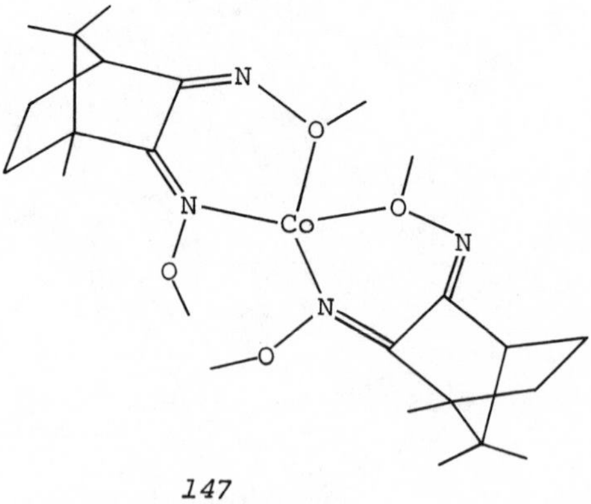

147

chelate ring, as shown in *147* (228). Only conjugated olefins having a terminal methylene undergo such cyclopropanation reactions. Neopentyl diazoacetate gave the highest optical yield (88% e.e., for *trans-148*).

$$Ph \atop H \Big\rangle C=CH_2 \quad + \quad N_2CHCO_2C_2H_5 \quad \longrightarrow$$

cis-148   trans-148

Catalytic cross-coupling between a Grignard reagent and an alkyl or aryl halide may produce optically active hydrocarbons. Coupling a 1-phenylethyl Grignard reagent with vinyl bromide yields, by catalysis with $NiCl_2 \cdot L_2$ (where L_2 is a chiral amino-phosphine or diphosphine), optically active 3-phenyl-1-butene in up to 63% e.e. (229). The process may be described as a slow reaction of the halide with the Grignard reagent, which is rapidly racemized via complex formation (see Sect. II-B-6). Interestingly the catalytic cross-coupling between 1-bromo-2-methylnaphthalene and 2-methyl-1-naphthylmagnesium bromide (two achiral molecules) produces optically active biarylatropisomers (230).

F. Asymmetric Oxidation

Catalytic asymmetric epoxidation of allylic alcohols has been performed with two different types of catalysts. The well-characterized molybdenum catalyst *149* is active in the epoxidation of allylic alcohols *150* with cumene hydroperoxide to give the epoxy alcohol *151* in 10 to 33% e.e. (231). On the other hand, using $VO(acac)_2$ and a chiral hydroxamic acid as a catalytic system, Sharpless obtained asymmetric inductions as high as 50% in the epoxidation of E-α-phenylcinnamyl alcohol with t-butyl hydroperoxide (232). Although the mechanism of these reactions is still under investigation, such high asymmetric inductions with allylic alcohols suggest

149

150

151

a coordination of the alcohol function to the metal during
the oxygen atom transfer step.

Asymmetric oxidation of sulfides to sulfoxides with
t-butyl hydroperoxide is catalyzed by $VO(acac)_2$ and $MoO_2(acac)_2$
in a benzene:chiral alcohol solvent (233). This result would
again indicate a complexation of an alcohol molecule to the
metal during the oxygen atom transfer step. The optical
yields obtained are comparable to those reported for the
asymmetric oxidation of sulfides with chiral peroxides (2).

G. Asymmetric Polymerization

Asymmetric polymerization of prochiral monomeric units
under the influence of a chiral catalyst is not usual (2).
Recently (234) polymers of a new type were prepared starting
from symmetric thiiranes such as cis-1,2-dimethylthiirane
and cyclohexene sulfide having two neighboring asymmetric
carbon atoms of opposite configuration. An optically active
initiator derived from the reaction of diethylzinc with (R)-
$(-)$-3,3-dimethyl-1,2-butanediol was used. The direction of
ring opening of episulfide is oriented by the chiral
catalyst, which preferentially attacks one of the asymmetric
carbon atoms, with inversion of configuration of the latter.
The resulting polymer is optically active due to the prevalence
of one type of configurational unit [e.g., $\Sigma(-R,R-) > \Sigma(-S,S-)$].
Optically active crystalline polymers were obtained. Crude
products can be separated by selective solubility of fractions
of different optical activity and crystallinity, as shown in
Table 7.

It is not possible at present to ascertain the optical
purity of the polymeric products so prepared. Using ^{13}C NMR
it was found that the polymer presents different types of
stereosequences. The signal corresponding to diisotactic
chains of the $-R,R-R,R-R,R-$ or $(S,S-S,S-S,S-)$ type was clearly
assigned, but the configuration of the prevailing chain is
still to be determined.

If the behavior of the catalyst with these monomers is
similar to that observed for methylthiirane (235), the optical
purity of the polymers might be at least of the order of 30%
(234).

IV. PHOTOCHEMICAL ASYMMETRIC SYNTHESIS

Creation of optically active compounds by photochemical
reactions was the focus of an old debate, and as early as 1874
Le Bel (236) had speculated that circularly polarized light
(CPL) might have led to the generation of optical activity on
earth. In spite of this early insight such photochemical asymmetric synthesis has only recently been achieved. A different

TABLE 7

Monomer	Time polymerization (hr)	Yield polymerization (%)	Fraction soluble in toluene at room temperature			Fraction insoluble in toluene, soluble in $CHCl_3$		
			%	(α_p) $(CHCl_3)$	m.p. (°C)	%	(α_p) $(CHCl_3)$	m.p. (°C)
cis-1,2-Dimethylthiirane	60	100	34	+24°	–	66	+66°	122
Cyclohexene sulfide	84	45	35	+3.8°a	65	10	+8.4b	80

a +20° in trichlorobenzene.
b +39° in trichlorobenzene.

type of enantioselective asymmetric photosynthesis can be
envisaged, in which the light is transferred from a chiral
sensitizer to a prochiral system. Still other methods involve
photoreactions of a molecule in which existing chiral centers
control the creation of an additional chiral unit. A special
case of asymmetric induction is photoreaction in a chiral medium
or in a chiral solvent.

Photochemical asymmetric syntheses, especially practical
ones leading to optically active compounds, are still poorly
developed. Most of the examples presented here are quite recent.

A. Photochemistry with CPL

CPL is a chiral physical reagent. It was only in 1929-1930
that Kuhn (237) succeeded in partially photoresolving a racemic
mixture with CPL. The method was recently extended to other
systems (238). It remained to be demonstrated that photosynthesis could be achieved with CPL. Most of the old reports
in this field, quoted in ref. 237, are certainly erroneous, as
pointed out in 1971 (239). The first unambiguous experiment was
the photocyclization of 1,2-diarylethylenes to give helicenes
(Fig. 22). A European group (240,241) and an American one
(242,243) published their results almost simultaneously and
reached similar conclusions. If, for example, diarylethylene
(152) (symbolized by [2] + [3]) is photolyzed at 290 to 370
nm, cyclization to 153 occurs, followed by dehydrogenation to
hexahelicene (154). The last step is very fast. Some I_2 is added
to the medium to help the oxidation. The [1] + [4] precursor
was used also. Both experiments led to optically active hexahelicene. With left CPL the specific rotations of isolated
hexahelicene were $[\alpha]_D$ = +7.9 ± 0.6° and +1.9 ± 0.5°, respectively (240). Using right CPL the sign of rotation was reversed. Highest specific rotations were obtained when [4] + [3] and
[4] + [4] precursors were irradiated with CPL (λ = 310 to 370
nm). The specific rotations of the corresponding helicenes were
34 ± 1.5° and 48.5 ± 1°, respectively (238b,241). Thus experimental proof of photoinduced asymmetric synthesis is established, since it was also demonstrated that the optical activity
was not the result of a partial photoresolution of helicene
(photolysis of dl-hexahelicene with left CPL gives a levoratory
hexahelicene very slowly). The postulated mechanism (240-244)
involves the photoselection of one enantiomeric conformer of
the cis-1,2-diarylethylene (155). This cannot be planar due to
steric hindrance and thus must be considered to be a labile
racemic mixture. The preferred excitation of one conformer is
related to its CD ($\Delta\varepsilon$). As in the beginning of a partial photoresolution (237,238,244) the optical purity of the product is
equal to 1/2 ($\Delta\varepsilon/\varepsilon$) = $g/2$, where g is the anisotropy factor of
the compound that absorbs the CPL. The anisotropy factor is
equal to the difference between absorption coefficients for a

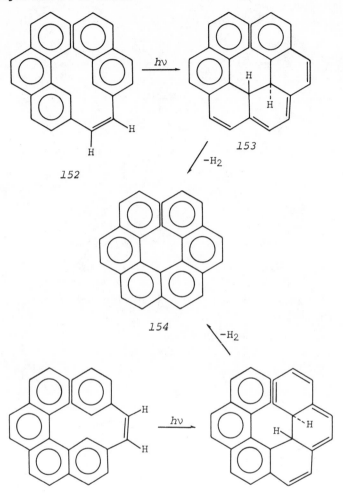

Fig. 22.   Asymmetric syntheses of hexahelicene using circularly polarized light (240,242).

given CPL divided by half the sum of these absorption coefficients. The (R) and (S) conformers give photoreactions whose
rates are strictly proportional to ε_R and ε_S. The ratio of the
two rate constants k_R/k_S is equal to $\varepsilon_R/\varepsilon_S$. It is easy to see
that a high enantiomeric excess requires a high k_R/k_S value,
that is, a high g factor.

After the excitation step (Fig. 23) cyclization occurs,
leading to a dihydrohelicene and then to the helicene. Racemization of conformer 155 is easy, either through cis-trans
photoisomerization or by simple rotation of aryl groups around
the vinylic bonds. Thus the starting material remains racemic,

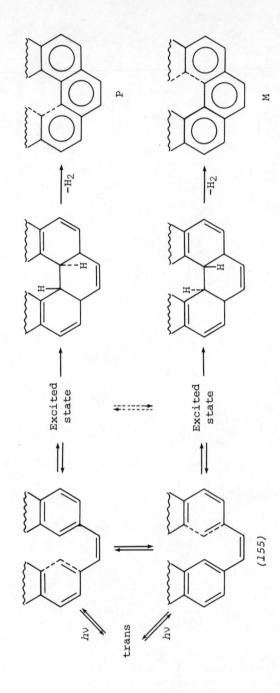

Fig. 23. Mechanism of asymmetric synthesis of helicenes.

and the optical yield does not change during the reaction (238b).

There is no racemization in the excited state; otherwise optically active helicenes could never be obtained. The same holds in the aromatization step. The optical yield for hexahelicene was calculated to be 0.2% (240). It may be estimated that in the photochemical synthesis of helicenes the best optical yields are in the range of 0.5 to 1%. Photocyclization of bis-1,2-diarylethylenes gives access to [10] through [14] helicenes. Several double photocyclizations were investigated with CPL, and slightly optically active helicenes were recovered. The optical process leading from 1,2-diarylethylene to chiral helicene can be summarized by the following sequence:

$$R \xrightarrow{h\nu} R^* \xrightarrow{\text{Cyclization}} R' \xrightarrow{(-H_2)} \text{(P) helicene}$$

$$S \xrightarrow{h\nu} S^* \xrightarrow{\text{Cyclization}} S' \xrightarrow{(-H_2)} \text{(M) helicene}$$

Whether this scheme represents an asymmetric synthesis is a matter of definition. As far as the conformers are concerned, it may be considered as a kinetic resolution coupled with a racemization (see Sect. II-B. for examples formally related to this case). The optically active helicenes are devoid of chiral centers, but are generated by a transfer of asymmetry from dihydrohelicenes which possess two chiral centers.

Recently a new asymmetric synthesis with CPL was realized leading to a chiral compound with a classical chiral center (246). Some of the results are summarized in Figure 24. N-arylenamines are nonplanar molecules, and photolysis with CPL selects one antipode. The true mechanism of the reaction is not fully understood (247), but it seems that the later steps involve concerted reactions. The result is a retention of some of the initial optical activity. Asymmetric synthesis with CPL is not of synthetic interest, but can help elucidate reaction mechanisms. It must be stressed that asymmetric induction was observed thanks to a "conformational effect" in the prochiral substrate. It might be hypothesized that a perfectly planar prochiral molecule could react preferentially on one of its faces after absorption of CPL, but such an effect has never been observed.

CPL is the only chiral physical reagent that has been shown to generate optical activity in chemical systems. A new absolute asymmetric synthesis was claimed (248) recently. Several reactions were carried out under the simultaneous influence of combined electric and magnetic fields. Specific rotations of the chiral products were extremely small. The theoretical basis of such experiments is contestable (249).

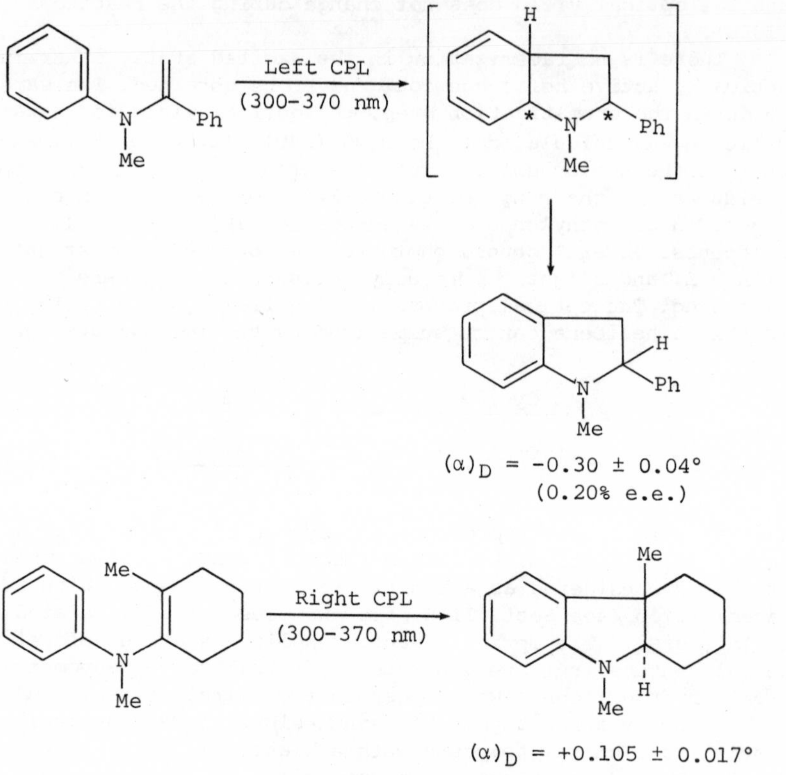

$(\alpha)_D = -0.30 \pm 0.04°$
(0.20% e.e.)

$(\alpha)_D = +0.105 \pm 0.017°$

Fig. 24. Photosynthesis of dihydroindoles (246).

B. Chiral Sensitizers

Many photochemical reactions take advantage of the preliminary excitation of a limited amount of sensitizer which can then transfer its energy to the substrate. This process may lead to asymmetric synthesis if the sensitizer is chiral. Since in many cases direct absorption of light cannot occur, the chiral sensitizer should play the role of a chiral catalyst. This method is a promising one, but until now there is no example of such an experiment. The only reports dealing with chiral sensitizers S* involve the photoactivation of a racemic mixture:

$$R \underset{k_S}{\overset{k_R}{\rightleftharpoons}} S$$

When $k_R \neq k_S$, the photoequilibrium is displaced toward one enantiomer. This can occur if there is a preferential energy transfer from the sensitizers to one enantiomer. Energy transfer can occur by several mechanisms. The most suitable for chiral recognition is one in which energy is transferred over short distances. An exciplex formation between the excited sensitizer and the acceptor in its ground state obeys this requirement. The few successful cases of asymmetric inductions using a chiral sensitizer seem to involve an exciplex mechanism. It was Hammond (250) who first demonstrated that optically active *trans*-1,2-diphenylcyclopropane is recovered from photo-isomerization of 1,2-diphenylcyclopropane in the presence of a singlet sensitizer (*N*-acetyl-α-naphthylethylamine). The e.e. was estimated (251,252) as 4%. 3-Methylindanone, a triplet sensitizer, is also able to induce optical activity (3% e.e.) in this system (252). An unsuccessful attempt was made to influence the decay modes of 1,2-diphenylcyclopropane excited with optically active solvents, the excitation being provided by singlet or triplet sensitizers (253).

The photochemical optical activation of *dl-p*-tolylmethyl-sulfoxide with *N*-acetyl-α-naphthylethylamine has been observed (252), the optical purity at the photostationary state being 4.5%. 1,4-Dimethylallene has been irradiated in the presence of the *t*-butyl ether of 21,22-dehydroneoergosterol as sensitizer, and an e.e. of 3.4% was attained (254). Conditions for a good asymmetric transfer were analyzed (255).

All the published data relating to chiral sensitizers indicate only a few percent e.e. in the recovered material. A substantial improvement was recently observed (256), a 12% e.e. being obtained when *dl-p*-tolylisopropylsulfoxide was irradiated in the presence of (+)-*N*-trifluoroacetyl-α-naphthylethylamine. Specific hydrogen bonds between sensitizer and sulfoxide could play a role in the process. Further developments may be expected in the near future in this new field.

C. Diastereoselective Photochemical Synthesis

Asymmetric synthesis, either enantioselective or diastereo-selective, has seldom been performed by photochemical reactions. It is, of course, possible to find many photoreactions of natural products which generate a new chiral center, but the concept of asymmetric induction leading to an operational asymmetric synthesis has never been seriously considered.

Factors influencing stereoselectivity in the photochemical reduction of some alkylcyclohexanones have been discussed (256). A hydroxycyclohexyl radical species is formed by hydrogen transfer from 2-propanol (the hydrogen donor). Asymmetric induction originates in this radical step and is very sensitive to the experimental conditions. An elegant asym-

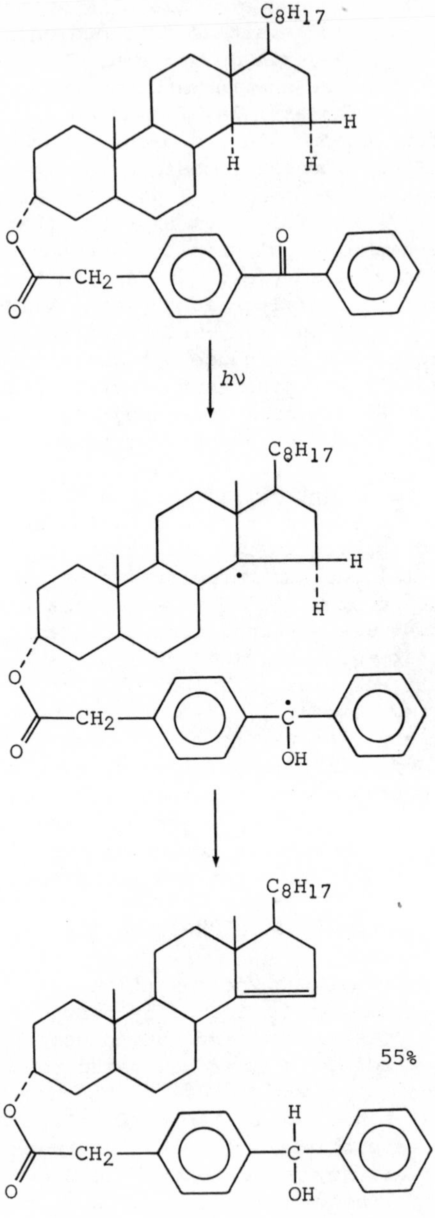

Fig. 25.

metric photochemical reduction of benzophenone esters by an
internal hydrogen transfer was described by Breslow (257). The
principle of this method is summarized in Figure 25. Two
diastereomers are formed in the ratio 55:45, implying that
saponification should release a chiral benzhydrol of 10% e.e.
Asymmetric induction was high in the by-product formed from
insertion in the C-7α-H bond of the steroid, with C-C coupling
between benzophenone and steroid occurring. Unfortunately the
asymmetric center in the chiral tertiary benzhydrol cannot be
removed from the steroid moiety.

One of the first examples that may be classified as a
photochemical asymmetric synthesis is the photoalkylation of
glycine derivatives by terminal olefins or toluene. Elad
demonstrated (258) that if glycine is part of a polypeptide
chain there is good control (up to 40% e.e.) in the creation
of the new chiral center (Fig. 26). A radical mechanism operates
after the first step of photoinitiation of the process.

$(\text{L-Ala-Gly-L-Ala})_n \xrightarrow[\text{PhCH}_3]{h\nu} (\text{L-Ala-L-Phe-L-Ala})_n$

$\downarrow$

L-phenylalanine
40% e.e.

$\text{TFa-L-Ala-Gly-L-Ala-OMe} \xrightarrow[\text{PhCH}_3]{h\nu} \text{Tfa-L-Ala-D-Phe-L-Ala-OMe}$

$\downarrow$

D-phenylalanine
20% e.e.

Fig. 26. Photoalkylation of some glycine derivatives
(259).

Asymmetric photosynthesis of a [6]helicene skeleton was
observed (259) when a menthyloxycarbonyl group was introduced
in the 1,2-diarylolefinic precursors shown in Figure 27. After
photocyclization and treatment with LAH a helicenic alcohol is
obtained with 5 to 6% e.e. When the menthyl group is replaced
by a 2-CH$_2$-[6]helicene group, the optical yield is increased
to 16% (260).

Photodimerization of thymine has been extensively investigated in connection with the photochemistry of nucleic acid
derivatives. If a frozen solution of thymidine (a combination
of thymine and a sugar) is photolyzed, four photoproducts
are obtained (261). After removal of the sugar the photodimers
of thymine were recovered, two of them being chiral. From the
data it is difficult to estimate the extent of the asymmetric
induction resulting from the sugar, but it appears quite high.
Reinvestigation in this field would be useful.

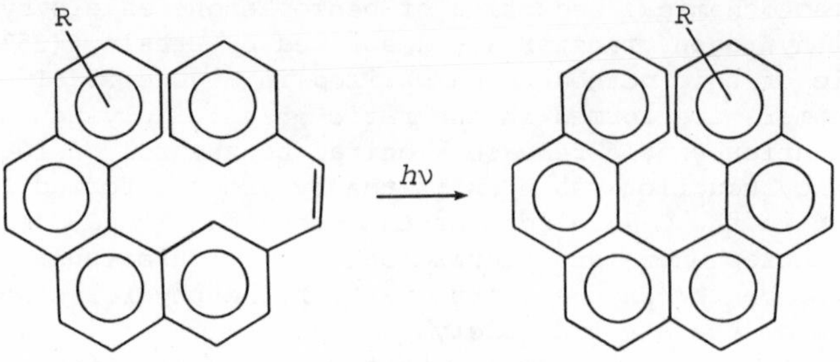

R = -CO$_2$-(-)-menthyl
Fig. 27

The best photochemical asymmetric synthesis at the present time is one published by Green and co-workers (262). Mannitol hexacinnamate, irradiated in benzene solution, undergoes a [2] + [2] photocycloaddition yielding dimers that are recoverable by transesterification with methanol. The main product is (+)-dimethyl-δ-truxinate (Fig. 28), which is obtained with 38-48% e.e. The carbohydrate skeleton acts as a template both to induce the dimerization (which does not occur for cinnamates in solution) and to orientate it toward formation of one enantiomer. Considerable improvement was obtained by working with D-mannitol-3,4-isopropylidene-1,2,5,6-tetra-δ-cinnamate: (+)-dimethyl-δ-truxinate was produced with 85% e.e. (263).

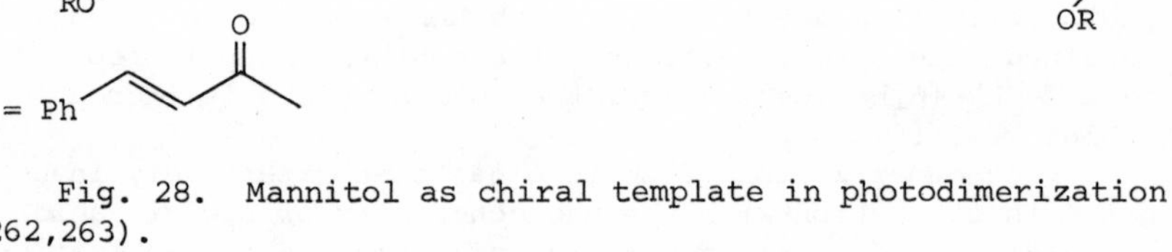

Fig. 28.   Mannitol as chiral template in photodimerization (262,263).

A special situation for a diastereoselective asymmetric synthesis is [2] + [2] photocycloaddition in the solid state, where the chiral crystal structure is the sole source of asymmetry (264). Asymmetric synthesis of a cyclobutane derivative was

carried out by irradiation of single mixed chiral crystals of
156 (15%) in 157. Selective excitation of the thienyl compound
(Fig. 29) through appropriate cutoff filters yielded the
mixed optically active dimer 158. It was demonstrated (265)
that its optical purity was at least 80%, showing that excited
156 reacts preferentially on one of its faces when included
in a chiral crystal. Photochemical polymerization of divinyl-
benzene derivatives that are allowed to crystallize into a
chiral crystal gives optically active polymers, apparently of
very high optical purity (166).

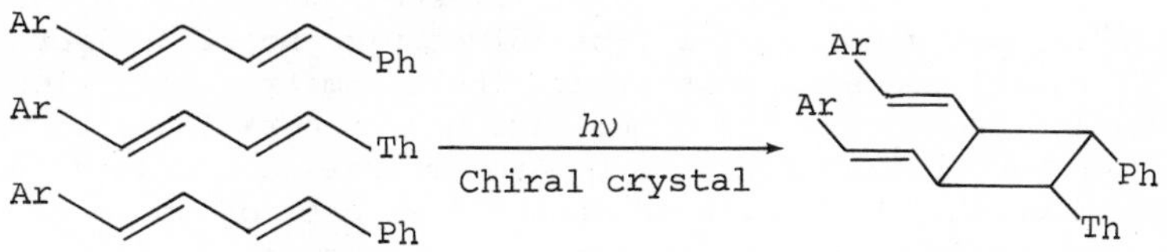

157   Ar = 2,6-C$_6$H$_3$Cl$_2$
156   Th = 2-thienyl

Fig. 29

   In the two preceding examples the result of the reaction
is a transfer of asymmetry, the crystal chirality becoming
molecular chirality (the crystal has to be dissolved to recover
the products). One of the crucial points is to control the
formation of the chiral crystal. A general discussion of
asymmetric synthesis of dimers and polymers via topochemical
reactions in chiral crystals may be found in ref. 267.
   It is known that some optically active molecules can
crystallize with inclusion of guest compounds. Thus several
penta-1,3-dienes have been included in deoxycholic acid. These
prochiral monomers were then polymerized by irradiation of the
crystals with γ-rays. After removal of deoxycholic acid optically active polymers were obtained (268). Here deoxycholic
acid plays the role of host matrix.
   A host reagent in an inclusion molecular complex should
react specifically with the host if the crystalline complex
is well defined. Such a situation was encountered with the 4:1
molecular complex of deoxycholic acid and di-t-butyl diperoxy-
carbonate. Photolysis at 25°C of the crystals leads to clean
5α-hydroxylation of the deoxycholic acid. This must reflect a
preferred orientation of the peroxycarbonate in the channel
structure of the complex (269).
   Only one example of asymmetric polymerization in liquid
crystals is known. A tertiary mixture of cholesteryl acrylate,
p-di-acroyloxybenzene, and p-acroyloxybenzylidene-p'-hexyloxya-

niline gives a cholesteric phase. Irradiation by UV light
initiates a three-dimensional polymerization, retaining the
cholesteric structure (270). This experiment would constitute
an effective asymmetric synthesis after removal of cholesteryl
moieties. Similar experiments with thermal polymerization are
also possible (271).

D. Enantioselective Photosynthesis

There is as yet no clear-cut example of enantioselective
synthesis. An asymmetric photopinacolization of acetophenone to
1,2,3-diphenyl-2,3-butanediol (6% e.e.) was obtained by Seebach
(272a) using DDB [(+)-1,4-bis(dimethylamino)-2,3-dimethoxybutane]
as chiral reagent and solvent. The asymmetric induction by DDB
must arise during the dimerization of the radicals formed by
hydrogen abstraction; DDB behaves more as a chiral solvent than
as a chiral reagent. More recently an e.e. of up to 29% has
been obtained by decreasing the temperature to -72°C (272b).
It is interesting to mention the great effectiveness of a
chiral solvent system-(+)-PhCHOHCF$_3$ in CFCl$_3$-in controlling the
photochemical ring closure of the nitrone of N-t-butyl-1,1-
diphenylimine into 1-t-butyl-2,2-diphenyloxaziridine. This
compound, in which the nitrogen atoms is the sole chiral center,
is obtained in good chemical yield with 31% e.e. (273).

V. ASYMMETRIC ELECTROCHEMISTRY

The future of this method is hard to evaluate, since there
are as yet only a few reports on asymmetric electrochemical
synthesis. The method is intrinsically attractive for enantioselective synthesis because the chiral auxiliary entity
introduced in the electrochemical system should be able to
control the production of a large number of chiral molecules.
A review and discussion of the stereochemistry of organic
electrode processes, especially asymmetric induction by use of
optically active supporting electrolytes, may be found in
ref. 274.

One of the first reports in this field was the 1968 work
of Horner and co-workers (275), who used a chiral supporting
electrolyte such as an ephedrinium salt in methanol. Phenylmethylcarbinol of 4.6% e.e. is thus obtained by reduction of
acetophenone at a mercury electrode. The origin of the asymmetric induction was discussed (274) by assuming the stereoselective formation of a carbanion at the chiral double layer
on the electrode. This would be the result of a preferred
approach of the acetophenone molecule to the absorbed
ephedrinium hydrochloride. Possible transition states for the
reduction of acetophenone were proposed.

Another approach was described by Grimshaw and collaborators in 1967 (276). A small amount of a chiral compound, which does not participate in current transport, is adsorbed on the electrode. With use of an alkaloid as chiral auxiliary moiety, some coumarins were reduced to chiral dihydrocoumarins in optical yields of up to 17%. More recently Kariv and co-workers (277) criticized the mechanism proposed in (276) and tested many alkaloids in the electroreduction of acetophenone at a mercury cathode. 1-Phenylethanol was formed with 14.9% e.e. in the presence of adsorbed quinine (see also similar work on acetophenone reduction in ref. 278). Better results were obtained in the reduction of acetylpyridines (279). 2-Acetylpyridine in the presence of a small amount of brucine yielded the corresponding alcohol with 31.2% e.e. Additional work led to asymmetric reduction of 2-acetylpyridine in as high as 47.5% e.e., which is the best optical yield yet achieved in asymmetric electrochemistry (280). 3-Acetylpyridine under the same conditions gives a racemic alcohol. A complete understanding of the reaction mechanism has not yet been achieved, but it was postulated that the crucial step for asymmetric induction occurs after the transfer of two electrons to the N-protonated substrate. Enamine intermediates would be formed and asymmetrically protonated in a reaction involving the chiral alkaloid. Detailed investigations of the reduction of phenylglyoxylic acid to mandelic acid were performed by Peltier and collaborators (281a) on a mercury electrode in the presence of several alkaloids. Strychnine is the best chiral inducer, mandelic acid being formed in excellent yield with about 20% e.e. To interpret the electrochemical data it was proposed that a delocalized carbanion is first formed in a chiral environment because of the alkaloid that is located in the neighborhood of the electrode. The details of interaction between this carbanion and the surrounding species are not understood but they are crucial for the protonation step.

Electrochemical reduction of phenylglyoxylic acid oxime leads to optically active phenylglycine (281b) in up to 17% e.e. It is interesting to note that the absolute configuration depends on the cathodic potential used.

The potential of asymmetric electrolysis is well illustrated by the small amount of alkaloid necessary to control asymmetric reduction. In the previous experiments 0.5% equivalent of the chiral inducer with respect to the ketone is enough to obtain good results.

Only one example of a chiral solvent effect in electrolysis has been described. A DAB-methanol-LiBr medium and Hg cathode allowed the formation of acetophenone pinacol with about 7% e.e. (273). This result was compared to that obtained in photopinacolization in DAB medium (272).

A completely new approach was envisaged by Miller and

co-workers (283), who constructed a chiral electrode. Graphite
was superficially oxidized into carboxylic groups, and (S)-
phenylalanine methyl ester was covalently bound to the surface.
Phenylglyoxylic acid was reduced in aqueous solution through use
of this chiral cathode. Mandelic acid was obtained in good
yield with 9.7% e.e. This type of experiment allows investigation of the mechanistic details of the electron transfer from
the electrode surface. It was demonstrated (283) that the
asymmetric reaction takes place on edge surfaces and not on
basal surfaces of the highly ordered pyrolitic graphite. In
addition the first asymmetric anodic reaction was obtained with
this chemically modified graphite electrode: p-tolymethylsul-
fide was oxidized to optically active p-tolymethylsulfoxide
[2.5% e.e. of (-)] from p-tolylmethylsulfide.

Chemically modified electrodes will open exciting perspec-
tives in asymmetric synthesis if high stereoselectivity can be
achieved.

VI. ASYMMETRIC SYNTHESIS WITH ENZYMES

A number of asymmetric syntheses have been described
involving purified enzymes, cell extracts, or growing cells
as catalysts. Only a few examples are cited here. We shall not
attempt to review the whole, rather extensive field but limit
ourselves to some results that appear promising or relevant to
organic synthesis.

Access to optically active D-hydroxynitriles has been
made easy by the use of D-hydroxynitrile lyase (284) in the
reaction of hydrocyanic acid with aldehydes. Chemical and
optical yields as high as 96% were obtained for the synthesis
of (D)-(+)-mandelonitrile. The process can be made continuous
by adsorption of the enzyme on a high-molecular-weight carrier
(e.g., ECTEOLA cellulose) (285).

Actively fermenting yeast has long been known to reduce
many "nonphysiological" compounds, especially ketone groups
(286), asymmetrically. There is at present some additional
work on reduction of ketones with other adjacent functional
groups. A number of α-diols $RCHOH-CH_2OH$ of known absolute
configuration have been prepared enantiomerically pure by
enzymatic reduction of α-ketols $RCOCH_2OH$ (287).

Other α-functionalized ketones such as α-chloroketones,
α-chloroesters, α-ketoacids, and α-ketoesters have been
investigated in reductions with actively fermenting yeast,
yielding the corresponding functionalized alcohols with
enantiomeric purities—when known—in excess of 90%. Yields
vary from 30 to 80% (288).

Chiral centers in key synthons for total synthesis of
optically active prostaglandins have been created by the use
of microbial enzymes (202). While chemical asymmetric reduction
of 136b to 137b was only partially successful, microbiological

reduction catalyzed by *Dipodascus uninucleatus* gave (R)-*137b*
and *Mucor rammanianus* yielded the enantiomer (S)-*137b*. Other
synthons such as *160* have been prepared in asymmetric form
from the unsaturated iodoketone *159*, with *Penicillium decumbers*
[to give (3S)-*160*] or *Aspergillus ustus* [to produce the (3R)
enantiomer], both in low (10 to 12%) yields. Asymmetric reduction
of ester *136a* was stereoselectively achieved by *D. uninucleatus*
to give *137a* in 48% yield. These microbiological reductions
afforded three chiral centers required for the synthesis of PGE$_1$
and PGE$_2$.

Takano (289) reported the asymmetric synthesis of lactone
164 by reduction of the achiral *cis*-3,5-diacetoxycyclopentene
161 with growing cultures of *Bacillius subtillis var. Niger*,
to give the chiral monoacetate *162*; transformation of this
latter into the rearranged compound *163* leads to lactone *164*,
which possesses the absolute configuration required for synthesis of prostaglandin F$_2$.

A process has been described by Tanabe (290) for addition
of ammonia to fumaric acid to give L-aspartic acid, using the
enzyme aspartase as a catalyst. A great variety of microorganisms are known for high fumarase activity, but in many
of them the enzyme, produced inside the cell, remains there.
Tanabe's procedure involves *Brevibacterium ammoniagenes* cells,
which produce aminoacylase extracellularly, so that the
extraction and separation of the desired enzyme is not required.

A number of problems had to be overcome, for example,
immobilization of the cells, enhancement of the aspartase
activity, and reduction of the activity of other enzymes that
are produced by the cells (leading to metabolization of fumaric

(3S)-*160*

(3R)-*160*

159

161   *162*   *163*   *164*

acid to succinic acid). These difficulties were solved by
treating the immobilized cells in a polyacrylamide gel lattice
with bile extract. Such a procedure would, in a month's
operation, using a 1000-liter column and a 200-liter/hr flow,
satisfy the present annual demand (1976) for (S)-malic acid.

$$R^3 \quad \backslash C=C-COOH \longrightarrow R^3 \quad \backslash CH-CH-COOH$$
$$R^2 \quad / \quad | \qquad \qquad R^2 \quad / \quad |$$
$$\qquad \quad R^1 \qquad \qquad \qquad \qquad R^1$$

165

$$D \quad \backslash \quad COOH \qquad HTO \qquad H$$
$$\quad C=C \quad \longrightarrow \quad T--C-CH-COOH$$
$$H \quad / \quad \backslash D \qquad \qquad D \quad | $$
$$\qquad \qquad \qquad \qquad \qquad D$$

166

Microorganisms have also been reported to act as catalysts
in asymmetric hydrogenation (291). *Clostridium klurjveri* allows
hydrogenation by hydrogen gas (atmospheric pressure) of a
number of α,β-unsaturated acids (*165*), and it has been shown,
insofar as the products were not new, that the reduction is
totally stereoselective. This procedure has provided what is
probably the simplest path to a chiral methyl group such as
in *166*.

Another example of a completely stereoselective enzymatic
reaction is the reduction of 2-methyl-1,2-di(3-pyridyl)-1-
propanone by *Botryodiplodia theobromac* Pat. to give (-)-2-
methyl-1,2-di(3-pyridyl)-1-propanol (*168*) with 99% e.e. The
same microorganism was found to convert isopropylaminomethyl
2-naphtyl ketone (*169*) to (-)-pronethalol (*170*) (292).

Enzymatic asymmetric epoxidation of 1,7-octadiene to
(R)-(+)-7,8-epoxy-1-octene (80% optical purity) is obtained

167   *168*

$$R-CO-CH_2-NH-i-Pr \longrightarrow R-CHOH-CH_2-NH-i-Pr$$

169   *170*

with a resting cell suspension of *Pseudomonas olevoraus,* a
well-known enzymatic system for ω-hydroxylation (293).

Asymmetric synthesis with enzymes seems to have a future
in some industrial processes, especially with the cell immobilization technique. It is not yet a standard method in
organic synthesis, but is almost irreplaceable when stereospecific labeling with deuterium or tritium is needed (294).

VII.   CONCLUSIONS

Asymmetric synthesis has evolved rapidly during recent
years. Less emphasis has been given to theoretical concepts and
mechanistic studies, most of the progress being registered in
synthetic chemistry. Methods have been devised for achieving
optical yields as high as 95%. Some stoichiometric reactions
with respect to the chiral auxiliary moiety are now highly
efficient (see Sects. II-A, B, and C). We have not discussed
immolative asymmetric syntheses.* However, the fact that the
inducing chiral center is destroyed is unimportant if the
chiral starting material is inexpensive with respect to the
product. An elegant illustration of this principle is the
beautiful two-step synthesis shown in Figure 30, which allowed

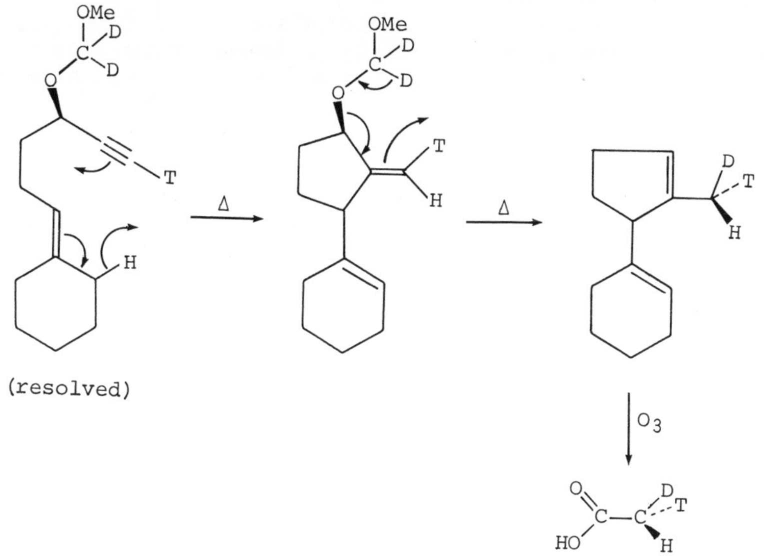

(resolved)

Fig. 30

*An immolative asymmetric synthesis is one in which the
existing auxiliary chiral center (or other chiral elements) is
destroyed in the course of generating a new one.

Arigoni (294) to obtain, on a preparative scale, optically
pure chiral acetic acid. Total synthesis of 11α-methylprogesterone
by Johnson (295) is an example of another useful type of asymmetric synthesis (Fig. 31) where a chiral center, which remains
in the product, controls formation of six other asymmetric
centers.

A trend in asymmetric synthesis is the utilization of
cheap natural products such as sugars, amino acids, hydroxy

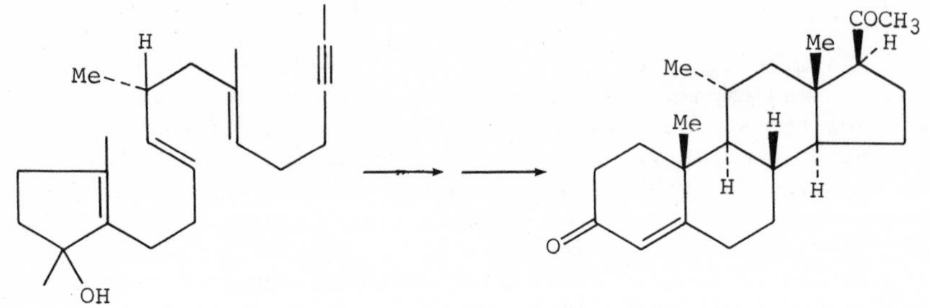

Fig. 31

acids, or terpenes as starting materials. Thus optically active
prostaglandins were synthetized from (S)-malic acid (297) or
carbohydrates (298). In these latter syntheses Stork uses 2,3-
isopropylidene-L-erythrose and D-glyceraldehyde, respectively
as the chiral precursors (Fig. 32). Asymmetric syntheses of

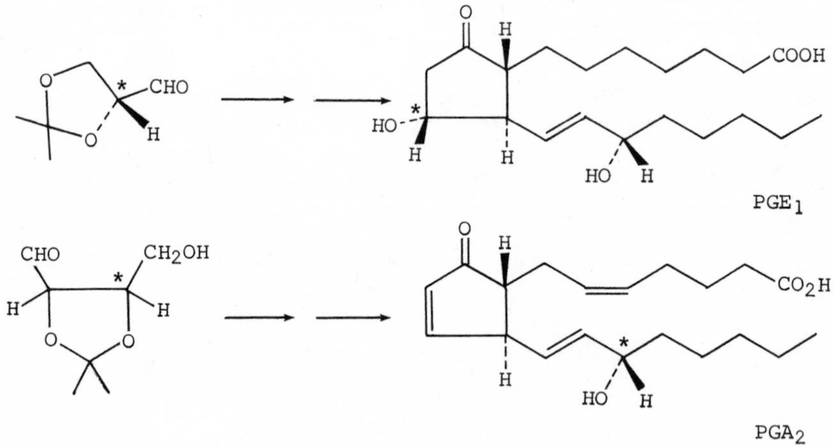

Fig. 32

avenaciolide (299) and biotin (300) were achieved from sugar derivatives. Chiral intermediates for the total synthesis of steroids with C/D rings having the desired stereochemistry were efficiently prepared form camphor (305).

Catalytic methods have improved notably in efficiency during the last five years. As a result industrial asymmetric processes are now possible based on methods of organic chemistry . Since, frequently, catalysts are expensive as well as sophisticated, numerous attempts are being made to prepare supported catalysts that can be recovered for reuse.

Only a few chiral heterogeneous catalysts are known (2,1b), the best-known instance being Raney nickel modified by tartaric acid. Unfortunately this type of catalyst is effective only for methyl acetoacetate and closely related compounds. It is interesting that, through careful study of the experimental conditions, optical yields, of up to 91% have been attained (303), the chiral catalyst being a nickel-palladium-Kieselguhr system modified with tartaric acid. The mechanism of asymmetric hydrogenation on nickel surfaces modified by tartaric acid was studied in vapor-phase reduction of methyl acetoacetate. The IR spectra of the adsorbed species and labeling experiments with deuterium suggested a model for the stereochemical control of the reduction based on weak interactions between the substrate and chiral modifier through hydrogen bonds (303).

A new approach involves attempts to support chiral catalysts on polymer beads. An insoluble DIOP analogue was thus synthesized on a support of polystyrene (204). With this ligand a supported rhodium catalyst was prepared that catalyzes hydrosilylation of acetophenone, the optical yields being the same as under homogeneous conditions. The catalyst can be reused after filtration. Unfortunately this supported complex does not catalyze the reduction of α-acetamido-α,β-unsaturated carboxylic acids; the hydrophilic character of the polymer seems responsible for this failure. Recently a cross-linked DIOP-bearing polymer was prepared by Stille (304) with free hydroxyl groups on some sidechains, increasing the hydrophilicity of this system. The resulting catalyst induced asymmetric synthesis of α-amino acids with the optical yield being the same as in solution.

It may be hoped than more supported catalysts will be operative in the future and will be carefully evaluated for efficiency and stability.

A revival of interest in mechanistic studies seems highly desirable in asymmetric synthesis to avoid purely empirical research, which tends to be slow and difficult. A better understanding of the reactions involved and the factors that control asymmetric induction would greatly help in devising appropriate structural features in chiral reactants and catalysts.

ACKNOWLEDGMENTS

We thank Drs. L. Salem, Nguyen Trong Anh, and Z. Welvart for many helpful discussions and critical reading of the manuscript. We are grateful to Drs. J. E. Baldwin, B. Bosnich, B. S. Green, A. Horeau, W. S. Knowles, L. L. Miller, D. Seebach, N. Spassky, B. M. Trost, A. Vasella, and J. P. Vigneron for kindly informing us of unpublished results in the area of asymmetric synthesis.

One of us (H. K.) thanks Professor A. I. Meyers for his hospitality at Colorado State University during the summer of 1976, and for providing facilities for writing a part of this chapter.

REFERENCES

1. W. Marckwald, *Chem Ber.*, 37, 1368 (1904).
2. J. D. Morrison and H. S. Mosher, *Asymmetric Organic Reactions*, Prentice-Hall, Englewoods Cliffs, N. J., 1971; paperback reprint, American Chemical Society, Washington D. C. 1976.
3. R. S. Cahn, C. K. Ingold, and V. Prelog, *Experientia,* 12, 81 (1956); R. S. Cahn, C. K. Ingold, and V. Prelog, *Angew. Chem., Int. Ed. Engl.,* 5, 385 (1966).
4. E. L. Eliel, *Stereochemistry of Carbon Compounds*, McGraw-Hill, New York, 1962.
5. H. C. Brown and S. Krishnamurthy, *J. Am. Chem. Soc.,* 94, 7159 (1972).
6. J. C. Fiaud, in *Fundamentals and Methods of Stereochemistry,* Vol. 3, H. B. Kagan, Ed., Georg Thieme Verlag, 1977.
7. E. J. Corey and H. E. Ensley, *J. Am. Chem. Soc.,* 97, 6908 (1975).
8. J. Mathieu and J. Weil-Raynal, *Bull. Soc. Chim. Fr.,* 1968, 1211.
9. D. R. Boyd and M. A. McKervey, *Q. Rev.,* 22, 95 (1968).
10. T. D. Inch, *Synthesis,* 1970, 466.
11. H. J. Schneider and R. Haller, *Pharmazie,* 28, 417 (1973).
12. H. E. Radunz, *Chem. Ztg.,* 97, 592 (1973).
13. J. W. Scott and D. Valentine, Jr., *Science,* 184, 943 (1974); for a complement see also D. Valentine, Jr., and J. W. Scott, *Synthesis,* 1978, 329.
14. E. L. Eliel, *Tetrahedron,* 30, 1503 (1974).
15. J. Mathieu, R. Bucourt, and J. Weill-Raynal, *Chimia,* 27, 217 (1874).
16. Y. Izumi, *Angew. Chem.,* 83, 956 (1971); *Angew. Chem., Int. Ed. Engl.,* 10, 871 (1971); for a new classification of asymmetric reactions, see Y. Izumi and A. Tai, *Stereo-*

Differentiating Reactions, Kodansha, Tokyo, and Academic Press, New York, 1977.

17. E. Ruch and I. Ugi, "The Stereochemical Analogy Model," in *Topics in Stereochemistry,* Vol. 4, N. L. Allinger and E. L. Eliel, Eds., Wiley-Interscience, New York, 1969, p. 99.

18. E. L. Eliel and N. L. Allinger, *Conformational Analysis,* Interscience, New York, 1965.

19. E. Anders, E. Ruch, and I. Ugi, *Angew. Chem.,* $\underline{85}$, 16 (1973); *Angew. Chem., Int. Ed. Engl.,* $\underline{12}$, 25 (1973).

20. L. Gruber, J. Tömösközi, and L. Ötvös, *Tetrahedron Lett.,* $\underline{1973}$, 811.

21. H. J. Schneider and R. Haller, *Tetrahedron,* $\underline{29}$, 2509 (1973).

22. J. P. Vigneron, M. Dhaenens, and A. Horeau, *Tetrahedron,* $\underline{33}$, 497 (1977).

23. M. Malissard, S. Sicsic, Z. Welvart, A. Chiaroni, C. Riche, and C. Pascard, *Bull. Soc. Chim. Fr.,* $\underline{1974}$, 1459, and references cited therein.

24. L. Salem, *J. Am. Chem. Soc.,* $\underline{95}$, 94 (1973).

25. W. G. Dauben, G. J. Fonken, and D. S. Noyce, *J. Am. Chem. Soc.,* $\underline{78}$, 2579 (1956).

26. G. Chauviere and Z. Welvart, *Bull. Soc. Chim. Fr.,* $\underline{1970}$, 774. R. P. Bell, *The Proton in Chemistry,* Cornell University Press, New York, 1959, p. 183; 2nd ed., Chapman and Hall, 1973, p. 250; More O'Ferral and J. Kouba, *J. Chem. Soc., B,* $\underline{1967}$, 985; W. A. Pryor and K. G. Kneipp, *J. Am. Chem. Soc.,* $\underline{93}$, 5584 (1971).

27. M. Cherest, H. Felkin, and N. Prudent, *Tetrahedron Lett.,* $\underline{1968}$, 2201, 2205.

28. P. Geneste, P. Lamaty, and J. P. Roque, *Tetrahedron Lett.,* $\underline{1970}$, 5007.

29. E. C. Ashby and S. A. Nody, *J. Am. Chem. Soc.,* $\underline{98}$, 2010 (1976).

30. D. H. Wertz and N. L. Allinger, *Tetrahedron,* $\underline{30}$, 1579 (1974).

31. J. P. Vigneron, H. B. Kagan, and A. Horeau, *Tetrahedron Lett.,* $\underline{1968}$, 5681. J. P. Vigneron, H. B. Kagan, and A. Horeau, *Bull. Soc. Chim. Fr.,* $\underline{1972}$, 3836.

32. H. Thompson and R. E. Naipawer, *J. Am. Chem. Soc.,* $\underline{95}$, 6379 (1973).

33. D. Varech and J. Jacques, *Tetrahedron Lett.,* $\underline{45}$, 4443 (1973).

34. J. M. Lefour, personal communication.

35. W. T. Wipke and P. Gund, *J. Am. Chem. Soc.,* $\underline{96}$, 299 (1975). W. T. Wipke and P. Gund, *J. Am. Chem. Soc.,* $\underline{98}$, 8107 (1976).

36. H. B. Bürgi, J. D. Dunitz, and E. Shefter, *J. Am. Chem. Soc.,* $\underline{95}$, 5065 (1973).

37. H. B. Bürgi, J. M. Lehn, and G. Wipf, *J. Am. Chem. Soc.*, 96, 1956 (1974).

38. G. Chauviere and Z. Welvart, *Bull. Soc. Chim. Fr.*, 1970, 774; see also reference 53.

39. E. Toromanoff, "Steric Course of the Kinetic 1,2 Addition of Anions to Conjugated Cyclohexenones," in *Topics in Stereochemistry*, Vol. 2, N. L. Allinger and E. L. Eliel, Eds., Wiley-Interscience, New York, 1967, p. 157.

40. J. E. Baldwin, *J. Chem. Soc., Chem. Commun.*, 1976, 738.

41. R. E. Carter, B. Nilsson, and K. Olsson, *J. Am. Chem. Soc.*, 97, 6155 (1975). R. E. Carter and P. Stilbs, *ibid.*, 98, 7515 (1976).

42. A. Liberles, A. Greenberg, and J. E. Eilers, *J. Chem. Educ.*, 50, 676 (1973).

43. N. D. Epiotis, *J. Am. Chem. Soc.*, 95, 3087 (1973).

44. E. J. Corey, K. B. Becker, and R. K. Varma, *J. Am. Chem. Soc.*, 94, 8618 (1972).

45. G. J. Karabatsos, *J. Am. Chem. Soc.*, 89, 1367 (1967).

46. V. Prelog, M. Wilhem, and D. Bruce Bright, *Helv. Chim. Acta*, 37, 221 (1954).

47. J. Weill-Raynal and J. Mathieu, *Bull. Soc. Chim. Fr.*, 1969, 115.

48. R. Parthasarathy, J. Ohrt, A. Horeau, J. P. Vigneron, and H. B. Kagan, *Tetrahedron*, 26, 4705 (1970).

49. H. K. Spencer and R. K. Hill, *J. Org. Chem.*, 41, 2485 (1976).

50. B. Denis, J. Ducom, and J. F. Fauvarque, *Bull. Soc. Chim. Fr.* 1972, 990.

51. D. J. Faulkner and M. R. Petersen, *J. Am. Chem. Soc.*, 95, 553 (1973).

52. D. Cabaret, personal communication.

53. N. L. Allinger and M. T. Tribble, *Tetrahedron Lett.*, 1971, 3259.

54. N. S. Crossley, A. C. Darby, H. B. Henbest, J. Mc Cullough, B. Nicholls, and M. F. Stewart, *Tetrahedron, Lett.*, 1961, 398.

55. G. Belluci, F. Marioni and A. Marsili, *Tetrahedron*, 28, 3393 (1972).

56. (a) J. P. Guetté, J. Capillon, M. Perlat, and M. Guetté, *Tetrahedron Lett.*, 1974, 2409.
    (b) J. Capillon, Thesis, Paris, 1975.

57. (a) M. Cherest, M. Felkin, M. Tacheau, J. Jacques, and D. Varech, *J. Chem. Soc., Chem. Commun.*, 1977, 372.
    (b) A. Vasella, *Helv. Chim. Acta,* 60, 426 (1977); A. Vasella, to be submitted to *Helv. Chim. Acta.*

58. Nguyen Trong Anh, O. Eisenstein, J. M. Lefour, and M. E. Tran Huu Dau, *J. Am. Chem. Soc.*, 95, 6146 (1973).

59. S. Inagaki and K. Fukui, *Chem. Lett.*, 1974, 509.

60. J. Klein, *Tetrahedron*, 30, 3349 (1974). E. C. Ashby and J. R. Boone, *J. Org. Chem.*, 41, 2890 (1976).

61. Nguyen Trong Anh and O. Eisenstein, *Tetrahedron Lett.*, 1976, 155; Nguyen Trong Anh and O. Eisenstein, *Nouv. J. Chim.*, 61, 1 (1977).

62. D. J. Cram and F. A. Abd Elhafez, *J. Am. Chem. Soc.*, 74, 5828 (1952).

63. J. Huet, Y. Maroni-Barnaud, Nguyen Trong Anh, and J. Seyden-Penne, *Tetrahedron Lett.*, 1976, 159.

64. U. Berg, R. Gallo, and J. Metzger, *J. Am. Chem. Soc.*, 98, 1260 (1976).

65. B. Caro and G. Jaouen, *J. Chem. Soc., Chem. Commun.*, 1976, 655.

66. E. J. Corey, R. S. McCaully, and H. S. Sachdev, *J. Am. Chem. Soc.*, 92, 2476 (1970).

67. D. Cabaret and Z. Welvart, *J. Organomet. Chem.*, 80, 199 (1974); D. Cabaret and Z. Welvart, *J. Organomet. Chem.*, 78, 295 (1974).

68. P. Audoye, A. Gaset, and A. Lattes, *J. Appl. Chem. Biotechnol.*, 25, 19 (1975).

69. T. Hiyama, T. Mishima, H. Sawada, and H. Nozaki, *J. Am. Chem. Soc.*, 97, 1626 (1975); *J. Am. Chem. Soc.*, 98, 641 (1976).

70. A. Horeau, *Tetrahedron Lett.*, 1969, 3121; A. Horeau and J. P. Guetté, *Tetrahedron*, 30, 1923 (1974).

71. W. H. Pirkle and D. L. Sikkenga, *J. Org. Chem.*, 40, 3430 (1975), and references cited therein.

72. M. D. Creary, D. W. Lewis, D. L. Wernick, and G. M. Whitesides, *J. Am. Chem. Soc.*, 96, 1038 (1975).

73. C. M. Thong, M. Marraud, and J. Neel, *C. R. Acad. Sci. Paris, C*, 281, 691 (1976).

74. E. Gil-Av and D. Nurok, *Adv. Chromatogr.*, 10, 99 (1974).

75. C. H. Lochmuller and R. W. Souter, *J. Chromatogr.*, 113, 283 (1975).

76. G. R. Sullivan, J. A. Dale, and H. S. Mosher, *J. Org. Chem.*, 38, 2143 (1974).

77. D. Valentine, Jr., K. K. Chan, C. G. Scott, K. K. Johnson, K. Toth, and G. Saucy, *J. Org. Chem.*, 41, 62 (1976).

78. H. Wynberg and B. Feringa, *Tetrahedron*, 1976, 2831.

79. J. C. Fiaud and A. Horeau, *Tetrahedron Lett.*, 1972, 2565.

80. M. S. Patel and M. Worsley, *Can. J. Chem.*, 48, 1881 (1970).

81. G. M. L. Cragg, *Organoboranes in Organic Synthesis*, Marcel-Dekker, New York, 1973; H. C. Brown, *Boranes in Organic Chemistry*, Cornell University Press, Ithaca, New York, 1972, p. 285.

82. J. J. Partridge, N. K. Chadha, and M. R. Uskokovic,

J. Am. Chem. Soc., <u>95</u>, 7171 (1973).

83. J. J. Partridge, N. K. Chadha, and M. R. Uskokovic,
     J. Am. Chem. Soc., <u>95</u>, 532 (1973).

84. H. C. Brown, N. R. de Lue, G. W. Kabalka, and H. C.
     Hedgecock, Jr., *J. Am. Chem. Soc.*, <u>98</u>, 1290 (1976).

85. R. Andrisano, A. S. Angeloni, and S. Marzocchi, *Tetra-
     hedron*, <u>29</u>, 913 (1973).

86. R. Haller and H. J. Schneider, *Chem. Ber.*, <u>106</u>, 1312
     (1973).

87. S. Yamaguchi, H. S. Mosher, and A. Pohland, *J. Am. Chem.
     Soc.*, <u>94</u>, 9254 (1972).

88. S. Yamaguchi and H. S. Mosher, *J. Org. Chem.*, <u>38</u>, 1870
     (1973).

89. C. J. Reich, G. R. Sullivan, and H. S. Mosher, *Tetra-
     hedron Lett.*, <u>1973</u>, 1505.

90. K. Kabuto and H. Ziffer, *J. Org. Chem.*, <u>40</u>, 3467 (1975).

91. J. P. Guetté and A Horeau, *Bull. Soc. Chim. Fr.*, <u>1967</u>,
     1747.

92. I. Jacquet and J. P. Vigneron, *Tetrahedron Lett.*, <u>1974</u>,
     2065.

93. S. I. Yamada, M. Kitamoto, and S. Terashima, *Tetrahedron*,
     <u>36</u>, 3165 (1976).

94. D. Seebach and H. Daum, *Chem. Ber.*, <u>107</u>, 1748 (1974).

95. D. Seebach and H. Meyer, *Angew. Chem.*, <u>86</u>, 40 (1974);
     Angew. Chem., Int. Ed. Engl., <u>13</u>, 77 (1974).

96. O. Cervinka and J. Fusek, *Coll. Czech. Chem. Commun.*,
     <u>38</u>, 441 (1973).

97. O. Cervinka, P. Malon, and H. Prochazkova, *Coll. Czech.
     Chem. Commun.*, <u>39</u>, 1869 (1974).

98. R. J. D. Evans, S. R. Landor, and J. P. Regan, *J. Chem.
     Soc., Perkin 1*, <u>1974</u>, 557.

99. D. Arigoni and E. L. Eliel, "Hydrogen Isotopes at Non-
     cyclic Positions," in *Topics in Stereochemistry*, Vol. 1,
     N. L. Allinger and E. L. Eliel, Eds., Interscience,
     New York, 1969, p. 127.

100. S. R. Landor, O. O. Sonola, and A. R. Tatchell, *J. Chem.
     Soc., Perkin 1*, <u>1974</u>, 1902.

101. L. D. Tomina, E. I. Klabunovskii, Y. I. Petrov, E. D.
     Lubuzh, and E. M. Cherkasova, *Izv. Akad. Nauk SSSR, Ser.
     Khim.*, <u>1972</u>, 2506; *Chem. Abstr.*, <u>78</u>, 71238e (1973).

102. D. Nasipuri and P. R. Mukherjee, *J. Indian Chem. Soc.*, _
     <u>51</u>, 171 (1974).

103. D. Cabaret and Z. Welvart, *J. Organomet. Chem.*, <u>80</u>, 185
     (1974).

104. G. M. Giongo, F. di Gregorio, N. Palladino, and W. Marconi,
     Tetrahedron Lett., <u>1973</u>, 3195.

105. G. Giacomelli, R. Menacagli, and L. Lardicci, *J. Org.
     Chem.*, <u>38</u>, 2370 (1973); *J. Org. Chem.*, <u>39</u>, 1757 (1974);
     J. Am. Chem. Soc., <u>97</u>, 4009 (1975).

106. L. Lardicci and G. Giacomelli, *J. Chem. Soc., Perkin 1,*
     1974, 337.
107. R. Bentley and Ed., *Molecular Asymmetry in Biology,* Vol.
     2, Academic Press, New York, London, 1970, pp. 1-67.
108. Y. Ohnishi, T. Numakunai, and A. Ohno, *Tetrahedron Lett.,*
     44, 3813 (1975).
109. Y. Ohnishi, M. Kagami, and A. Ohno, *J. Am. Chem. Soc.,*
     97, 4768 (1975).
110. K. Nishiyama, N. Baba, J. Oda, and Y. Inouye, *J. Chem.
     Soc., Chem. Commun.,* 1976, 101.
111. S. Brandange, S. Josephson, and S. Vallen, *Acta Chem.
     Scand.,* B31, 179 (1977).
112. T. Sugita, Y. Yamasaki, O. Itoh, and K. Ichikawa, *Bull.
     Chem. Soc. Jpn.,* 47, 1945 (1974).
113. M. F. Grundon, D. Stewart, and W. E. Watts, *J. Chem.
     Soc., Chem. Commun.,* 1973, 573.
114. C. R. Johnson and C. W. Schroeck, *J. Am. Chem. Soc.,* 95,
     7418 (1973).
115. S. Musierowicz, A. Wroblewski, and H. Krawczyk, *Tetra-
     hedron Lett.,* 1975, 437.
116. J. S. Zweig, J. L. Luche, E. Barreiro, and P. Crabbé,
     Tetrahedron Lett., 1975, 2355.
117. T. A. Whitney and A. W. Langer, Jr., *Adv. Chem. Ser.,*
     130, 270 (1974).
118. D. Seebach, personal communication.
119. A. Davidson, W. C. Krusell, and R. C. Michaelson, *J. Organo-
     met. Chem.,* 72, C7 (1974).
120. G. Saucy and R. Borer, *Helv. Chim. Acta,* 54, 2034 (1971);
     J. W. Scott, R. Borer, and G. Saucy, *J. Org. Chem.,* 37,
     1659 (1972); M. Rosenberger, A. J. Diggan, R. Borer, R.
     Muller, and G. Saucy, *Helv. Chim. Acta,* 55, 2663 (1972);
     N. Cohen, B. L. Banner, R. Borer, R. Mueller, R. Yang,
     M. Rosenberger, and G. Saucy, *J. Org. Chem.,* 37, 3385
     (1972); N. Cohen, B. L. Banner, J. F. Blount, M. Tsai,
     and G. Saucy, *J. Org. Chem.,* 38, 3229 (1973).
121. J. H. Dopper, B. Greijdanus, D. Oudman, and H. Wynberg,
     J. Chem. Soc. Chem. Commun., 1975, 972.
122. K. Harada, T. Okawara, and K. Matsumoto, *Bull. Chem.
     Soc. Jpn.,* 46, 1865 (1973).
123. I. Ojima, S. Yamada, and Y. Nagai, *Chem. Lett.,* 1975,
     737.
124. R. Urban and I. Ugi, *Angew. Chem.,* 87, 67 (1975); *Angew.
     Chem., Int. Ed. Engl.,* 14, 61 (1975).
125. R. Urban, G. Eberle; D. Marquarding, D. Rehn, H. Rehn,
     and I. Ugi, *Angew. Chem.,* 88, 644 (1976); *Angew. Chem.,
     Int. Ed. Engl.,* 15, 627 (1976).
126. K. Harada, T. Iwasaki, and T. Okawara, *Bull. Chem. Soc.
     Jpn.,* 46, 1901 (1973).
127. S. I. Yamada and S. I. Hashimoto, *Tetrahedron Lett.,*
     1976, 997.

128. S. I. Yamada, N. Ikota, and K. Achiwa, *Tetrahedron Lett.*, 1976, 1001.

129. B. W. Bycroft and G. R. Lee, *J. Chem. Soc., Chem. Commun.*, 1975, 988.

130. K. Achiwa and S. I. Yamada, *Tetrahedron Lett.*, 1974, 1799.

131. A. I. Meyers and E. D. Mihelich, *Angew. Chem.*, 88, 321 (1976); *Angew. Chem., Int. Ed. Engl.*, 15, 270 (1976).

132. A. I. Meyers, G. Knaus, K. Kamata, and M. E. Ford, *J. Am. Chem. Soc.*, 98, 567 (1976), and references cited therein.

133. A. I. Meyers and C. E. Whitten, *Heterocycles*, 4, 1687 (1976).

134. K. Hiroi and S. I. Yamada, *Chem. Pharm. Bull.*, 21, 47, 51 (1973).

135. G. Otani and S. I. Yamada, *Chem. Pharm. Bull.*, 21, 2130 (1973).

136. M. Kitamoto, K. Hiroi, S. Terashima, and S. I. Yamada, *Chem. Pharm. Bull.*, 22, 459 (1974).

137. S. I. Yamada, M. Shibassaki, and S. Terashima, *Tetrahedron Lett.*, 1973, 381.

138. A. I. Meyers, D. R. Williams, and M. Druelinger, *J. Am. Chem. Soc.*, 98, 3032 (1976).

139. A. I. Meyers and K. Kamata, *J. Org. Chem.*, 39, 1603 (1974).

140. D. Enders and H. Eichenauer, *Angew. Chem.*, 88, 579 (1976); *Angew. Chem., Int. Ed. Engl.*, 15, 549 (1976).

141. S. I. Yamada, T. Oguri, and T. Shiori, *J. Chem. Soc., Chem. Commun.*, 1976, 136.

142. B. M. Trost and T. J. Dietsche, *J. Am. Chem. Soc.*, 95, 8200 (1973).

143. B. M. Trost and P. E. Strege, *J. Am. Chem. Soc.*, 99, 1649 (1977).

144. M. Mikolajczyk and J. Drabowicz, *J. Chem. Soc., Chem. Commun.*, 1974, 547.

145. M. Mikolajczyk, J. Drabowicz, J. Omelanczuk, and E. Fluck, *J. Chem. Soc., Chem. Commun.*, 1975, 382.

146. R. Bucourt, L. Nedelec, J. C. Gasc, and J. Weill-Raynal, *Bull. Soc. Chim. Fr.*, 1967, 561, and references cited therein.

147. H. Matsushita, M. Noguchi, and S. Yoshikawa, *Chem. Lett.*, 1975, 1313.

148. H. Matsushita, N. Nogichi, M. Saburi, and S. Yoshikawa, *Bull. Chem. Soc. Jpn.*, 48, 3715 (1975).

149. L. Duhamel, *C. R. Acad. Sci., Ser. C.*, 282, 125 (1976).

150. R. E. Pincock and K. R. Wilson, *J. Am. Chem. Soc.*, 93, 1291 (1971). R. E. Pincock, R. R. Perkins, A. S. Ma, and K. R. Wilson, *Science*, 174, 1018 (1971).

151. A. Fischli, K. Klaus, H. Mayer, P. Schoenholzer, and R. Rueegg, *Helv. Chim. Acta*, 38, 564 (1975).

152. J. C. Clark, G. H. Phillips, and M. R. Steer, *J. Chem. Soc., Perkin 1*, 1976, 476.

153. S. Sifniades, W. J. Boyle, Jr., and J. F. Van Peppen, *J. Am. Chem. Soc.*, 98, 3738 (1976).

154. U. Eder, G. Sauer, and R. Wiechert, *Angew. Chem.*, 83, 492 (1971); *Angew. Chem., Int. Ed. Engl.*, 10, 496 (1971).

155. G. Z. Hajos and D. R. Parrish, German Patent, 2, 102, 623; *Chem. Abstr.*, 76, 59072x (1972).

156. Z. G. Hajos and D. R. Parrish, *J. Org. Chem.*, 39, 1615 (1974).

157. S. Danishefsky and P. Cain, *J. Am. Chem. Soc.*, 97, 5282 (1975).

158. U. Obenius and G. Bergson, *Acta Chem. Scand.*, 26, 2546 (1972).

159. B. Langström and G. Bergson, *Acta Chem. Scand.*, 27, 3118 (1973).

160. H. Wynberg and R. Helder, *Tetrahedron Lett.*, 1975, 4057.

161. (a) T. Hiyama, H. Sawada, M. Tsukanaka, and H. Nozaki, *Tetrahedron Lett.*, 1975, 3013.

162. J. C. Fiaud, *Tetrahedron Lett.*, 1975, 3495.

163. S. Colonna and R. Fornaster, *Synthesis*, 1975, 531.

164. J. Balcells, S. Colonna, and R. Fornaster, *Synthesis*, 1976, 266.

165. J. P. Masse and E. R. Parayre, *J. Chem. Soc., Chem. Commun.*, 1976, 438.

166. R. Helder, J. C. Hummelen, R. W. P. M. Laane, J. S. Wiering, and H. Wynberg, *Tetrahedron Lett.*, 1976, 1831.

167. L. Markó and B. Heil, *Catal. Rev.*, 8, 269 (1975). E. R. James, *Homogeneous Hydrogenation*, Wiley-Interscience, New York, R. E. Harmon, S. K. Gupta, and D. J. Brown, *Chem. Rev.*, 1973, 7321; J. Benes and J. Hetflejs, *Chem. Listy*, 68, 916 (1974).

168. J. D. Morrison, W. F. Masler, and S. Hataway, *Catalysis in Organic Synthesis*, Academic Press, New York, 1976.

169. J. D. Morrison, W. F. Masler, and M. K. Neuberg, in *Advances in Catalysis*, Vol. 25, D. D. Eley, H. Pines, and P. B. Weisz, Eds.,

169. F. J. McQuillin *Homogeneous Hydrogenation in Organic Chemistry*, D. Reidel, Dordrecht-Holland, Boston 1976.

170. A. P. G. Kielboom and F. van Rantwijk, *Hydrogenation and Hydrogenolysis in Synthetic Organic Chemistry*, Delft University Press, 1977.

171. I. Ojima, K. Yamamoto, and M. Kumada, in *Aspects of Homogeneous Catalysis*, Vol. 3, Ugo Ed., D. Reidel, Dordrecht-Holland and Boston, 1977, p. 186.

172. B. Bogdanovic, B. Henc, A. Lösler, B. Meister, H. Pauling, and G. Wilke, *Angew. Chem.*, 85, 1013 (1973); *Angew. Chem., Int. Ed. Engl.*, 12, 954 (1973).

173. L. Horner and H. Stegel, *Phosphorus*, 1, 199, 209 (1972).

174. W. S. Knowles, M. J. Sabacky, and B. D. Vineyard, *Chem. Technol.*, 1972, 591.

175. A. M. Aguiar, C. J. Morrow, J. D. Morrison, R. E.
      Burnett, W. F. Masler, and N. S. Bhacca, *J. Org. Chem.*,
      <u>41</u>, 1545 (1976).

176. H. B. Kagan and T. P. Dang, *J. Am. Chem. Soc.*, <u>94</u>, 6429
      (1972).

177. W. S. Knowles, M. J. Sabacky, B. D. Vineyard, and D. J.
      Weinkauff, *J. Am. Chem. Soc.*, <u>97</u>, 2569 (1975).

178. G. Gelbard, H. B. Kagan, and R. Stern, *Tetrahedron*, <u>32</u>,
      233 (1976).

179. D. Sinou and H. B. Kagan, *J. Organomet. Chem.*, <u>114</u>, 325
      (1976).

180. A. P. Stoll and R. Süess, *Helv. Chim. Acta*, <u>57</u>, 2487
      (1974).

181. T. Hayashi, T. Mise, S. Mitachi, K. Yamamoto, and M.
      Kumada, *Tetrahedron Lett.*, <u>1976</u>, 1133.

182. H. B. Kagan, *Pure Appl. Chem.*, <u>43</u>, 401 (1975).

183. H. B. Kagan, manuscript in preparation, to be submitted
      to *Acc. Chem. Res.*

184. T. P. Dang, J. C. Poulin, and H. B. Kagan, *J. Organomet.
      Chem.*, <u>91</u>, 39 (1975).

185. Rhone-Poulenc, French Patent No. 73.18319 (1973).

186. S. Brunie, J. Mazan, N. Langlois, and H. B. Kagan, *J.
      Organomet. Chem.*, <u>114</u>, 225 (1976).

187. H. B. Kagan, N. Langlois, and T. P. Dang, *J. Organomet.
      Chem.*, <u>90</u>, 353 (1975).

188. R. Glaser, *Tetrahedron Lett.*, <u>1975</u>, 2127.

189. B. D. Vineyard, W. S. Knowles, M. J. Sabacky, G. L.
      Bachman, and D. J. Weinkauff, *J. Am. Chem. Soc.*, in
      press.

190. M. D. Fryzuk and B. Bosnich, *J. Am. Chem. Soc.*, <u>99</u>,
      6262 (1977).

191. J. C. Poulin, C. Detellier, and H. B. Kagan, unpublished
      results.

192. P. Abley and F. J. McQuillin, *J. Chem. Soc.*, *C*, <u>1971</u>,
      844.

193. V. A. Pavlov, S. I. Klabunovskii, G. S. Barycheva, L. N.
      Kaigorodeva, and Y. S. Airapetov, *Izv. Akad. Nauk SSR,
      Ser. Khim.*, <u>1975</u>, 2374; *Bull. Acad. Sci. USSR, Div. Chem.
      Sci.*, <u>24</u>, 2262 (1975).

194. (a) M. Tanaka and I. Ogata, *J. Chem. Soc., Chem. Commun.*,
      <u>1975</u>, 735.
      (b) T. Hayashi, M. Tanaka, and I. Ogata, *Tetrahedron
      Lett.*, <u>1977</u>, 295.
      (c). R. H. Grubbs and R. A. de Vries, *Tetrahedron Lett.*,
      <u>1977</u>, 1875.

195. M. Fiorini, G. M. Giongo, F. Marcati, and W. Marconi, *J.
      Mol. Catal.*, <u>1</u>, 451 (1975-1976).

196. B. R. James, D. K. W. Wang, and R. F. Voigt, *J. Chem.
      Soc., Chem. Commun.*, <u>1975</u>, 574.

197. S. Takeuchi, Y. Ohgo, and J. Yosshimura, *Chem. Lett.*,
     <u>1973</u>, 265.
198. A. Levi, G. Modena, and G. Scorrano, *J. Chem. Soc., Chem.
     Commun.*, <u>1975</u>, 6.
199. (a) P. Bonvicini, A.Levi, G. Modena, and G. Scorrano,
     J. Chem. Soc., Chem. Commun., <u>1972</u>, 1188.
     (b) M. Tanaka, Y. Watanabe, T. Mitsudo, H. Iwane, and Y.
     Takegami, *Chem. Lett.*, <u>1973</u>, 239.
200. B. Heil, S. Törös, S. Vastag, and L. Markó, *J. Organomet.
     Chem.*, <u>94</u>, C47 (1975).
201. J. Solodar, *Chem. Technol.*, <u>1975</u>, 421.
202. C. J. Sih, J. B. Heather, R. Sood, G. Peruzotti, L. F.
     Hsu Lee, and S. S. Lee, *J. Am. Chem. Soc.*, <u>97</u>, 865 (1975).
203. Y. Ohgo, Y. Natori, S. Takeuchi, and J. Yoshimura, *Chem.
     Lett.*, <u>1974</u>, 709, 1327, and references quoted.
204. W. Dumont, J. C. Poulin, T. P. Dang, and H. B. Kagan,
     J. Am. Chem. Soc., <u>95</u>, 8295 (1973).
205. R. J. P. Corriu and J. J. E. Moreau, *J. Organomet. Chem.*,
     <u>64</u>, C51 (1974).
206. T. Hayashi, H. Ohmizo, S. Baba, K. Shinohara, K. Kagusa,
     K. Yamamoto, and M. Kumada, 22nd Symposium on Organo-
     metallic Chemistry, Japan, 1974.
207. I. Ojima and Y. Nagai, *Chem. Lett.*, <u>1974</u>, 223.
208. T. Hayashi, K. Yamamoto, and M. Kumada, *Tetrahedron Lett.*,
     <u>1974</u>, 4405.
209. (a) I. Ojima, T. Kogure, and Y. Nagai, *Chem. Lett.*,
     <u>1975</u>, 985.
     (b) I. Ojima, T. Kogure, and Y. Nagai, *Tetrahedron Lett.*,
     <u>1974</u>, 1899; I. Ojima and Y. Nagai, *Chem. Lett.*, <u>1975</u>,
     191.
210. J. F. Peyronel, J. C. Fiaud, and H. B. Kagan, unpublish-
     ed results.
211. N. Langlois, T. P. Dang, and H. B. Kagan, *Tetrahedron
     Lett.*, <u>1973</u>, 4865.
212. R. J. P. Corriu and J. J. E. Moreau, *J. Organomet.
     Chem.*, <u>85</u>, 19 (1975).
213. R. J. P. Corriu and J. J. E. Moreau, *Nouv. J. Chim.*, <u>1</u>,
     71 (1977).
214. R. J. P. Corriu and J. J. E. Moreau, *Bull. Soc. Chim.
     Fr.*, <u>1975</u>, 901.
215. P. Pino, G. Consiglio, G. Botteghi, and C. Salomon, in
     Adv. Chem. Ser., <u>132</u>, 295, *Homogeneous Catalysis* (1974).
216. C. Salomon, G. Consiglio, and P. Pino, *Angew. Chem.*, <u>85</u>,
     663 (1973); *Angew. Chem., Int. Ed. Engl.*, <u>12</u>, 669 (1973).
217. I. Ogata and Y. Ikeda, *Chem. Lett.*, <u>1972</u>, 487.
218. M. Tanaka, Y. Ikeda, and I. Ogata, *Chem. Lett.*, <u>1975</u>,
     1115.
219. G. Consiglio and P. Pino, *Helv. Chim. Acta*, <u>59</u>, 642
     (1976), and references quoted.

220. G. Consiglio, *Helv. Chim. Acta,* 59, 124 (1976).
221. B. Bogdanovic, B. Henc, B. Meister, H. Pauling, and G. Wilke, *Angew. Chem.,* 84, 1070 (1972); *Angew. Chem., Int. Ed. Engl.,* 11, 1023 (1972).
222. M. Hidai, H. Ishiwatari, H. Yagi, E. Tanaka, K. Onozawa, and Y. Uchida, *J. Chem. Soc., Chem. Commun.,* 1975, 170.
223. B. M. Trost and P. E. Strege, *J. Am. Chem. Soc.,* 99, 1649 (1977).
224. H. Kagan, A. Hibon de Gournay, M. Larcheveque, and J. C. Fiaud, *J. Organomet. Chem.,* in press (1978).
225. German Patent, 2,240,257; *Chem. Abstr.,* 78, 159,048d (1973).
226. T. Aratani, Y. Yoneyoshi, and T. Nagase, *Tetrahedron Lett.,* 1975, 1707.
227. Y. Tatsuno, A. Konishi, A. Nakamura, and S. Otsuka, *J. Chem. Soc., Chem. Commun.,* 1974, 588.
228. S. Otsuka, personal communication.
229. T. Hayashi, M. Tajika, K. Tamao, and M. Kumada, *J. Am. Chem. Soc.,* 98, 3718 (1976).
230. K. Tamao, A. Minato, N. Miyake, T. Matsuda, Y. Kiso, and M. Kumada, *Chem. Lett.,* 1975, 133.
231. S. I. Yamada, T. Mashiko, and S. Terashima, *J. Am. Chem. Soc.,* 99, 1988 (1977).
232. R. C. Michaelson, R. E. Palermo, and K. B. Sharpless, *J. Am. Chem. Soc.,* 99, 1990 (1977).
233. F. Di Furia, G. Modena, and R. Curci, *Tetrahedron Lett.,* 50, 4637 (1976).
234. N. Spassky, A. Momtaz, and M. Reix, private communication, to be published.
235. A. Deffieux, M. Sepulchre, N. Spassky, and P. Sigwalt, *Makromol. Chem.,* 175, 339 (1974).
236. J. A. Le Bel, *Bull. Soc. Chim. Fr.,* 22, 337 (1874).
237. W. Kuhn and E. Knopf, *Naturwissenschaften,* 18, 183 (1930).
238. (a) H. B. Kagan, G. Balavoine, and A. Moradpour, *J. Mol. Evol.,* 4, 41 (1974).
     (b) G. Balavoine, A. Moradpour, and H. B. Kagan, *J. Am. Chem. Soc.,* 96, 5152 (1974).
239. P. Boldt, W. Thielecke, and H. Luthe, *Chem. Ber.,* 104, 353 (1971).
240. A. Moradpour, J. F. Nicoud, G. Balavoine, H. B. Kagan, and G. Tsoucaris, *J. Am. Chem. Soc.,* 93, 2353 (1971).
241. H. B. Kagan, A. Moradpour, J. F. Nicoud, G. Balavoine, R. H. Martin, and J. P. Cosyn, *Tetrahedron Lett.,* 1971, 2479.
242. W. J. Bernstein, M. Calvin, and O. Buchardt, *J. Am. Chem. Soc.,* 94, 494 (1972).
243. W. J. Bernstein, M. Calvin, and O. Buchardt, *Tetrahedron Lett.,* 1972, 2195.

244. O. Buchardt, *Angew. Chem.*, 86, 222 (1974); *Angew. Chem., Int. Ed. Engl.*, 13, 179 (1974).

245. A. Moradpour, H. B. Kagan, B. Maes, G. Morren, and R. H. Martin, *Tetrahedron*, 31, 2139 (1975).

246. J. F. Nicoud and H. B. Kagan, *Isr. J. Chem.*, 15, 78 (1976/1977).

247. O. L. Chapman, G. L. Eian, A. Bloom, and J. Clardy, *J. Am. Chem. Soc.*, 93, 2918 (1971).

248. P. Gerike, *Naturwissenschaften*, 62, 38 (1975).

249. P. G. De Gennes, *C. R. Acad. Sci., Ser. B*, 270, 891 (1970); C. A. Mead, A. Moscowitz, H. Wynberg, and F. Meuwese, *Tetrahedron Lett.*, 1977, 1603.

250. G. S. Hammond and R. S. Cole, *J. Am. Chem. Soc.*, 90, 2957 (1968).

251. G. Ouannes, R. Beugelmans, and G. Roussi, *J. Am. Chem. Soc.*, 95, 8472 (1973).

252. G. Balavoine, S. Jugé, and H. B. Kagan, *Tetrahedron Lett.*, 1973, 4159.

253. A. Faljoni, K. Zinner, and R. G. Weiss, *Tetrahedron Lett.*, 1974, 1127.

254. C. S. Drucker, V. G. Toscano, and R. G. Weiss, *J. Am. Chem. Soc.*, 95, 6482 (1973).

255. T. J. Wilkinson, P. S. Mariano, and G. I. Glover, *Proceedings of the Meeting on Photochemistry*, Nashville, June 1974.

256. J. C. Micheau, N. Paillous, and A. Lattes, *Tetrahedron*, 31, 441 (1975).

257. R. L. Wife, D. Prezant, and R. Breslow, *Tetrahedron Lett.*, 1976, 517.

258. D. Elad and J. Sperling, *J. Am. Chem. Soc.*, 93, 967 (1971).

259. Y. Cochez, J. Jespers, V. Libert, K. Mislow, and R. H. Martin, *Bull. Soc. Chim. Belg.*, 84, 1033 (1975).

260. Y. Cochez, R. H. Martin and J. Jespers, *Isr. J. Chem.*, 15, 29 (1976/1977).

261. D. Weinblum and H. E. Johns, *Biochim. Biophys. Acta*, 114, 450 (1966).

262. B. S. Green, Y. Rabinsohn, and M. Rejtö, *J. Chem. Soc., Chem. Commun.*, 1975, 313.

263. B. S. Green, A. T. Hagler, Y. Rabinsohn, and M. Rejtö, *Isr. J. Chem.*, 15, 124 (1976/1977).

264. A. Elgavi, B. S. Green, and G. M. J. Schmidt, *J. Am. Chem. Soc.*, 95, 2058 (1973).

265. B. S. Green, A. T. Hagler, Y. Rabinsohn and M. Rejtö, *Isr. J. Chem.*, 15, 124 (1976/1977).

266. L. Addadi, M. D. Cohen, and M. Lahav, *J. Chem. Soc., Chem. Commun.*, 1975, 471.

267. L. Addadi, M. Cohen, and M. Lahav, *Proceedings of the Nato Institute on Optically Active Polymers*, Forges les Eaux, France (1975).

268. G. Audisio and A. Salvani, *J. Chem. Soc., Chem. Commun.,* 1976, 481.
269. N. Friedman, M. Lahav, L. Leiseronitz, R. Popovitz-Biro, C. P. Tang, and Z. Zaretzkii, *J. Chem. Soc., Chem. Commun.,* 1975, 864.
270. L. Liebert, L. Strzelecki, and D. Vacogne, *Bull. Soc. Chim. Fr.,* 1975, 2073.
271. L. Liebert and L. Strzelecki, *Bull. Soc. Chim. Fr.,* 1973, 597, 603.
272. (a) D. Seebach and H. Daum, *J. Am. Chem. Soc.,* 93, 2795 (1971).
     (b) D. Seebach, personal communication.
273. D. R. Boyd and D. C. Neil, *J. Chem. Soc., Chem. Commun.,* 1977, 51.
274. L. Eberson and L. Horner, in *Organic Electrochemistry,* Baizer, Ed., Marcel-Dekker, New York, 1973, p. 869.
275. L. Horner, D. Degner, and D. Skaletz, *Tetrahedron Lett.,* 1968, 5889.
276. R. N. Gourley, J. Grimshaw, and P. G. Millar, *J. Chem. Soc., C,* 17, 2318 (1970).
277. E. Kariv, H. A. Terni, and E. Gileadi, *J. Electrochem. Soc.,* 120, 639 (1973).
278. J. Kopilov, S. Shatzmiller, and E. Kariv, *Electrochim. Acta,* 21, 535 (1976).
279. I. A. Titova, I. M. Levinson, V. G. Mairanovskii, and A. B. Ershler, *Elektrokhimiya,* 9, 424 (1973).
280. J. Kopilov, E. Kariv and L. L. Miller, *J. Am. Chem. Soc.,* 99, 3450 (1977).
281. M. Jubault, E. Raoult, and D. Peltier, *C. R. Acad. Sci., C,* 277, 583 (1973); *Electrochim. Acta,* 19, 865 (1974).
     (b) M. Jubault, E. Raoult, J. Armand, and L. Boulared, *J. Chem. Soc., Chem. Commun.,* 1977, 250.
282. B. F. Watkins, J. R. Behling, E. Kariv, and L. L. Miller, *J. Am. Chem. Soc.,* 97, 3549 (1975).
283. L. L. Miller, personal communication.
284. W. Becker, H. Freund, and E. Pfeil, *Angew. Chem.,* 77, 1139 (1965); *Angew. Chem., Int. Ed. Engl.,* 4, 1079 (1965).
285. W. Becker and E. Pfeil, *J. Am. Chem. Soc.,* 88, 4299 (1966).
286. R. Bentley, in *Molecular Asymmetry in Biology,* Vol. 1, Academic Press, New York, 1970, p. 41.
287. J. P. Guetté and N. Spassky, *Bull. Soc. Chim. Fr.,* 1972, 4217.
288. D. D. Ridley and M. Stralow, *J. Chem. Soc., Chem. Commun.,* 1975, 400.
289. S. Takano, K. Tanigawa, and K. Ogasawara, *J. Chem. Soc., Chem. Commun.,* 1976, 189.
290. Tanabe Seiyaku Co., Japan Patent, 74, 25, 189; *Chem. Abstr.,* 81, 8979h (1974).

291. H. Simon, B. Rambeck, H. Hashimoto, H. Günther, G.
     Nonhynek, and N. Neumann, *Angew. Chem.*, <u>86</u>, 675 (1974);
     Angew. Chem., *Int. Ed. Engl.*, <u>13</u>, 608 (1974).
292. R. Howe and R. H. Moore, *J. Med. Chem.*, <u>14</u>, 287 (1971).
293. S. W. May and R. D. Schwartz, *J. Am. Chem. Soc.*, <u>96</u>,
     4031 (1974).
294. C. A. Townsend, T. Scholl, and D. Arigoni, *J. Chem. Soc.*,
     Chem. Commun., <u>1975</u>, 921.
295. W. S. Johnson and G. E. Dubois, *J. Am. Chem. Soc.*, <u>98</u>,
     1039 (1976).
296. K. G. Paul, F. Johnson, and D. Favara, *J. Am. Chem. Soc.*,
     <u>98</u>, 1285 (1976).
297. G. Stork and S. Raucher, *J. Am. Chem. Soc.*, <u>98</u>, 1584
     (1976).
298. G. Stork and A. Takahashi, *J. Am. Chem. Soc.*, <u>99</u>, 1275
     (1977).
299. H. Ohrui and S. Emoto, *Tetrahedron Lett.*, <u>1975</u>, 3657.
300. H. Ohrui and S. Emoto, *Tetrahedron Lett.*, <u>1975</u>, 2765.
301. Y. Izumi, S. Tatsumi, and M. Imaida, *Bull. Chem. Soc.
     Jpn*, <u>42</u>, 2373 (1969).
302. Y. Orito, S. Niwa, and S. Imai, *Yuki Gosei Kagaku*, <u>34</u>
     (9), 672 (1976); *Chem. Abstr.*, <u>85</u>, 123,295n (1977).
303. I. Yasumori, Y. Inoue, and K. Okabe, in *Catalysis
     Heterogeneous and Homogeneous*, Elsevier, Amsterdam,
     1975, p. 41.
304. N. Takaishi, H. Imai, C. A. Bertelo, and J. K. Stille,
     J. Am. Chem. Soc., <u>98</u>, 5401 (1976).
305. R. V. Stevens and F. C. A. Gaeta, *J. Am. Chem. Soc.*, <u>99</u>,
     6105 (1977).
306. N. S. Zefirov, *Tetrahedron*, <u>33</u>, 3193 (1977).

Chiral Lanthanide Shift Reagents

Glenn R. Sullivan*

Department of Chemistry,
California Institute of Technology
Pasadena, California

I.	Background	288
II.	Theory	288
III.	Introduction and Uses of Chiral LSRs	289
IV.	Choice of Chiral LSR	305
	A. β-Diketonate Ligand	305
	B. Lanthanide Ion	306
V.	Experimental Techniques	307
	A. The Typical LSR Experiment	307
	B. Choice of Solvent	308
	C. Substrate Concentration	309
	D. Magnetic "Fish"	309
	E. Sample Filtration or Centrifugation	309
	F. Tuning the Homogeneity of the Magnetic Field	309
VI.	Special Cases and Techniques	310
	A. Diastereotopic Groups	310
	B. Simplification by Spin-Spin Decoupling	312
	C. Effect of LSR-Substrate Ratio	314
	D. Effect of Lowering the Sample Temperature	315
	E. Nuclei Other Than ^{1}H	317
	F. Alteration of Functionality	319
	G. Requirement for Both Enantiomers	320
	H. Polyfunctional Molecules	321
	I. Recovery of Substrate	323
VII.	Correlation of Absolute Configuration	324
	Acknowledgments	325
	References	325

*Present address: Nicolet Technology Corp., 145 E. Dana St., Mountain View, California.

I. BACKGROUND

The use of chiral lanthanide shift reagents (chiral LSRs)
for the determination of enantiomeric purities by NMR spectroscopy has several advantages over other procedures commonly
employed for such determinations. Classical methods are normally
considered experimentally cumbersome (1). Pirkle's method (2) of
using a chiral solvent in NMR is simple to apply; however, the
amount of shift difference between enantiomers is generally
small, not infrequently too small to be usable. Another approach,
introduced by Mislow and Raban (3) and studied extensively by
Mosher (4), the conversion of chiral alcohols and amines to
diastereomeric esters and amides with an enantiomerically pure
chiral acid chloride, followed by NMR analysis, is limited to
these two functionalities. A further drawback is the fact that
derivatives of the compound must be isolated without altering
the original ratio of enantiomers. Gerlach (5) extended Mosher's
method by adding an achiral LSR to increase the shift difference
between the diastereomeric esters. Still other approaches have
been the use of gas chromatography, liquid chromatography, and
high-pressure liquid chromatography (6,7), usually after chemical conversion of enantiomers to diastereomers.*

Determination of enantiomeric purity by chiral LSRs is
applicable to a much broader range of compounds than some of
these other techniques and is normally simpler. Furthermore the
magnitudes of the shift differences observed between enantiomers
in the presence of chiral LSRs ($\Delta\Delta\delta$) are often much larger than
for the other NMR methods.

II. THEORY

The shifts induced by paramagnetic ions can arise from
both contact and dipolar interactions. The contact shift arises
from the Fermi-contact interaction, a "through bond" interaction that occurs only if there is a finite probability of
finding an unpaired electronic spin on the atomic s orbital
of the nucleus being observed. The dipolar shift arises from a
"through space" dipole-dipole interaction between the unpaired
electron and the nucleus being observed. This dipolar interaction was first described in detail by McConnell and Robertson
(8). A more complete discussion of contact and dipolar interactions can be found in several reviews of achiral LSRs (9-13).

Under normal conditions the equilibrium between the substrate and the LSR is rapid on the NMR time scale:

*The chiral auxiliary reagent used for this conversion
must itself be enantiomerically pure, which is sometimes a
drawback in these methods.

$$(S)\text{-substrate} \cdot (R)\text{-LSR} \xrightarrow{k_S} \left\{ \begin{matrix} (R)\text{-substrate} \\ (S)\text{-substrate} \end{matrix} \right\} \;+\; 2(R)\text{-LSR}$$

(A)

$$\xrightarrow{k_R} (R)\text{-substrate} \cdot (R)\text{-LSR} \qquad\qquad [1]$$

(B)

Therefore only a single time-averaged spectrum results from the average of complexed and uncomplexed substrate molecules. Rapidly equilibrating complexes are formed by an enantiomerical-ly pure chiral LSR* binding to each of two enantiomers. These complexes are diastereomeric and can have different averaged chemical shifts. This difference in shifts may have at least two causes. First, the equilibrium constants (k_R, k_S) may be different for diastereomeric complexes, thereby causing larger shifts for the complex having the larger binding constant (14). Second, the two diastereomeric complexes (A, B) formed may differ in their geometry, thus causing a difference in the induced shift for corresponding signals in the two complexes.

III. INTRODUCTION TO AND USES OF CHIRAL LSRs

Whitesides and Lewis (15) reported the first chiral LSR, tris(3-*t*-butylhydroxymethylene-*d*-camphorato)europium(III) (*1*).

1. R = *t*-Butyl
2. R = $CF_2CF_2CF_3$
3. R = CF_3

They showed that *1* was quite effective in separating the signals of the enantiomers of α-phenylethylamine (Fig. 1) and of several other amines. This LSR was not found useful for functionalities other than amines, however.

It was only a short time before Fraser (16) and Goering (17) independently introduced the chiral LSR tris(3-hepta-fluorobutyryl-*d*-camphorato)europium(III), Eu(hfbc)₃ (2), and tris(3-trifluoroacetyl-*d*-camphorato)europium(III), Eu(facam)₃ (*3*). Both were shown to be effective for many different

*The chiral LSR itself (arbitrarily assumed in eq. [1] to have *R* configuration) is actually a mixture of four diastereomers in rapid equilibrium by virtue of the chirality of the octa-hedral europium.

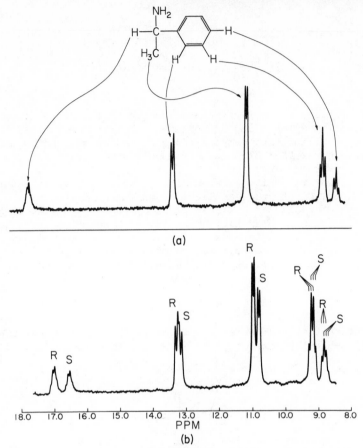

Fig. 1.   Spectra of solutions prepared from (a) (S)-α-
phenylethylamine (10 µl) and (b) a mixture of (R)- and (S)-α-
phenylethylamine (7 and 5 µl, respectively), in 0.3 ml of a
carbon tetrachloride solution of 1 (∿0.15M). The chemical shift
scale applies only to the spectrum of the mixture; that of the
pure (S) isomer was displaced slightly to lower field due to
differences in concentrations of the samples. Reprinted with
permission from J. Am. Chem. Soc., 92, 6979 (1970) (15). Copyright by the American Chemical Society.

functional groups. Though 2 and 3 have been by far the most
widely used, other chiral LSRs have been reported (14,16,18,19).
Of these, tris(d,d-dicampholylmethanato)europium(III), Eu(dcm)₃
(4), introduced by Whitesides (14) though not widely used,*

*The primary reasons for lack of widespread use of 4 are
probably the facts that it is not yet commercially available and
that it is slightly more difficult to prepare than 2 and 3.

bears particular mention. These chiral LSRs are discussed more fully in Sect. IV-A.

4

Through 1976 the enantiomeric purities of more than 170 different compounds have been determined by use of chiral LSRs (6,13-87). Some of these compounds and their pertinent shift data are collected in Table 1 to provide an overview of the types of structures for which chiral LSRs have been success-fully utilized. It is apparent that virtually any chiral mole-cule containing a functional group capable of binding to an achiral LSR is a possible candidate for enantiomeric purity determination with a chiral LSR. Although most of the compounds in Table 1 are monofunctional, a number of fairly complex species are also included. Generally the closer the function-ality to the source of chirality, the better are the chances of success.

The data in Table 1 may be illustrated by two simple examples. Figure 1a shows the spectrum of (S)-α-phenylethyl-amine in the presence of 1. This spectrum appears much as it would in the presence of an achiral LSR. That is, the individual resonances are shifted dramatically downfield and are well separated, allowing easy assignment. The spectrum of a mixture of (R)- and (S)-α-phenylethylamine in the presence of 1 is shown in Figure 1b. The extent of the shifts induced for the different types of protons is similar to that of the S isomer alone. However, each type of proton now displays two signals, one for each isomer. For the CHNH$_2$ and the CH$_3$ signals, the separation is sufficient to allow direct determination of the enantiomeric purity. The $\Delta\Delta\delta$ values listed in Table 1 for these protons are the shift differences between the signals of enantiotopic groups (13). The effective separation of the signals can be increased significantly by selectively spin-spin decoupling from neighboring protons (see Sect. VI-B).

In practice one does not normally achieve separation of all the signals of a pair of enantiomers. The experiment may be considered successful, however, as soon as analytically ade-quate separation of any one signal has been achieved.

The first spectrum in Figure 2 shows the signals of dl-2-phenyl-2-butanol in the presence of the achiral LSR tris(2,2,6,6-

TABLE 1

Separation of the ^{1}H NMR Signals of Enantiomers in the Presence of Chiral LSRs

Compound	Observed atom	Solvent	LSR[a]	Molar ratio[b]	$\Delta\Delta\delta$[c] (ppm)	Reference
Alcohols						
1-Phenyl-2,2-dimethyl-1-propanol	CH C(CH3)3 ortho protons	CCl4	3	0.5	0.17 0.05 0.07	20
1-Phenyl-2,2-dimethyl-1-propanol	CH C(CH3)3 ortho protons	CDCl3	2	0.52	0.09 0.14 0.22	21
2-Butanol	CH CHCH3 CH2CH3	CDCl3	2	0.52	0.00 0.06 0.00	21
1,1,1-Trifluoro-2-propanol	CH CH3	CDCl3	2	0.58	0.26 0.06	21
1-Phenyl-1-butanol	CH CH2CH3 ortho protons	CDCl3	2	0.57	0.44 0.03 0.00	21
1-Phenyl-1-propanol	CH3 CH	CCl4	3	0.5	0.02 0.29	20
1-Phenylethanol	CH3 CH ortho protons	CCl4	2	0.4	0.05 0.07 0.02	16
CH3CH2CH(CH3)CHaHbOH (2-Methyl-1-butanol)	Ha Hb	CCl4	2	0.4	0.05 0.11	16

Compound	Proton	Solvent	n			Ref.
1-(ortho-Methylphenyl)ethanol	CHOH	CCl₄	3	0.18	0.08	22
2,2-Dimethyl-1-propanol-1-d	CHDOH	CDCl₃	2	0.5	0.11	23
Benzyl-α-d alcohol	CHDOH	CDCl₃	2	0.5	0.15	23
CH₃(SnMe₃)CHCH₂CH₂OH	Me₃	CS₂	3	0.71	0.08	24
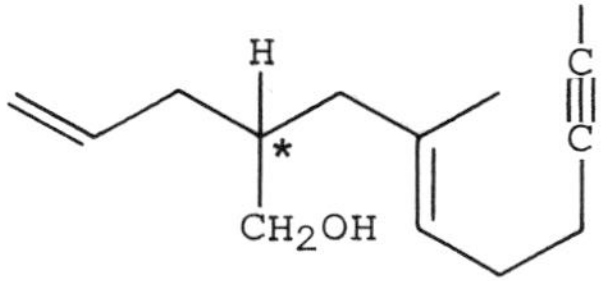 (structure below)	CH₂OH	CDCl₃	2	≈1.5	0.13ᵈ	25
1,3-Di-t-butylpropargyl alcohol	CHOH / 1-t-butyl	CCl₄	4	≈1.0	2.50 / 2.43	14
1,3-Di-t-butylpropargyl alcohol	CHOH / 1-t-butyl	CCl₄	3	≈1.0	0.59 / 0.09	14
Cyclohexylmethylcarbinol	CHCH₃	CCl₄	4	≈1.0	1.22	14
Cyclohexylmethylcarbinol	CHCH₃	CCl₄	3	≈1.0	0.07	14
2-Methyl-endo-norbornanol	2-CH₃	CCl₄	3	0.5–1.0	0.20	26
dl-Menthol	4-CH₃	CCl₄	3	0.5–1.0	0.07	26
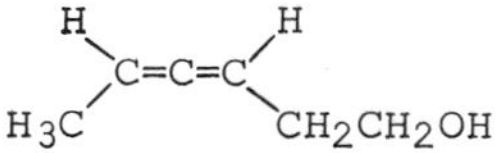 (structure below)	CH₂OH	CDCl₃	2	1.1	0.30, 0.32	27
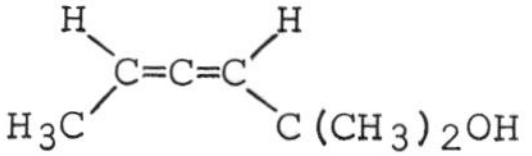 (structure below)	(CH₃)₂OH	CDCl₃	2	1.2	0.08, 0.14	27

Amines

Compound	Proton	Solvent	n			Ref.
1-Phenylethylamine	CHCH₃ / CHCH₃ / ortho protons	CDCl₃	4	≈1.0	0.66 / 4.42 / 0.15	14
1-Phenylethylamine	CHCH₃ / CHCH₃	CCl₄	2	0.5–1.0	0.17 / 0.22	26

Structures:

$$-C\equiv C-$$

with

$$\underset{\underset{\displaystyle CH_2OH}{|}}{\overset{\displaystyle H}{\underset{\displaystyle *}{C}}}$$

$$\underset{H_3C}{}\overset{H}{C}=C=C\overset{H}{\underset{\displaystyle}{}}CH_2CH_2OH$$

$$\underset{H_3C}{}\overset{H}{C}=C=C\overset{H}{\underset{\displaystyle}{}}C(CH_3)_2OH$$

293

TABLE 1
Continued

Compound	Observed atom	Solvent	LSR[a]	Molar ratio[b]	ΔΔδ[c] (ppm)	Reference
1-Phenylethylamine	CHCH3	CCl4	3	1.0-2.0	0.18	26
	CHCH3				0.92	
1-Phenylethylamine	CHCH3	e	1	0.5	0.55	15
N-Methyl-1-phenylethylamine	CHCH3	CDCl3	4	≈1.0	0.21	14
	N-CH3				1.46	
	ortho protons				1.45	
N,N-Dimethyl-1-phenylethylamine	CHCH3	CDCl3	4	≈1.0	0.26	14
	N-(CH3)2				0.85	
C6H5CH2CH(NH2)CH3 (Amphetamine)	CHCH3	CDCl3	3	0.2	0.1	28
	CHCH3	e	1	e	0.7	14
Esters						
2-Butyl acetate	COCH3	CCl4	4	≈1.0	0.29	14
	α-CH3				0.35	
	CH				0.37	
2-Butyl acetate	α-CH3	CCl4	3	0.5-1.0	0.23	26
Ethyl 2-methylbutanoate	2-CH3	CCl4	4	≈1.0	0.21	14
Methyl 2-methyl-2-phenylbutanoate	OCH3	CCl4	2	1.0-2.0	0.22	26
	CH3				0.13	
	CH2CH3				0.20	
1-Acetoxy-1-phenyl-2-butene	CCH3	CCl4	3	0.5-1.0	0.18	26
	COCH3				0.10	
3-Acetoxy-1-octyne	e	e	3	e	e	29

Compound	Group	Solvent				Ref
Methyl 2-phenylpropionate	CHCH₃	CS₂	4	1.2	0.015	6
	CCH₃	d	3	e	e	30
Ketones						
Camphor	CH₃	CCl₄	4	≈1.0	0.12	14
	CMeₐMeb				0.75	
	CMeₐMeb				0.14	
Camphor	CH₃	CCl₄	3	≈1.0	0.03	14
	CMeₐMeb				0.12	
	CMeₐMeb				0.00	
3-Methyl-3-phenyl-2-pentanone	COCH₃	CCl₄	2	1.0–2.0	0.13	26
	CCH₃				0.10	
	CH₂CH₃				0.00	
3-Methyl-2-pentanone	COCH₃	CCl₄	4	≈1.0	0.12	14
	CHCH₃				0.27	
	CH₂CH₃				0.00	
Ethers and Epoxides						
cis-1-Methoxy-2-methylcyclohexane	OCH₃	CCl₄	4	≈1.0	0.98	14
2-Methoxy-2-phenylbutane	OCH₃	CCl₄	2	1.0–2.0	0.02	26
2-Butyl methyl ether	OCH₃	CS₂	4	≈1.0	0.21	14
	CHCH₃				0.12	
cis-β-Methylstyrene oxide	β-CH₃	CCl₄	3	0.6	0.27	17
Styrene oxide	α-H	CCl₄	2	0.4	0.31	16
	trans-β-H				0.75	
	cis-β-H				0.15	
Other Functionalities						
sec-Butylformamide	CHCH₃	CCl₄	4	≈1.0	0.30	14
	CH₂CH₃				0.22	

295

TABLE 1
Continued

Compound	Observed atom	Solvent	LSR[a]	Molar ratio[b]	ΔΔδ[c] (ppm)	Reference
sec-Butylformamide	CHC$\underline{H_3}$ CH$_2$C$\underline{H_3}$	CCl$_4$	3	≈1.0	0.10 0.05	14
Benzyl-α-*d* azide	α-H	CDCl$_3$	2	1.0	0.08	31
2-Nitrobutane	CHC$\underline{H_3}$ CH$_2$C$\underline{H_3}$	CS$_2$	4	0.5	0.15[f] 0.00[f]	14
2-Cyanobutane	CHC$\underline{H_3}$ CH$_2$C$\underline{H_3}$	CS$_2$	4	1.5	0.05[g] 0.29[g]	14
2-Phenylpropionaldehyde	C$\underline{H}$O C$\underline{H}$CH$_3$	CCl$_4$	2	0.4	0.03 0.01	16
[structure: 2-(1-methylsulfonylethyl)-4-nitrotoluene] CHC$\underline{H_3}$ / CH-SO$_2$CH$_3$ / CH$_3$ / NO$_2$	CHC$\underline{H_3}$ C$\underline{H}$CH$_3$ SO$_2$C$\underline{H_3}$ aryl-C$\underline{H_3}$	CDCl$_3$	2	0.14	0.12 0.13 0.12 0.12	32
2-Butanethiol	CHC$\underline{H_3}$ CH$_2$C$\underline{H_3}$	CS$_2$	4	1.5	0.14[f] 0.00[f]	14
[structure: t-Butyl, H, C=C=C, PO(C$_6$H$_5$)$_2$, C$_6$H$_5$] allenic *H* t-butyl *H*	allenic *H* t-butyl *H*	CCl$_4$	1	0.36	0.07 0.02	33
Multifunctional 2-Methoxypropionic acid	OC$\underline{H_3}$	CDCl$_3$	2	e	e	34

Compound	Proton	Solvent				Ref.
Methyl 3-phenyl-3-hydroxy-pentanoate	ortho protons $CH_2CO_2CH_3$	CCl_4	3	0.8	1.0 0.4	35
(structure)	C-2 proton	$CDCl_2$	2	0.13	≈0.03	36
(structure) OCH_3	OCH_3 CO_2CH_3	e	2	e	e e	37
$CH_3OCH_2CH_2CH_2CH(CH_2CH_3)CH_2CO_2CH_3$	OCH_3 CO_2CH_3 OCH_3 CH CH_3	e	3	e	e e	38
Alanine methyl ester		CCl_4	3	0.5	0.05 0.26 0.06	38
(structure) OCH_3 C_4H_9 OH	e	e	3	e	e	39

297

TABLE 1
Continued

Compound	Observed atom	Solvent	LSR[a]	Molar ratio[b]	$\Delta\Delta\delta^{c}$ (ppm)	Reference
	1,1'-N-methyl 3,3'-N-methyl	CDCl$_3$	3	e	0.04 0.07	40
	24-OCOC$\underline{H}_3$	CCl$_4$	2	e	e	40
	one of the OC$\underline{H}_3$	CCl$_4$	3	1.07	0.32	18

one of the OCH_3 CCl$_4$ 3 1.07 0.26 18

C_{11}-aromatic proton CDCl$_3$ 3 0.07 0.05^d 41
C_1–OCH_3 0.05

e CDCl$_3$ 3 0.07 e 41

299

TABLE 1
Continued

Compound	Observed atom	Solvent	LSR[a]	Molar ratio[b]	$\Delta\Delta\delta c$ (ppm)	Reference
	e	CDCl₃	3	0.07	e	41
	CH₃ PH	Toluene d_8	3	e	e e	42

ortho protons	CDCl₃	1	0.6	d,e	14
meta protons				d,e	
all five OC$\underline{H}_3$	CDCl₃	4	0.8	d,e	14
C$\underline{H}_3^a$	CDCl₃	2	0.62	0.65	43
C$\underline{H}_3^b$				1.82	
C$\underline{H}_3$				1.84	
C$\underline{H}$					

TABLE 1
Continued

Compound	Observed atom	Solvent	LSR[a]	Molar ratio[b]	$\Delta\Delta\delta$[c] (ppm)	Reference
$CH_3O_2C(CH_2)_5\overset{*}{C}H(C_6H_5)CO_2CH_3$	CHCO_2C$\underline{H_3}$	CCl_4	2	0.90	0.12	44
Methyl 3-hydroxyoctadecanoate	CO_2C$\underline{H_3}$	$CDCl_3$	3	0.25	0.06	45
	C$\underline{H}$OH				0.34	
	3,3'-C$\underline{H_3}$	$CDCl_3$	3	e	0.02	46
	6,6'-C$\underline{HH}$				0.05	
	C$\underline{H_3}$	e	3	e	e	47
	C$\underline{H_3}$				e	
	$\underline{H}$	CCl_4	2	0.53	0.15	48

CH₃ structure with O epoxide, C₆H₅, and CO₂C₂H₅, H

CH_3 ⸻ $CO_2C_2H_5$
C_6H_5 $\underline{H}$
O

| 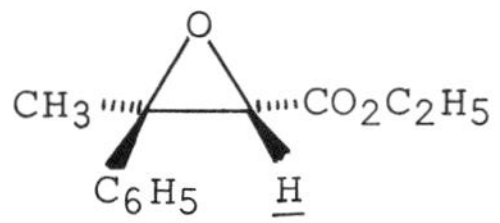 $\underline{H}$ | CCl_4 | 3 | e | 0.10 | 49 |
| $\underline{CH_3}$ | | | | 0.21 | |

---

[a] As defined in the text, 2 = Eu(hfbc)₃, 3 = Eu(facam)₃, and 4 = Eu(dcm)₃.

[b] Molar ratio of chiral LSR/substrate.

[c] $\Delta\Delta\delta$ represents the difference in shift induced between signals of two enantiotopic groups.

[d] This compound is discussed more fully in the text.

[e] Details were not given, but it was reported that the $\Delta\Delta\delta$s obtained were large enough to make an enantiomeric purity determination on the sample.

[f] The spectrum was obtained at −75°C.

[g] The spectrum was obtained at −50°C.

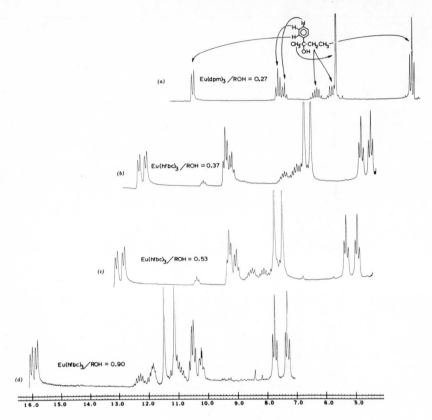

Fig. 2.   Spectra of *dl*-2-phenyl-2-butanol in CCl₄ in the
presence of 0.1*M* Eu(dpm)₃ (*a*) and in the presence of 0.2*M*
Eu(hfbc)₃ at three different Eu(hfbc)₃-substrate molar ratios
(*b–d*). Reprinted with permission from *J. Am. Chem. Soc.*, <u>96</u>,
1495 (1974) (26). Copyright by the American Chemical Society.

tetramethyl-3,5-heptanedionate)europium(III), Eu(dpm)₃. The
signals are shifted by varying amounts and can be readily
assigned to their respective protons. The remaining curves in
Figure 2 show the effect of three different amounts of *2* on the
NMR spectrum of *dl*-2-phenyl-2-butanol. The signals of the aromatic ortho protons and the CH₂CH₃ of the two enantiomers are
well separated, and the enantiomeric purity of a nonracemic
mixture can easily be determined from either pair of signals.
The CCH₃ signals are also well separated, but they overlap
other signals, preventing their use for enantiomeric purity
determination in these spectra. In such situations of overlap
use of a different lanthanide ion can sometimes be helpful

(see Sect. IV-B). Other signals in the spectra (Fig. 2) may have been separated, but their complexity and the overlapping of the patterns render them useless unless additional special techniques (see Sect. VI) are employed.

IV. CHOICE OF CHIRAL LSR

A. β-Diketonate Ligand

As yet there is no way to predict which β-diketonate ligand will give the best results for a particular compound. The three previously mentioned chiral LSRs [2, Eu(hfbc)$_3$; 3, Eu(facam)$_3$; 4, Eu(dcm)$_3$] frequently give good results. Compounds 2 and 3 have been shown to be useful for many substrates; however, 4 seems to induce larger ΔΔδs than 2 or 3 for the limited number of compounds for which it has been employed. Table 2 gives data comparing 4 to 2 and 3 for some of the available cases. Although these data were not obtained using the same conditions and concentrations, the differences observed for the ΔΔδ values are so large that they appear to arise from more than just different experimental conditions. Therefore, 4 seems to be the best first choice, followed by 2 and 3 in close order.

TABLE 2

Comparison of the Separating Ability of Chiral LSR[a]

Compound	Observed atom	ΔΔδ (ppm)		
		4	3	2
1,3-Di-t-butylpropargyl alcohols	CHOH	2.50	0.59	
	1-t-butyl	2.43	0.09	
Cyclohexylmethylcarbinol	CHCH$_3$	1.22	0.07	
Camphor	CH$_3$	0.12	0.03	
	CMe$_a$Me$_b$	0.75	0.12	
	CMe$_a$Me$_b$	0.14	0.00	
sec-Butylformamide	CHCH$_3$	0.30	0.10	
	CH$_2$CH$_3$	0.22	0.05	
1-Phenylethylamine	CHCH$_3$	0.66	0.18	0.17
	CHCH$_3$	4.42	0.92	0.22

[a]These data are all taken from Table 1. Because the shifts were not all recorded using the same conditions and concentrations, they can be used only for a rough comparison of these reagents.

To obtain the desired result the LSR should be dry. The chiral LSRs made from camphor derivatives are normally not easily sublimed* (unlike the common achiral LSRs) and are usually dried under vacuum over phosphorus pentoxide for one or two days. Since these chiral LSRs are hydroscopic, their exposure to air should be minimized (14).

B. Lanthanide Ion

Since achiral LSRs were introduced, every lanthanide(III) ion has been tested for applicability. Although many of these have been used for specific purposes, Eu(III), Pr(III), and Yb(III) appear to be the most generally useful. Since the primary purpose of using a chiral LSR is to separate the signals of a pair of enantiomers and to observe their relative intensities, it is important to choose a lanthanide that induces only a small amount of line broadening. For this reason Yb(III)

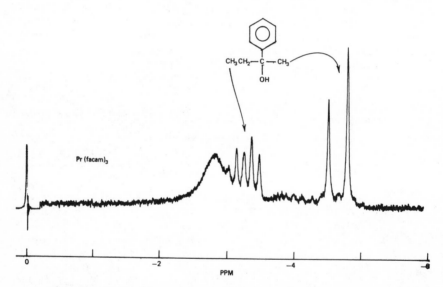

Fig. 3. Upfield portion of the spectrum of partly resolved (excess (R) isomer) 2-phenyl-2-butanol in CCl₄ in the presence of 0.3M Pr(facam)₃. Reprinted with permission from *J. Am. Chem. Soc.*, 96, 1495 (1974) (26). Copyright by the American Chemical Society.

*These derivatives are readily sublimed when prepared from a fresh NaOH solution, to form the complex β-diketone and the Eu(III). However, in this author's experience commercially obtained chiral LSRs are not easily sublimed.

has not proven suitable in chiral LSRs for proton NMR. Since
Eu(III) causes the least line broadening, it has been by far
the most commonly used lanthanide ion.

Pr(III), though more prone to causing line broadening, can
sometimes be a useful alternative to Eu(III), since it has the
redeeming feature of shifting most signals to higher fields
[Eu(III) generally shifts signals to lower fields]. This can
be helpful in special cases where Eu(III) causes some signals
to overlap. An example of the use of Pr(facam)$_3$ (the praseodymium analogue of 3) is shown in Figure 3. The α- and β-methyl
signals of 2-phenyl-2-butanol in the presence of Pr(facam)$_3$
(26) have line widths comparable to those given in the presence
of Eu(hfbc)$_3$ (2) (Fig. 2). Therefore Pr(facam)$_3$ is about as
useful as Eu(hfbc)$_3$ in separating these signals. Unfortunately
the aromatic proton absorptions downfield of TMS (not shown
in Fig. 3) are reported to be broad and lacking in fine
structure, which is often the case in the presence of Pr(III).
When the ^{13}C spectrum is observed rather than the proton
spectrum, Pr(III) and Yb(III) are sometimes more effective for
determining enantiomeric purities than Eu(III) (see Sect. VI-E).

V. EXPERIMENTAL TECHNIQUES

Although one cannot choose the most effective chiral LSR
in advance, an experiment can be planned and executed in such
a manner as to afford the highest probability of success. In
the following sections several of the details important to a
successful enantiomeric purity determination are discussed.

A. The Typical LSR Experiment

Although there are many ways of carrying out a chiral LSR
experiment, all involve obtaining a series of spectra in which
the molar ratio of chiral LSR to substrate is varied. From the
series of spectra obtained, the individual signals can be
traced from their original positions and usually their assignments can be made. The requisite change in molar ratios (LSR
to substrate) can be made in a number of ways, two of which
are described here.

1. The easiest and most common technique is to prepare a
solution of the substrate in the desired solvent, to which
solid portions of chiral LSR are then added incrementally. A
series of spectra are thus obtained. These incremental additions
can be made simply by using a small spatula to deliver a few
milligrams of LSR from a bottle to the sample tube.

2. Another method is to start with a substrate sample in
solution as in the preceding technique, but to add the LSR
incrementally by syringe as a concentrated solution. This

method does cause some dilution of the substrate, but offers
the advantage of allowing the concentrated chiral LSR solution
to be filtered to remove any precipitate that may be present
before it is introduced into the substrate solution (see Sect.
V-E). It has also been found useful for handling the submilli-
gram quantities of LSR often necessary for dilute samples when
a time-averaging method is employed.

B. Choice of Solvent

Though most determinations of enantiomeric purity reported
in the literature have been conducted in CCl_4 and $CDCl_3$, several
other solvents are suitable for use with chiral LSRs (14,9,88,
89). Most of the common NMR solvents are shown in Table 3 along
with their effects on the $\Delta\Delta\delta$ and chemical shifts (δ) for
2-butanol and 3-methyl-2-pentanone in the presence of 4. Although
the largest $\Delta\Delta\delta$s are obtained in pentane, that solvent's
practical use is severely limited by its interfering resonance
as well as by the insolubilities of many substrates in it.
Generally CCl_4 and $CDCl_3$ are the most convenient solvents, even
though others may be useful in some cases. It should be emphasized that, for best results, whatever solvent is chosen
should be dry.

TABLE 3

Solvent Effects on $\Delta\Delta\delta$ for 2-Butanol and
3-Methyl-2-pentanone in the Presence of Eu(dcm)$_3$ (4) (14)

Solvent[a]	$CH_3CH\underline{O}HCH_2CH_3$		$CH_3COCH(C\underline{H}_3)CH_2CH_3$	
	$\Delta\Delta\delta$ (ppm)	δ (ppm)[b]	$\Delta\Delta\delta$ (ppm)	δ (ppm)[b]
n-C_5H_{12}	1.05	15.78	0.30	9.32
CCl_2FCF_2Cl	0.86	15.55	0.28	8.75
$CFCl_3$	0.80	15.17	0.23	8.06
CCl_4	0.76	15.00	0.23	7.80
CS_2	0.69	14.47	0.16	7.30
C_6D_6	0.67	13.57	0.25	7.05
$CDCl_3$	0.63	12.18	0.11	4.18
CD_2Cl_2	0.46	9.83	0.10	3.67

[a]Solutions were made by adding 100 mg of Eu(dcm)$_3$ (4) to
350 µl of solvent and 7.5 µl of 2-butanol or 10 µl of 3-methyl-
2-pentanone.

[b]Chemical shifts in parts per million downfield of TMS at
32°C.

C. Substrate Concentration

The concentration of substrate should be kept as low as is compatible with having adequate signal strength. Normally the more concentrated the sample, the broader will be the signals and therefore the poorer the results. It is not unusual to have to add chiral LSR to substrate in a 1:1 or even greater molar ratio. Therefore the amount of chiral LSR required can be quite large if the sample is too concentrated in substrate. Convenient substrate concentrations are 0.1 to 0.25M.

D. Magnetic "Fish"

After the addition of a large amount of chiral LSR, there sometimes appears a small amount of insoluble material that becomes magnetized in the strong spectrometer field. After the sample has been spun in the magnetic field for a few minutes, this material may aggregate, giving macroscopic particles. These magnetic "fish" can then be detected by moving a small bar magnet from side to side of the sample tube and watching the particles "swim" about. These fish can cause gross deterioration of the resolution of the spectrum as well as causing the lock signal to fluctuate as the particles drift around in the sample tube. These particles can be easily removed by dipping a small magnet (encased in glass) into the sample and pulling them out.

E. Sample Filtration or Centrifugation

Even after removing any magnetic fish present, samples often have a fine white precipitate suspended in the solution after large amounts of LSR are added. This precipitate may cause cloudiness and must be removed for optimum spectral resolution. This precipitate can be removed by filtration through a plug of cotton (14), filtration through a millipore filter incorporated in a hypodermic syringe (90), or by centrifugation to give a completely transparent solution.

F. Tuning the Homogeneity of the Magnetic Field

Even a completely clear solution of substrate plus chiral LSR may give a poor NMR spectrum if the spectrometer is not shimmed correctly for this particular sample (which is now highly concentrated in paramagnetic metal ions). The amount of retuning necessary can sometimes be reduced by allowing the sample to spin in the magnet for a few minutes before attempting to tune on it (91). It should be noted that it is preferable to tune the field on some internal signal such as TMS,

CHCl$_3$, or the deuterated solvent signal if appropriate, since tuning on an external capillary may lead to inadequate resolution. Tuning on the paramagnetic sample will often require changing the off-axis shim controls as well as the major-axis controls. Difficulties are sometimes reduced by tuning the instrument on samples initially containing only small amounts of chiral LSR and then working up to larger amounts.

VI. SPECIAL CASES AND TECHNIQUES

The information provided so far should enable the reader to conduct a normal chiral LSR experiment with satisfactory results. There are, however, instances in which an awareness of special cases or techniques may be helpful.

A. Diastereotopic Groups

Chiral LSR analyses of compounds containing diastereotopic groups* involve a special problem since these groups can give two distinct NMR signals even in the absence of LSRs that may be shifted differently by either chiral or achiral LSRs. Thus if there are diastereotopic groups on two enantiomers being shifted by a chiral LSR, four distinct signals (two for each isomer) may be seen. Though the presence of diastereotopic groups may make the analysis more difficult, it does not necessarily preclude it. Figure 4A shows the spectrum of the geminal methyl signals of a *dl*-mixture of *5* in the presence of *2* (27). Figure 4B shows the effect of *2* on a partially optically active sample

$$\begin{array}{c} H \\ | \\ \end{array} C=C=C \begin{array}{c} H \\ | \\ \end{array}$$
$$H_3C \qquad\qquad C(CH_3)_2OH$$

5

of *5*. In this spectrum the two signals from the diastereotopic geminal methyl groups of one of the enantiomers are much smaller than those of the other. To obtain the enantiomeric purity, it is necessary only to compare the integrals of one of the larger signals and one of the smaller signals.

*We need not be concerned with enantiotopic groups here, since it follows from symmetry considerations that a chiral molecule cannot contain enantiotopic ligands. It should be mentioned, however, that enantiotopic groups in achiral molecules, such as the benzyl protons in $C_6H_5CH_2OH$, display distinct signals in the presence of chiral shift reagents.

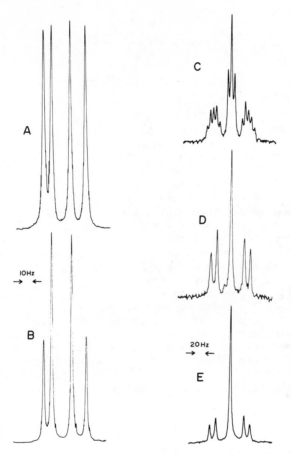

Fig. 4. (A-B) NMR spectra of the geminal methyls of 5 at a molar ratio of 1:2. (A) dl mixture; (B) 5 (33% enantiomeric excess). (C-E) NMR spectra of H_a, $H_{a'}$, H_b, and $H_{b'}$ of 7 at a molar ratio of Eu(hfbc)$_3$ to 7 of 1:1. (C) dl mixture with no decoupling; (D) 7a (16% enantiomeric excess) with C-2 protons decoupled; (E) 7b (13% enantiomeric excess) with C-2 protons decoupled. Reprinted with permission from J. Am. Chem. Soc., 97, 2919 (1975) (27). Copyright by the American Chemical Society.

When the diastereotopic groups are geminal protons, there is the further complication that the two protons will be coupled to each other. For example, the CH$_2$OH group of dl-6 (92) in the presence of a chiral LSR may give 16 lines arising from four different chemical shifts. That is, for one of the enantiomers the two diastereotopic geminal proton signals should

6

be separated and coupled to each other, giving an *AB* pattern
further split by coupling to the vicinal methine proton. Therefore each geminal proton should appear as the *A* part of an *ABX*
pattern, and thus as a doublet of doublets. And since each
geminal proton should give a four-line pattern, each of the
enantiomers should give an eight-line spectrum, and if the
signals of the two enantiomers are separated by the chiral
LSR, a 16-line spectrum can result.

The upper portion of Figure 5 shows the observed signals
of the CH₂OH of *dl-6* in the presence of *2* (25). This complex
pattern may well consist of 16 lines, with some being over-
lapped. This spectrum by itself is too complicated to be useful for analysis. However, the lower curves in Figure 5 are the
spectra of the same diastereotopic geminal protons for purified *d-6* and *l-6* recorded separately. In these spectra, each
individual geminal proton can indeed be observed as a doublet
of doublets, giving eight lines for each enantiomer. Therefore,
in spite of the complex appearance of the spectrum of the *dl*
mixture, the use of chiral LSRs can provide a good estimate of
the enantiomeric purity, once the signals are assigned by
examination of the spectra of the pure (or highly enriched)
enantiomers.

B. Simplification by Spin-Spin Decoupling

Many times complex multiplets may arise from spin-spin
coupling and the desired separation of signals of enantiotopic
groups can often be greatly enhanced by decoupling (22,27).
Because many of the signals will be separated only after the
addition of a chiral LSR, it is often easiest to carry out
decoupling at this stage. The following example shows how
decoupling can simplify a complex signal pattern (27).

Addition of a 1.1 molar equivalent of *2* to a racemic
mixture of *7* produces signals for H_a, H_a', H_b, and H_b', as
shown in Figure 4*C*. The pattern is complicated by the fact

7a   7b

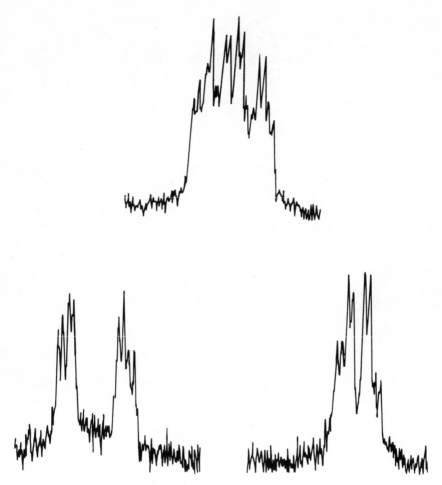

Fig. 5. Proton NMR spectra of CH_2OH signals of 6 in the presence of Eu(hfbc)$_3$ in CDCl$_3$. Approximate chemical shift is $\delta = 19$ ppm. The upper spectrum is a *dl* mixture of 6. The lower two spectra are of each of the two enantiomers separately (25).

that these two protons in each of the enantiomers are dia-stereotopic (Sect. VI-A). In the presence of *2* the signals for H$_b$ and H$_b$′ are shifted differently, yielding the *AB* part of an *ABX$_2$* pattern appearing on either side of the central triplet in Figure 4*C*. H$_a$ and H$_a$′ are each shifted by the same amount, yielding a simple triplet appearing in the center of Figure 4*C*. These patterns are simplified considerably by decoupling the adjacent methylene protons, which have all been shifted equally, giving the patterns shown in Figures 4*D* and 4*E*. From

these simplified spectra the enantiomeric purities can be
obtained with fair accuracy.

C. Effect of LSR-Substrate Ratio

As one adds more and more chiral LSRs, the spectral
resolution generally becomes progressively worse. This is not
a serious problem as long as the signals for the enantiomers
are still separated. However, at some stage the signals will
not separate any further and may actually move back together.
At this point further addition of chiral LSR is contraindi-
cated. Figure 6 (see also Fig. 2) shows a dramatic example of

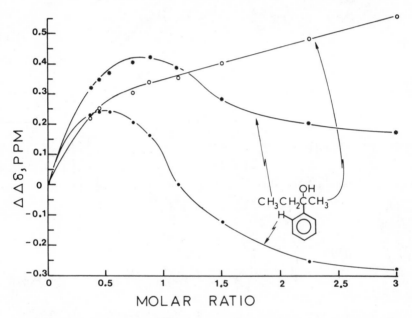

Fig. 6. Plots of ΔΔδ vs. molar ratio for the designated
protons of 2-phenyl-2-butanol in the presence of 0.3M Eu(hfbc)$_3$.
Reprinted with permission from *J. Am. Chem. Soc.*, 96, 1495
(1974) (26). Copyright by the American Chemical Society.

how ΔΔδ can change in sign upon the addition of increasing
amounts of chiral LSR. In the case of 2-phenyl-2-butanol the
ΔΔδ of the ortho proton signal reaches a positive maximum of
about 0.24 ppm, collapses back to 0.0, then reaches a negative
maximum of about -0.28 ppm.

Since ΔΔδ is an unpredictable function of the LSR-substrate
ratio, each individual compound may require a different amount
of chiral LSR. An example where only a very small amount of
chiral LSR is required is given in Sect. VI-H.

D. Effect of Lowering the Sample Temperature

Even though most of the studies involving chiral LSRs have been at room temperature, lower temperatures may offer important advantages (14). Changes in temperature often cause dramatic changes in the amount of shift induced in a spectrum. Figure 7

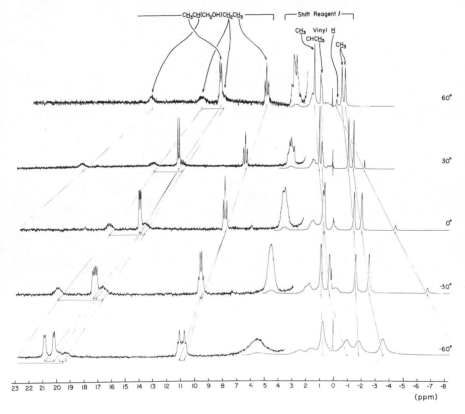

Fig. 7.   Increase in the magnitudes of $\Delta\Delta\delta$ for 2-methyl-1-butanol with decreasing temperature in the presence of Eu(dcm)$_3$. The concentrations of substrate and Eu(dcm)$_3$ in CS$_2$ in these spectra are 0.3 and 0.15M, respectively. Reprinted with permission from *J. Am. Chem. Soc.*, 96, 1038 (1974) (14). Copyright by the American Chemical Society.

shows the effect of temperature on 2-methyl-1-butanol in the presence of 4. Both the amount of shift and the $\Delta\Delta\delta$ increase substantially as the temperature is lowered. At temperatures above -30°C the enantiomeric purity could not be determined from this sample, but at -60°C it could be determined readily.

Table 4 shows the effect of temperature on $\Delta\Delta\delta$ induced by *4*. These data show that lowering the temperature may be very useful in increasing $\Delta\Delta\delta$ and can be used successfully for a wide variety of functional groups. In several of the cases given, enantiomeric purity can be measured only at temperatures lower than ambient.

TABLE 4

Temperature Dependence of $\Delta\Delta\delta$ Induced by Eu(dcm)$_3$ *(4)* (14)

Substrate[a]	Observed signal	$\Delta\Delta\delta$ (ppm) at 25°C	Maximum[b] useful $\Delta\Delta\delta$
2-Butyl acetate	COC$\underline{H}_3$	0.13	0.65 (-25°C)
	CHC$\underline{H}_3$	0.18	0.47 (-25°C)
Methyl 3,7-dimethyl-octanoate	CH$_2$C$\underline{H}_3$	0.03	0.13 (-25°C)
	CHC$\underline{H}_3$	0.10	0.60 (-50°C)
	OC$\underline{H}_3$	0.07	0.07 (25°C)
2-Butyl methyl ether	OC$\underline{H}_3$	0.21	0.21 (25°C)
	CHC$\underline{H}_3$	0.12	0.98 (-25°C)
	CH$_2$C$\underline{H}_3$.	c	0.45 (-50°C)
Camphor	C$\underline{H}_3$	0.00	0.14 (0°C)
	C(C$\underline{H}_3)_2$	0.39	1.06 (-25°C)
	C(C$\underline{H}_3)_2$	0.06	0.28 (-25°C)
2-Methyl-1-butanol	CHC$\underline{H}_3$	0.02	1.06 (-75°C)
	CHC$\underline{H}_3$	0.02	0.60 (-75°C)
2-Nitrobutane	CHC$\underline{H}_3$	0.00	0.15 (-75°C)
	CH$_2$C$\underline{H}_3$	0.00	0.00 (-50°C)
2-Cyanobutane	CHC$\underline{H}_3$	0.00	0.05 (-50°C)
	CH$_2$C$\underline{H}_3$	0.03	0.29 (-50°C)
2-Butanethiol	CHC$\underline{H}_3$	0.00	0.14 (-75°C)
	CH$_2$C$\underline{H}_3$	0.00	0.00 (-75°C)

[a]The concentration of substrate was ca. 0.3M in CS$_2$ (dried over 3 Å molecular sieves). The concentration of *4* was ca. 0.15M for 2-butyl acetate and 2-nitrobutane; ca. 0.3M for 2-butyl methyl ether, camphor, and 2-methyl-1-butanol; and ca. 0.45M for 2-cyanobutane, methyl 3,7-dimethyloctanoate, and 2-butanethiol.

[b]In some cases larger $\Delta\Delta\delta$s were observed at lower temperatures, but the signals became broadened, making the separation ineffective.

[c]The substrate resonance was not observed due to interfering resonances from *4*.

E. Nuclei Other Than [1]H

Although the large majority of the work with chiral LSRs has been carried out with ^{1}H NMR signals, other nuclei such as ^{19}F and ^{13}C may be used. Fraser (53) and Williamson (93) have studied the effect of chiral LSRs on the ^{13}C signals of a number of compounds shown in Tables 5 and 6. The results reveal two important points: first, ^{13}C NMR is quite useful

TABLE 5

Effect of Eu(hfbc)$_3$ and Pr(hfbc)$_3$ on the
[13]C Signals of Several Substrates[a] (53)

Substrate	Observed ^{13}C signal	Eu(hfbc)$_3$[b]		Pr(hfbc)$_3$[b]	
		$\Delta\Delta\delta$ (ppm)	$\Delta\delta$ (ppm)	$\Delta\Delta\delta$ (ppm)	$\Delta\delta$ (ppm)
1-Phenylethanol	$\underline{C}H_3$	<0.05	2.1	<0.05	-6.8
	$\underline{C}H$	0.11	6.7	0.28	-13.5
	aromatic C-1	<0.05	0.9	0.17	-6.2
1-Phenylethylamine	$\underline{C}H_3$	0.18	2.9	<0.08	-6.8
	$\underline{C}H$	0.34	9.2	0.20	-12.6
	aromatic C-1	<0.05	6.5	<0.05	-3.8
Styrene oxide	$\underline{C}H_2$	0.40	6.4	0.57	-10.9
	$\underline{C}H$	0.33	6.0	<0.08	-10.5
	aromatic C-1	<0.05	1.4	0.16	-4.9
trans-2-Methylcyclo-hexanol	$\underline{C}H_3$	<0.05	1.7	0.12	-7.5
	C-2	<0.05	1.4	<0.05	-15.6
	C-1	<0.05	6.9	0.38	-21.5
2-Methylpiperidine	$\underline{C}H_3$	0.18	2.1	0.56	-6.2
	C-2	0.34	4.4	0.49	-8.7
	C-6	0.13	5.0	0.48	-11.0
	C-3	<0.05	-0.3	0.16	-2.7
	C-5	<0.05	0.8	0.16	-3.2
	C-4	<0.05	0.9	<0.08	-3.7

[a]Each sample (0.25 mmol) was dissolved in 0.4 ml CCl$_4$-C$_6$D$_6$ (4:1 by volume) containing 3 or 4 drops TMS, then 0.05 mmol of Eu(hfbc)$_3$ or Pr(hfbc)$_3$ was added.

[b]For each chiral LSR $\Delta\delta$ represents the average shift (ppm) of the indicated nucleus in the two enantiomers. Positive values denote downfield shifts. The values of $\Delta\Delta\delta$ given as upper limits were estimated from the line widths compared with that of TMS.

TABLE 6
Effect of Yb(facam)$_3$ on the ^{13}C Signals of
1-Phenylethanol and 2-Phenyl-2-butanol[a] (93)

Substrate	Observed ^{13}C signal	$\Delta\Delta\delta$	$\Delta\delta$[b]
1-Phenylethanol[c]	C̲H$_3$	0.18	16.0
	C̲H	0.78	33.9
	aromatic C-1	0.44	15.0
	ortho carbons	0.36	9.5
	meta carbons	0.11	4.5
	para carbon	0.11	3.8
2-Phenyl-2-butanol[d]	C-1	1.01	13.9
	C-2	1.80	26.8
	C-3	0.74	12.9
	C-4	0.51	7.2
	aromatic C-1	0.93	11.9
	ortho carbons	0.43	6.4
	meta carbons	0.12	2.6
	para carbon	0.15	3.2

[a]Both substrates were ca. 1.0M in carbon disulfide containing 10% TMS.

[b]These values represent the average $\Delta\delta$ of the indicated nucleus in the two enantiomers.

[c]The values for this substrate are the ones that were actually observed at an LSR-substrate ratio of 0.24.

[d]The values for this substrate are the ones observed at an LSR-substrate ratio of 0.35.

in the determination of enantiomeric purity of a variety of compounds; second, using Pr(III) or Yb(III) as the lanthanide is sometimes more effective than using Eu(III).

The only difficulty with using ^{13}C for a determination of enantiomeric purity is in obtaining quantitative intensities from the ^{13}C spectrum. The difficulty arises from differences in relaxation times and nuclear Overhauser enhancements. This problem has been discussed elsewhere for general cases (94). However, since the two signals being quantitatively compared in enantiomeric purity determinations are for enantiotopic carbon atoms (diastereotopic in the chiral LSR-substrate complex), the difficulties from these two sources should be minimal.

F. Alteration of Functionality

Sometimes it is desirable to alter the functional groups
present in the compound under study. Thus a functional group
can be altered to become more capable of binding a chiral LSR
or masked in such a way as to be no longer effective in binding.
Examples of the first type of alteration include the
following:

Acid → Ester
Acid → Ketone
Acid → Alcohol
Ketone → Alcohol
Aldehyde → Alcohol
Ester → Alcohol
Alcohol → Ester
Alcohol → Ketone

The most useful of these transformations is the conversion of
acids, ketones, and esters to alcohols which interact strongly
with chiral LSRs thus giving larger induced shifts ($\Delta\delta$) than the
other oxygen functions. Another quite useful alteration is the
conversion of an alcohol to an acetate (26). This can be bene-
ficial when the parent alcohol fails to give a useful $\Delta\Delta\delta$. The
binding site of the acetate (C=O) is different from that of the
alcohol (OH), and is expected to produce different shifts upon
the addition of a chiral LSR. The acetate has the added advantage of adding a singlet methyl resonance to the spectrum,
which may be helpful in determining enantiomeric purity. Another
useful transformation is the conversion of an acid to its methyl
ester. This adds a singlet methyl resonance and provides an
ester binding site which is often superior to that of the free
acid.
The second type of alteration, which masks a particular
functional group so that it is no longer effective in binding
the chiral LSR (95), is useful when there is more than one
binding site, and when the site remote from the chiral center
monopolizes the binding. In this case the bound chiral LSR may
not be close enough to the chiral center to give a significant
diastereomeric interaction. Some examples of this type of
masking are:

Alcohol → Trifluoroacetyl ester
Ketone → Ethylene thioketal
Amine → Trifluoroacetamide

Though it may be advantageous in such cases to block the
functional group or groups more distant from the chiral center,
such blocking is not invariably required, as is shown in
Sect. VI-H.

G. Requirement for Both Enantiomers

If one has an enantiomerically pure sample, there is no
way to determine whether or not a chiral LSR experiment has
succeeded. Thus, if only one set of signals is observed in the
presence of a chiral LSR, either the sample is optically pure
or the signals of the enantiomers were not successfully
separated. For this reason it is necessary to have a sample
that is either racemic or only partially resolved to first
determine whether the chiral LSR does indeed separate the
signals of the enantiomers. Observing induced shifts of the
separate enantiomers in two different experiments, in contrast,
can be misleading, and should not be relied upon. Cases have
been encountered (25) in which some separately observed signals
in the (R) isomer appeared to have larger induced shifts ($\Delta\delta$)
than those in the (S) isomer (relative to signals that are
equally shifted in both isomers). However, upon mixing the
enantiomers in different proportions and adding the chiral LSR,
the (S) isomer had the larger induced shift, showing that the
two enantiomers must be in the same sample to display their
true relative positions.

One may wonder why separation cannot be tested by use of
a racemic mixture of the chiral LSR enantiomers, such as would
be easily available from d- and l-camphor. If the rapid exchange
equilibria taking place in the solution are examined in detail,
it is seen that by using a racemic chiral LSR, one obtains a
spectrum that cannot show separation of the substrate enantiomers. That is, since a substrate molecule will be bound to
many different chiral LSR molecules during the period of the
NMR observation, the observed signal of the substrate will
simply result from an average of binding with the two chiral
LSR enantiomers and thus appear as one signal.* Therefore to

*This may best be seen by considering the following
equilibria, both of which are rapid on the NMR time scale:

$$(R)\text{-substrate} \cdot (S)\text{-LSR} \; \underset{k_{RR}}{\overset{k_{RS}}{\rightleftharpoons}} \; (R)\text{-substrate} + \begin{Bmatrix} (R)\text{-LSR} \\ (S)\text{-LSR} \end{Bmatrix}$$

$$\rightleftharpoons \; (R)\text{-substrate} \cdot (R)\text{-LSR} \quad [a]$$

$$(S)\text{-substrate} \cdot (S)\text{-LSR} \; \underset{k_{SR}}{\overset{k_{SS}}{\rightleftharpoons}} \; (S)\text{-substrate} + \begin{Bmatrix} (R)\text{-LSR} \\ (S)\text{-LSR} \end{Bmatrix}$$

$$\rightleftharpoons \; (S)\text{-substrate} \cdot (R)\text{-LSR} \quad [b]$$

Now it follows from symmetry principles that $k_{RR} = k_{SS}$ and k_{RS}
$= k_{SR}$; in addition the shift of (R)-substrate·(R)-LSR is the
same as that of (S)-substrate·(S)-LSR, and that of (R)-sub-
strate·(S)-LSR is the same as that of (S)-substrate·(R)-LSR.
As a result the average shift corresponding to the (rapidly
established) equilibrium [a] is the same as that corresponding

test the chiral LSR method one must have a sample containing
both substrate enantiomers. This is a disadvantage relative to
the methods mentioned in Sect. I, involving the preparation of
diastereomeric esters and amides.

H. Polyfunctional Molecules

In many instances molecules containing more than one
binding site can be examined as satisfactorily as any others.
Two examples are shown in Figures 8 and 9. The spectrum of 8
was observed in the presence of 1 and 4 (Fig. 8). Either of
the chiral LSRs gave a $\Delta\Delta\delta$ adequate for the determination of
enantiomeric purity. It is interesting to note that 1 induced
separation of the aromatic proton signals, whereas 4 separated
the methoxy signals. This difference in action between two
chiral LSRs emphasizes the importance of trying different
chiral LSRs if one fails to work. Similar results were obtained
for 9 in the presence of two chiral LSRs (Fig. 9). In other
cases some alteration of one or more of the functional groups
may be necessary, as mentioned in Sect. VI-F.

The approximate order of binding capabilities of some
functional groups are (88):

primary amine > hydroxyl > ketone > aldehyde >
ether > ester > nitrile

For example, in a compound containing both hydroxyl and ester
groups the LSR will complex almost exclusively with the hydroxyl.
For this reason some polyfunctional molecules will behave as if
they were monofunctional.

Some types of polyfunctional molecules need to be examined
in the presence of a chiral LSR under special conditions. This
is exemplified by the enantiomeric purity determination of some
isoquinoline alkaloids (e.g., glaucine, 10) (41). When the

---

to equilibrium [b]; that is, (R)-substrate and (S)-substrate
will display exactly the same shifts in the presence of a
racemic shift reagent.

This may also be the place to point out that if the shift
reagent is *partially* resolved, that is, if the situation is
intermediate between that shown here and that shown on p. 289,
there will be a *partial* differential shift $\Delta\Delta\delta$ as between (R)-
substrate and (S)-substrate; in other words, a shift less than
the observable maximum with enantiomerically pure LSR but
greater than zero. While this point is of little practical
importance for chiral shift reagents (which are almost invariably used in enantiomerically pure form), a corresponding
argument applies to chiral solvents, for which it may be of
practical significance.

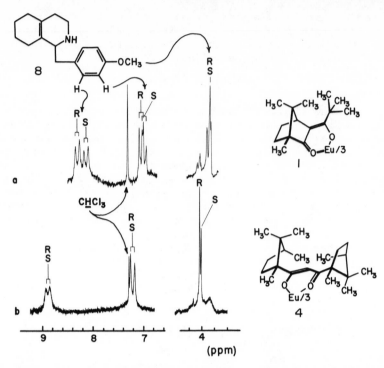

Fig. 8. Resolved methoxy and aromatic resonances observed
(100 MHz) for the enantiomers of amine 8 in the presence of
chiral LSR 1 (upper trace) and 4 (lower trace); chemical shifts
are referenced to SiMe₄ (T = 31°). The (R) isomer of 8 is present in higher concentration than the (S) isomer. (a) Concentration of substrate 0.5M and that of 1 ca. 0.3M in CDCl₃; (b)
concentration of substrate 0.5M and that of 4 ca. 0.2M in CDCl₃.
Reprinted with permission from J. Am. Chem. Soc., 96, 1038
(1974) (14). Copyright by the American Chemical Society.

spectrum of 10 is observed in the presence of 0.5 to 1.0 molar
ratio of LSR to substrate, the signals are prohibitively broad.
However, small amounts of a chiral LSR (LSR-substrate ratio of
0.07) are sufficient for determination of the enantiomeric
purity (41). In such cases the absolute induced shifts (Δδ) are
very small, while the ΔΔδ's are fairly large. The same problem
and solution have been found for some tertiary amines (96). This
technique may or may not be widely applicable to multifunctional
· molecules, but it emphasizes the need for trying a variety of
different conditions before abandoning a difficult problem.

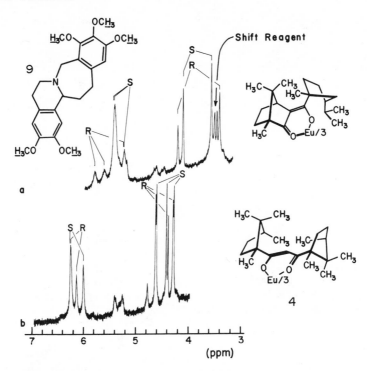

Fig. 9. Resolved methoxy resonances observed (100 MHz) for the enantiomers of amine 9 in the presence of the un- numbered LSR (upper trace) and 4 (lower trace); chemical shifts are referenced to SiMe₄ (T = 31°). The (S) isomer of 9 is present in higher concentration than the (R) isomer, (a) Con- centration of substrate 0.5M and that of the chiral LSR ca. 0.3M in CDCl₃; (b) concentration of substrate 0.5M and that of 4 ca. 0.4M in CDCl₃. Reprinted with permission from J. Am. Chem. Soc., 96, 1038 (1974) (14). Copyright by the American Chemical Society.

I. Recovery of Substrate

It is sometimes desirable to recover the substrate after obtaining its spectrum in the presence of an LSR. Column chro- matography has been shown to be an effective method of separating substrates from LSR. Silica gel eluted with benzene has been effective in recovering 11 from Eu(facam)₃ (57) and in recovering a 3-hydroxysteroid from the achiral LSR tris- (1,1,1,2,2,3,3-heptafluoro-7,7-dimethyl-4,6-octanedionate)-

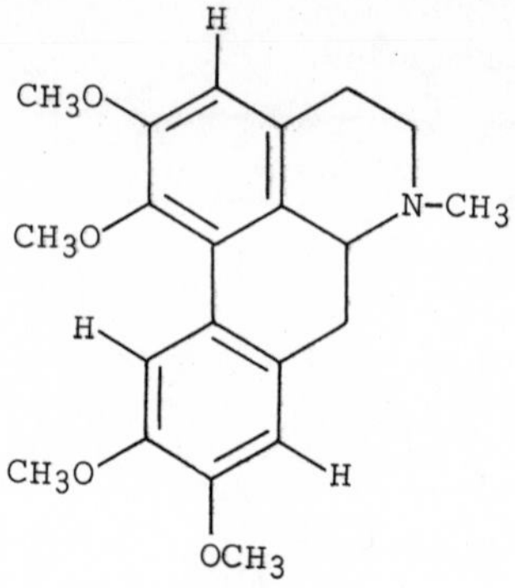

10

europium(III), Eu(fod)$_3$ (97). Another method used to recover a
series of chiral alcohols from Eu(fod)$_3$ involved adding *N*-
hexylamine to the alcohol Eu(fod)$_3$ solution and then passing
this mixture through a short column of silica gel using benzene
as the eluent (98). Still another system found useful in
separating alcohols from Eu(dcm)$_3$ (96) involves TLC on silica
gel using an eluent of petroleum ether (30 to 60°C)-ethyl ether
in a 97:3 ratio.

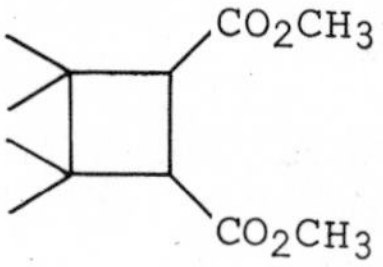

11

VII. CORRELATION OF ABSOLUTE CONFIGURATION

In addition to their usefulness in the direct determination of enantiomeric purities, chiral LSRs have also been shown
to be of value, albeit limted, in the correlation of absolute
configurations. This use of chiral LSRs for the correlation
of configuration in closely related compounds has been
demonstrated for a series of α-amino esters (38), a series of
1-deuterated primary alcohols (23), a series of alkyl aryl
carbinols (22), for four β-hydroxy-β-phenyl esters (35), and
for a series of secondary carbinols (21).

The correlation of configuration seems to be fairly
reliable when one group around the asymmetric carbon is varied
and the other three groups are held constant. However, even in
these cases exceptions have been found (21). Even more dis-
turbing than this is the fact that the sign of ΔΔδ can change

for some signals, depending on the amount of chiral LSR
present (see Sect. VI-C). For these reasons great care is
required in the applications of chiral LSRs to the correlation
of absolute configurations.

ACKNOWLEDGMENTS

I would like to thank the National Science Foundation
(Grant No. CH76-04760) for support through Professor J. D.
Roberts while this chapter was written at the California
Institute of Technology. Thanks are also due to the numerous
people who read the manuscript critically and who made many
helpful suggestions. Special thanks go to my wife Pat, for both
reading the chapter critically and proofreading it at its
successive stages of production.

REFERENCES

1. M. Raban and K. Mislow, *Top. Stereochem.*, 2, 199 (1967).
2. W. H. Pirkle, *J. Am. Chem. Soc.*, 88, 1837 (1966).
3. M. Raban and K. Mislow, *Tetrahedron Lett.*, 1965, 4249.
4. J. A. Dale, D. L. Dull, and H. S. Mosher, *J. Org. Chem.*,
    34, 2543 (1969); J. A. Dale and H. S. Mosher, *J. Am.
    Chem. Soc.*, 90, 3732 (1968), and 95, 512 (1973).
5. H. Gerlach and B. Zagalak, *J. Chem. Soc., Chem. Commun.*,
    1973, 274.
6. D. Valentine, Jr., K. K. Chan, C. G. Scott, K. K.
    Johnson, K. Toth, and G. Saucy, *J. Org. Chem.*, 41, 62
    (1976).
7. (a) R. Eberhardt, H. Lehner, and K. Schlögl, *Tetrahedron
    Lett.*, 1974, 4365.
    (b) Review: C. H. Lochmüller and R. W. Souter, *J. Chrom-
    atogr.*, 113, 283 (1975).
8. H. M. McConnell and R. E. Robertson, *J. Chem. Phys.*, 29,
    1361 (1958).
9. A. F. Cockerill, G. L. O. Davies, R. C. Harden, and D.
    M. Rackham, *Chem. Rev.*, 73, 553 (1973).
10. J. Reuben, *Prog. Nucl. Magn. Reson. Spectrosc.*, 9, Part 1
    (1973).
11. B. D. Flockhart, *C. R. C. Crit. Rev. Anal. Chem.*, 5, 69
    (1976).
12. R. von Ammon and R. D. Fischer, *Angew. Chem., Int. Ed.
    Engl.*, 11, 675 (1972); *Angew. Chem.*, 84, 737 (1972).
13. O. Hofer, *Top. Stereochem.*, 9, 111 (1975).
14. M. D. McCreary, D. W. Lewis, D. L. Wernick, and G. M.
    Whitesides, *J. Am. Chem. Soc.*, 96, 1038 (1974).
15. G. M. Whitesides and D. W. Lewis, *J. Am. Chem. Soc.*, 92,
    6979 (1970).

16. R. R. Fraser, M. A. Petit, and J. K. Saunders, *J. Chem. Soc., Chem. Commun.*, 1971, 1450.

17. H. L. Goering, J. N. Eikenberry, and G. S. Koermer, *J. Am. Chem. Soc.*, 93, 5913 (1971).

18. H. L. Goering, A. C. Backus, C. Chang, and D. Masilamani, *J. Org. Chem.*, 40, 1533 (1975).

19. V. M. Potapov, V. G. Bakhmutskaya, I. G. Ilina, G. I. Vinnik, and E. G. Rukhadze, *Zh. Obshch. Khim.*, 45, 2101 (1975); *J. Gen. Chem. USSR.*, 45, 2071 (1975).

20. K. Yamamoto, T. Hayashi, and M. Kumada, *Bull. Chem. Soc. Jpn.*, 47, 1555 (1974).

21. G. R. Sullivan, D. Ciavarella, and H. S. Mosher, *J. Org. Chem.*, 39, 2411 (1974).

22. O. Cervinka, P. Malon, and P. Trska, *Coll. Czech. Chem. Commun.*, 38, 3299 (1973).

23. C. J. Reich, G. R. Sullivan, and H. S. Mosher, *Tetrahedron Lett.*, 1973, 1505.

24. A. Rahm and M. Pereyre, *Tetrahedron Lett.*, 1973, 1333.

25. G. R. Sullivan, unpublished results.

26. H. L. Goering, J. N. Eikenberry, G. S. Koermer, and C. J. Lattimer, *J. Am. Chem. Soc.*, 96, 1495 (1974).

27. A. Claesson, L. Olsson, G. R. Sullivan, and H. S. Mosher, *J. Am. Chem. Soc.*, 97, 2919 (1975).

28. R. V. Smith, P. W. Erhardt, D. B. Rusterholz, and C. F. Barfknecht, *J. Pharm. Sci.*, 65, 412 (1976).

29. J. L. Luche, E. Barreiro, J. M. Dollat, and P. Crabbe, *Tetrahedron Lett.*, 1975, 4615.

30. K. S. Y. Lau, R. W. Fries, and J. K. Stille, *J. Am. Chem. Soc.*, 96, 4983 (1974).

31. C. Fisher, H. S. Mosher, and G. R. Sullivan, unpublished results.

32. S. J. Campbell and D. Darwish, *Can. J. Chem.*, 54, 193 (1976).

33. F. Lefevre, M. Marin, and M. Capmau, *C. R. Acad. Sci. Paris, Ser. C.*, 275, 1387 (1972).

34. A. I. Meyers and G. Knaus, *Tetrahedron Lett.*, 1974, 1333.

35. E. B. Dongala, A. Solladie-Cavallo, and G. Solladie, *Tetrahedron Lett.*, 1972, 4233.

36. J. D. McKinney, H. B. Matthews, and N. K. Wilson, *Tetrahedron Lett.*, 1973, 1895.

37. A. I. Meyers and C. E. Whitten, *Tetrahedron Lett.*, 1976, 1947.

38. K. Ajisaka, M. Kamisaku, and M. Kainosho, *Chem. Lett.*, 1972, 857.

39. H. Meyer and D. Seebach, *Liebigs Ann. Chem.*, 1975, 2261.

40. J. Redel, N. Bazely, Y. Calando, F. Delbarre, P. A. Bell, and K. Kodicek, *J. Steroid Biochem.*, 6, 117 (1975).

41. N. A. Shaath and T. O. Soine, *J. Org. Chem.*, 40, 1987 (1975).

42. D. Houalla, M. Sanchez, and R. Wolf, *Org. Magn. Reson.*, <u>5</u>, 451 (1973).

43. A. F. Cockerill, G. L. O. Davies, R. G. Harrison, and D. M. Rackham, *Org. Magn. Reson.*, <u>6</u>, 669 (1974).

44. J. Bus and O. Korver, *Rec.*, *J. R. Neth. Chem. Soc.*, <u>94</u>, 254 (1975).

45. M. G. Kienle, G. Cighetti, and E. Santaniello, *Bioorg. Chem.*, <u>4</u>, 64 (1975).

46. T. H. Koch, J. A. Olesen, and J. DeNiro, *J. Org. Chem.*, <u>40</u>, 14 (1975).

47. K. Mori, *Tetrahedron*, <u>31</u>, 1381 (1975).

48. B. Stridsberg and S. Allenmark, *Acta Chem. Scand.*, *B.*, <u>30</u>, 219 (1976).

49. J. M. Domagala, R. D. Bach, and J. Wemple, *J. Am. Chem. Soc.*, <u>98</u>, 1975 (1976).

50. G. M. Whitesides and D. W. Lewis, *J. Am. Chem. Soc.*, <u>93</u>, 5914 (1971).

51. H. L. Goering and Chiu-Shan Chan, *J. Org. Chem.*, <u>40</u>, 3276 (1975).

52. M. K. Kainosho, W. H. Pirkle, and S. D. Beare, *J. Am. Chem. Soc.*, <u>94</u>, 5924 (1972).

53. R. R. Fraser, J. B. Stothers, and C. T. Tan, *J. Magn. Reson.*, <u>10</u>, 95 (1973).

54. P. Y. Johnson, I. Jacobs, and D. J. Kerkman, *J. Org. Chem.*, <u>40</u>, 2710 (1975).

55. C. Mioskowski and G. Solladie, *Tetrahedron*, <u>29</u>, 3669 (1973).

56. A. Mannschreck, V. Jonas, H. Bodecker, H. Elbe, and G. Kobrich, *Tetrahedron Lett.*, <u>1974</u>, 2153.

57. G. L. Goe, *J. Org. Chem.*, <u>38</u>, 4285 (1973).

58. H. Nozaki, K. Yoshino, K. Oshima, and Y. Yamamoto, *Bull. Chem. Soc. Jpn.*, <u>45</u>, 3496 (1972).

59. R. R. Fraser, M. A. Petit, and M. Miskow, *J. Am. Chem. Soc.*, <u>94</u>, 3253 (1972).

60. K. Jankowski and A. Rabczenko, *J. Org. Chem.*, <u>40</u>, 960 (1975).

61. V. Schurig, *Tetrahedron Lett.*, <u>1976</u>, 1269.

62. D. L. Reger, *Inorg. Chem.*, <u>14</u>, 660 (1975).

63. D. Tatone, T. C. Dich, R. Nacco, and Botteghi, *J. Org. Chem.*, <u>40</u>, 2987 (1975).

64. K. Okamoto, T. Kinoshita, Y. Takemura, and H. Yoneda, *J. Chem. Soc., Perkin Trans. 2*, <u>1975</u>, 1426.

65. D. P. G. Hamon, G. F. Taylor, and R. N. Young, *Tetrahedron Lett.*, <u>1975</u>, 1623.

66. Jean-Louis Fourrey and J. Moron, *Tetrahedron Lett.*, <u>1976</u>, 301.

67. K. Mori, *Tetrahedron*, <u>32</u>, 1101 (1976).

68. P. G. Duggan and W. S. Murphy, *J. Chem. Soc., Perkin Trans. 1*, <u>1976</u>, 634.

69. J. Stackhouse, R. J. Cook, and K. Mislow, *J. Am. Chem. Soc.*, 95, 953 (1973).

70. C. A. Maryanoff, B. E. Maryanoff, R. Tang, and K. Mislow, *J. Am. Chem. Soc.*, 95, 5839 (1973).

71. K. S. Y. Lau, P. K. Wong, and J. K. Stille, *J. Am. Chem. Soc.*, 98, 5832 (1976).

72. A. Sinnema, F. van Rantwijk, A. J. De Koning, A. M. van Wijk, and H. van Bekkum, *Tetrahedron*, 32, 2269 (1976).

73. J. Bus, C. M. Lok, and A. Groenewegen, *Chem. Phys. Lipids*, 16, 123 (1976).

74. M. Le Plouzennec, F. Le Moigne, and R. Dabard, *J. Organomet. Chem.*, 111, C38 (1976).

75. G. Seitz and G. Kromeke, *Arch. Pharm.*, 309, 930 (1976).

76. R. Haller and H. J. Bruer, *Arch. Pharm.*, 309, 367 (1976).

77. S. Miura, S. Kurozumi, T. Toru, T. Tanaka, M. Kobayashi, S. Matsubara, and S. Ishimoto, *Tetrahedron*, 32, 1893 (1976).

78. D. Seebach, V. Ehrig, and M. Teschner, *Liebigs Ann. Chem.*, 1976, 1357.

79. H. Hikino, K. Mohri, Y. Hikino, S. Arihara, and T. Takemoto, *Tetrahedron*, 32, 3015 (1976).

80. A. H. Conner and J. W. Rowe, *Phytochemistry*, 15, 1949 (1976).

81. V. P. Zaitsev, V. M. Potapov, V. M. Demyanovich, and L. D. Soloveva, *Zh. Org. Khim.*, 12, 2326 (1976).

82. H. Matsushita, M. Noguchi, and S. Yoshikawa, *Bull. Chem. Soc. Jpn.*, 49, 1928 (1976).

83. J. A. Berson, P. B. Dervan, R. Malherbe, and J. A. Jenkins, *J. Am. Chem. Soc.*, 98, 5937 (1976).

84. T. Schmidt, C. Fedtke, and R. R. Schmidt, *Z. Naturforsch.*, *C*, 31, 252 (1976).

85. B. Stridsberg and S. Allenmark, *Acta Chem. Scand.*, *B*, 30, 219 (1976).

86. P. Reisberg, I. A. Brenner, and J. I. Bodin, *J. Pharm. Sci.*, 65, 592 (1976).

87. D. B. Repke and W. J. Ferguson, *J. Heterocycl. Chem.*, 13, 775 (1976).

88. J. K. M. Sanders and D. H. Williams, *J. Am. Chem. Soc.*, 93, 641 (1971).

89. J. Bouquant and J. Chuche, *Tetrahedron Lett.*, 1973, 493.

90. G. M. Whitesides, private communication.

91. M. R. Willcott, III, R. E. Davis, and R. W. Holder, *J. Org. Chem.*, 40, 1952 (1975).

92. Synthesized by W. S. Johnson, B. Ganem, and R. Muller, unpublished results.

93. K. L. Williamson, C. P. Beeman, and J. G. Magyar, in manuscript.

94. G. A. Gray, *Anal. Chem.*, 47, 557A (1975).

95. D. R. Crump, J. K. Sanders, and D. H. Williams, *Tetrahedron Lett.*, 1970, 4949.

96. R. Elsenbaumer, private communication.
97. W. L. Tan, private communication.
98. N. H. Anderson, B. J. Bottino, A. Moore, and J. R. Shaw, *J. Am. Chem. Soc.*, <u>96</u>, 603 (1974).

SUBJECT INDEX

Ab initio calculations, see
    Quantum mechanical
    calculations
Absolute configuration, 151,
    324
Acetic-d,t acid, chiral, 270
Acetophenone, 205, 210, 211,
    237, 243, 245, 264, 265
  hydrosilylation of, 271
Acetophenone pinacol, 265
3-Acetoxy-1-octyne, 294
1-Acetoxy-1-phenyl-2-butene,
    294
2-Acetylpyridine, 265
Alanine, 218, 250
  methyl ester, 297
Aldol condensation, 234, 235
Alkaloids, 321
Alkoxy aluminum dichloride,
    210
2-Alkoxy-1,3-dioxolane, 64
α-Alkylalkanoic acids, 220
Alkylation, 224, 228, 237, 249
N-Alkyl-N-methylephedrinium
    salt, 236
N-Alkylpyrrolidine, 70
2-Allyl-1,3,4-cyclopentane-
    trione, 208
2-Allylphenol, oxymercuration
    of, 213
α-Amino acids, asymmetric
    synthesis of, 199, 202,
    203, 217, 219, 227, 239,
    242
  asymmetric synthesis with,
    218, 234, 270
  enantiomeric purity of, 203
Amino alcohols, 205, 208
α-Amino caprolactam, 233

N-(2'-Aminoethyl)-1,2-diamino-
    ethane, chelate
    complexes of, 116
α-Aminoketones, 201
α-Aminonitriles, 203
2-Amino-3-phenylpropane, 218
Aminophosphine, 251
Amphetamine, 294
Androsterone, 24, 25
Angle bending strain, 5
Arsolane, 78
3-Aryl-1,2,3-oxathiazolidin-
    2-oxide, 78
Aspartic acid, 116, 185, 186,
    267
Aspergillus ustus, 267
Asymmetric alkylation, 213,
    215
Asymmetric electrochemistry,
    264
Asymmetric Grignard reagent,
    210
Asymmetric hydroboration, 203
Asymmetric hydrogenation, 238
Asymmetric induction, 176,
    179, 182, 184, 191, 192,
    212, 229, 230
  ab initio, 195
  models of, 179, 182, 191,
    196, 197, 247
  rules of, 191, 199
  solvent dependent, 200, 243
  temperature effect, 184,
    201, 208, 220
  see also Asymmetric
    synthesis
Asymmetric oxidation, 251
Asymmetric phase transfer
    catalysis, 238

Asymmetric polymerization, 252, 263
Asymmetric reduction, of conjugated double bonds, 244
  of ketones, 193, 198, 243
  of α,β-unsaturated carboxylates, 243
Asymmetric synthesis, of alkyl iodides, 205
  of α-amino acids, 186, 202, 203, 217, 219
  of atrolactic acid, 181, 191
  with chiral oxazolines, 186, 199, 221
  of dipeptides, 220
  with enzymes, 269
  of isoxalidine-ribosides, 195
  photochemical, 252, 257, 261
  of polypeptides, 218
  of prostaglandin, 190
  of steroids, 215, 229, 232, 261
  from sugars, 270
  from terpenes, 270
  see also Asymmetric induction.
Atrolactic acid, 181, 191
Atropaldehyde, 231
Avenaciolide, 271

Bacillius subtillis, 267
Benzaldehyde, 236
Benzocyclopentene, 49
Benzophenones, 261
  reduction of, 193
N-Benzoyl-valine-t-butylamide, 218
Benzyl-α-d-alcohol, 293
Benzyl-α-d-azide, 296
3-O-Benzyl-1,2-O-cyclohexylidene-α-D-glucofuranose, 208
5-Benzyloxymethylcyclopentadiene, 217
Bicyclo[5.3.0]decane, 20
Bicyclo[2.2.1]heptane, 56
Bicyclo[2.1.1]hexane, 58, 59
Bicyclo[3.1.0]hexane, 19, 49, 50

Bicyclo[3.3.0]hexan-3-one, 51
Bicyclo[4.3.0]nonane, 20
Bicyclo[3.3.0]octane, 20, 51
Bidentate complexes with six-membered chelate rings, 107
2,2'-Bi-1,3-dioxolane, 63, 64
2,2'-Bi-1,3-dithiolane, 68
Biotin, 271
Bis(aspartato)cobalt(lll), 116, 118, 119, 162
Bis(bidentate) complexes, 113, 114
Bis(1,2-diaminocyclohexane)-platinum(ll), 115
Bis(diethylenetriamine)-cobalt(lll), 116
Bisectional(b), 3
Bis(2-methyl-1,4,7-triaza-cyclononane)cobalt(lll), 123
Bis(prolinato)palladium(ll), 115
Bis{tribenzo(b,f,j)-(1,5,9)-triazacyclododecahexaene}cobalt(lll), 122
Bis(1,1,1-tris(aminoethyl)-ethane)cobalt(lll), 125
2-Bornene, 248
Botryodiplodia theobromac Pat., 268
Brevibacterium ammoniagenes, 267
Bromocyclopentane, 35
1-Bromo-2-methylnapthalene, 251
Brucine, 265
2-Butanethiol, 296, 316
2-Butanol, 245, 292
1-Butene, 248
2-Butene, 205, 248
2-Butyl acetate, 294, 316
sec-Butylformamide, 295, 296, 305
t-Butyl hydroperoxide, 238, 251
2-Butyl methyl ether, 295, 316

2-*t*-Butyl-1,3-oxathiolane, 70

C_2, 3, 6, 7, 11, 14
C_s, 3, 6, 7, 11, 14
Camphor, 295, 305, 316
Camphorquinone, 250
2-Carboethoxycyclohexanone, 249
2-Carbomethoxyindanone, 236
Carbomethoxymethylcyclopenta-
    diene, 203, 204
Carbon-13, NMR, 317, 261
Carbonato(3,8)-dimethyltri-
    ethylenetetramine)-
    cobalt(III), 131, 132
2-Carboxymethyl-1-indanone, 235
Catalysis, asymmetric, 177, 178
Catalyst, chiral, 178, 238, 271
  molybdenum, 251
  platinum, 248
  rhodium, 247
  supported, 271
Cervinka's rule, 208
Charge control, in stereo-
    specific additions, 197
Charge transfer complex, 193,
    194
Chelate rings, six-membered,
    conformations of, 107
Chelating diphosphines, 239
Chemical shifts, causes of,
    289
Chiral catalysts, 178, 238,
    271
Chiral electrode, 266
Chiral enamines, 224, 229
Chiral Grignard
    reagents, 193
Chiral inducer, 265
Chirality, crystal, 263
  molecular, 263
Chiral methyl group, 268
Chiral oxazolines, 186
Chiral phosphines, 228, 229,
    238
Chiral sensitizers, 258, 259
Chiral shift reagents, 202
Chiral solvent, 202, 264

1-Chloro-2-bromocyclopentane,
    37, 38, 39
Chlorocyclopentane, 34, 35
Chloro-1,13-diamino-2,4,10-
    triazatridecanecobalt-
    (III), 142
α-Chloroesters, 266
α-Chloroketones, 266
2-Chloropropanal, 195
Cholesteryl acrylate, 263
Circular dichroism, 151
  of bis-bidentate cobalt(III)
    complexes, 156
  of complexes containing
    multidentate ligands,
    160, 161
  of transition metal com-
    plexes, 151
  of tris-diamine complexes,
    152, 153, 154
Circularly polarized light,
    252, 254, 255
Citronellol, 249
Claisen rearrangement, 191
Clostridium Klurjveri, 268
Complexes involving multi-
    dentate ligands, 116
Conformational Rule, 51
Conformational transmission,
    54
Contact shift, 288
Cram's Rule, 191, 196, 197
    199
Cumene hydroperoxide, 251
Cuprate reagents, 215
Curtin-Hammett principle, 182,
    196
2-Cyanobutane, 296, 316
Cyclitols, 73
α-Cyclocitral, 224
1,2-Cyclohexanediol, 242
Cyclohexanones, reduction of,
    185
Cyclohexene sulfide, 253
Cyclohexenones, 188, 214, 215
Cyclohexyl(2-isopropoxyphe-
    nyl)methylphosphine, 243

Cyclohexylmethylcarbinol, 293, 305
1,3-Cyclooctadiene, 248
Cyclopentane, conformational analysis of, 2, 3, 4, 6 7, 11, 12, 14, 29, 31
  dialkyl, 37
  electron diffraction of, 30
  energy of, 5, 7
  entropy of, 2, 3
  force field calculations of, 12, 13
  nuclear magnetic resonance of, 30
  phase angle(f) of, 4
  pseudorotation of, 5, 10, 11, 12, 14, 30, 31
  pucker of, 4, 8
  quantum mechanical calculations on, 13, 14, 77
  Raman spectrum of, 30, 33
  R-value method with, 32
  strain energy of, 4
  symmetry number of, 4
  torsional angles in, 6, 9, 10, 11, 12, 13, 31
1,1-Cyclopentanedicarboxylic acid, 40, 42
1,2-Cyclopentanedicarboxylic acid, 40, 41, 42
1,2-Cyclopentanediol, 243
Cyclopentanes, bridged, 26
  deuterated, 32
  1,1-disubstituted, 36
  1,2-disubstituted, 37
  1,3-disubstituted, 15, 18, 19, 42
  fused, 21
  maximally puckered, 23, 53
  monosubstituted, 5, 14, 15, 33
  quantum mechanical calculations on, 77
Cyclopentanol, 36
Cyclopentanols, 1,3-disubstituted, 44
  oxidation rates of, 43

2-substituted, 37
Cyclopentanone, 27
  conformation of, 5, 6, 46, 48
  force field calculations on, 27
Cyclopentene, 27, 49, 77
Cyclopentyl halides, C-X frequencies in, 35
Cyclopentyl tosylates, acetolysis rates of, 45
Cyclopropanation, 250

DAB, 215
Darvon alcohol, 205, 206
DDB, 208, 264
Decoupling, 312
21,22-Dehydroneoergosterol, 259
Dehydropestalotin, 208
Deoxycholic acid, 263
Desoxybenzoin oxime, 209
Dewar benzenes, 217
Diacetoxycyclopentene, 267
Dialkylcyclopentanes, 37
1,10-Diamino-4,7-diazadecane, chelate complexes of, 137
1,8-Diamino-3,6-diazaoctane, chelate complexes of, 125
2,4-Diaminopentane, chelate complexes of, 109
2,3-Diaminopropionic acid, chelate complexes of, 119
1,14-Diamino-3,6,9,12-tetra-azatetradecane, 142
1,14-Diamino-3,6,9,12-tetraaza-tetradecanecobalt(lll), 142, 143, 144, 145, 161
1,11-Diamino-3,6,9-triazaunde-cane, chelate complexes of, 141
Diarylethylenes, 254
Diastereoselective photo-synthesis, 259

Diastereoselective syntheses,
    179, 182, 198, 201, 202
Diastereotopic groups, NMR of,
    310, 313
Dibromocarbene, 236
1,2-Dibromo-1-chlorocyclopen-
    tane, 40
Dibromocyclopentane, see 1,2-
    Dihalocyclopentanes;
    1,3-Dihalocyclopentanes
3,5-Dibromocyclopentene, 49
Dibromocyclopropane, 236
4,5-Dibromo-γ-valerolactone,
    62
1,3-Di-t-butylpropargyl
    alcohol, 293, 305
3,5-Dicarboalkoxy-2,6-di-
    methyl-1,4-dihydropyri-
    dine, 212
1,2-Dicarbomethoxycyclopen-
    tane, 37
Dichlorobis(ethylenediamine)-
    cobalt(111), 114, 159
Dichlorobis(N-methylethylene-
    diamine)cobalt(111), 114
Dichlorobis(propylenediamine)-
    cobalt(111), 114
Dichlorocarbene, 236
2β,3α-Dichloro-5α-cholestane,
    23, 24
Dichlorocyclopentane, see
    1,2-Dihalocyclopentanes;
    1,3-Dihalocyclopentanes
Dichlorocyclopropane, 236
Dichloro-1,10-diamino-4,7-di-
    azadecanecobalt(111), 138
Dichloro(1-methylamino-2-
    aminopropane)platinum-
    (11), 114
2,3-Dichlorotetrahydrofuran, 61
Dicyanobis(ethylenediamine)-
    cobalt(111), 114, 159
Diels-Alder reaction, 177,
    191, 198, 199, 217
Diethylenetriamine, 116
1,2-Dihalocyclopentanes, 16,
    37, 38, 39, 40

1,3-Dihalocyclopentanes, 17,
    18, 19, 38, 45
Dihydroindole, 258
2,3-Dihydro-2-methylbenzofu-
    ran, 213
1,3-Dihydroxycyclopentane, 42
Diisopinocampheylborane, 204
Dimenthyl(methyl)phosphine,
    249
1,4-Dimethylallene, 259
1,4-Dimethylamino-2,3-butane-
    diol, 207, 208
4-Dimethylamino-3-methyl-1,2-
    diphenyl-2-butanol, 205
3,3-Dimethyl-1,2-butanediol,
    252
1,2-Dimethylcyclopentane, 15,
    37
1,3-Dimethylcyclopentane, 15,
    16, 42
3,4-Dimethylcyclopentanone, 48
2,4-Dimethyl-1,3-dioxolane, 63
2,5-Dimethyl-2,4-hexadiene,
    249
2,6-Dimethylphenol, 207, 208
N,N-Dimethyl-1-phenylethyl-
    amine, 294
2,2-Dimethyl-1-propanol-1-d,
    293
5,7-Dimethyl-1,4,8,11-tetra-
    azaundecane, chelate
    complex of, 136
1,2-Dimethylthiirane, 252, 253
Dimethyl δ-truxinate, 262
Dinitrobis(ethylenediamine)-
    cobalt(111), 113, 159
Dinitrobis(propylenediamine)-
    cobalt(111), 114, 159
Dinitro-1,10-diamino-4,7-di-
    azadecanecobalt(111),
    138
Dinitro(3,8-dimethyltriethyl-
    enetetramine)cobalt(111),
    126, 127, 128, 130
Dinitro-1,3,8,10-tetramethyl-
    4,7-diazadecanecobalt-
    (111), 138

DIOP, 243, 245, 271
Dioxaarsolanes, 72
1,3-Dioxolane, 28, 63, 78
Dipeptides, asymmetric synthe-
    sis of, 220
1,2-Diphenylaminoethanol, 185
1,2-Diphenylcyclopentane, 37
1,2-Diphenylcyclopropane, 259
1,1-Diphenylethylene, 250
Diphenylmethyl alkyl ketone,
    205
Diphenylsilane, 245, 246
Diphosphine, 242, 251
Dipodascus uninucleatus, 267
Dipolar shift, 288
Dipole-dipole interaction, 17
1,2-Dithiolane-4-carboxylic
    acid, 68
1,3-Ditholanes, 68, 78
DOPA, 239

Enamine, chiral, 224, 225, 229,
    265
Enantiomeric excess, 202
Enantiomeric purity, deter-
    mination of, 291
Enantioselective synthesis,
    178, 179, 198, 202, 231,
    254
Enantiotopic groups, 310
Envelope, of cyclopentane, 3,
    6, 7, 11, 14
Enzymes, asymmetric synthesis
    with, 178, 266, 268, 269
Ephedrine, 264
Epoxidation, 237, 238, 251
  enzymatic asymmetric, 268
Erythrose, 270
Estrone, 215, 216, 235
Ethyl diazoacetate, 249, 250
Ethylene carbonate, 65
Ethylenediaminediacetato (pro-
    pylenediamine) cobaltate-
    (111), 140, 161
Ethylenediaminediacetic acid,
    chelate complexes of,
    139

Ethylenediaminetetraacetato-
    cobaltate (111), 149, 161
Ethylenediaminetetraacetic
    acid, 149
Ethylene thiourea, 68
α,β-Ethylenic esters, reduc-
    tion of, 210
Ethyl 2-methylbutanoate, 294
Eu (dcm) 3, 290, 315
Eu (dpm) 3, 304
Eu (facam) 3, 289
Eu (hfbc) 3, 289, 307, 314
Europium, as shift reagent,
    306. *See also under
    individual compounds*
Exciton circular dichroism, of
    tris-bidentate com-
    plexes, 163

Felkin model, for asymmetric
    induction, 191, 196,
    197
Five-membered rings, hetero-
    cyclic, 27, 59, 71, 72,
    78
Fluorine-19, NMR, 317
Fluorocyclopentane, 16, 34,
    35, 77
4-Fluoroproline, 67
Force field calculations, on
    cycloalkanes, 7, 20, 78
  on cyclopentanone, 27
  on heterocycles, 27
  on norbornanes, 26
  on steroids, 24
Furanose, 73, 74
Furanosides, 78

D-Galactano-γ-lactone, 62
Gauche-butane interaction, 185
Gem effect, 193
Geminal methyls, NMR spectra
    of, 311
Geminal protons, in NMR, 311,
    312
Germacyclopentane, 72
Germacyclopentanol, 78

Glaucine, 321
Glutamobis(ethylenediamine)co-
    balt(111), 135, 157
Glyceraldehyde, 270
Glycinatotriethylenetetramine-
    cobalt(111), 132
Glycosides, phase angles of,
    75
Grignard, additions to ketones,
    210
Grignard reductions of ketones,
    119

Half-chair, of cyclopentane,
    3, 6, 7, 11, 14
α-Halocyclopentanones, 48
Halogenocyclopentanes,
    conformations of, 16,
    33, 37, 39, 78
    infrared and Raman spectra
        of, 33, 34
    NMR of, 36
Hantzsch esters, 212
Helicene, 254, 256, 257, 261
Heterocycles, force field
    calculations on, 27
    quantum mechanical calcula-
        tions on, 27, 78
Heterocyclic five-membered
    rings, conformations
    of, 27, 59, 71, 72, 78
Heterogeneous reduction, 186
Hexahelicene, 254, 255, 262
Horeau method of kinetic
    resolution, 181
Hückel calculations, see
    Quantum mechanical
    calculations
Hydrotropaldehyde, 248
Hydrindane, 20, 22, 23, 52
Hydroformylation, 247, 248
Hydrogenation, asymmetric,
    238, 268
Hydrogen bonds, intramolecular,
    2
β-Hydrogen transfer, 210, 211,
    259

Hydrosilylation, 239, 244
    of acetophenone, 271
    of imines, 247
Hydroxy acids, asymmetric
    synthesis from, 270
2-Hydroxycyclopentanecarboxy-
    lic acid, 40
2-(Hydroxymethyl)-quinucli-
    dine, 235
Hydroxynitriles, 266
2-Hydroxy-A-norcholestanol, 53
β-Hydroxy-β-phenyl esters,
    absolute configurations
    of, 324
4-Hydroxyproline, 66

Imines, hydrosilylation of,
    247
Induction, asymmetric, see
    Asymmetric induction
Infrared spectrum, see under
    compound
2-Iodobutane, 205
Isoclinal, 3
Isoprene, telomerization of,
    249
Isopropylaminomethyl-2-naph-
    thyl ketone, 268
Isopropyldimenthylphosphine,
    249
Isotopic effects, in
    asymmetric synthesis,
    184, 197

Karabatsos model, of
    asymmetric induction,
    191, 196, 197
α-Keto acids, 219, 266
α-Keto esters, 191, 266
α-Ketols, 266
Ketones, additon of Grignards
    to, 210
    alkylation of, 226, 227
    asymmetric reduction of,
    198, 243
    photoreduction of, 259
    reduction of, 191, 215

reduction of by chiral
    Grignard reagents, 210
Kinetic anomeric effect, 195

Lanthanide shift reagents,
    chiral, 288, 305
    table of ^{1}H NMR signals
    with, 292
    effect of concentration on,
    315
    effect of solvent on, 308
    effect of temperature on,
    315
    line broadening with, 306,
    308, 315, 322
    polyfunctional molecules
    with, 321
    sample handling technique,
    307, 323
    substrate, functional group
    alteration, 319
    see also under individual
    rare earths
Lipoic acid, 68
Lissajous curves, 11, 12
Lysine, 233

Magnetic impurities, 309
Malic acid, 268, 270
Mandelic acid, 265, 266
Mannitol, 262
Meerwein-Ponndorf reactions,
    191
Menthol, 293
Menthyl benzoylformate, 212
Menthyldiphenylphosphine, 239,
    249
Menthyl mandelate, 212
Menthyl phenylpropynoate, 217
Mesembrine, 224, 225
1-Methoxy-2-methylcyclohexane,
    295
2-Methoxy-2-phenylbutane, 295
2-Methoxypropionic acid, 296
Methoxy-α-trifluoromethyl-
    phenylacetyl chloride
    (MTPA), 202, 204

Methyl acetoacetate, 243, 271
N-Methylalaninatobis(ethyl-
    enediamine)cobalt(lll),
    134
α-Methylbenzylamine, 243
N-(α-Methylbenzyl)-N-methyl-
    aminoalane, 210
N-α-Methylbenzylsalicylaldi-
    mine, 247
2-Methylbutanal, 196, 248
2-Methyl-1-butanol, 292, 315,
    316
2-Methylbutylaluminum, 210
Methyl-2-butylmagnesium, 211
Methyl cinnamate, 214
2-Methylcyclohexanol, 176,
    317
3-Methylcyclohexanone, 214,
    215
2-Methyl-2-cyclohexenol, 245
2-Methyl-2-cyclohexenone, 245
5-Methyl-1,3-cyclopentadiene,
    203
Methylcyclopentane, conforma-
    tion of, 5, 15, 36, 77
2-Methyl-1,3-cyclopentane-
    dione, 215, 234
Methyl 3,7-dimethyloctanoate,
    316
2-Methyl-1,2-di(3-pyridyl)-1-
    propanone, 268
N-Methylephedrine, 206, 214,
    215, 236
Methyl ethylene phosphate, 71
Methyl ethyl ketone, 245
2-Methyl-2-ethylsuccinic
    acid, 202
3-Methyl-4-hexenoate, 228
8-Methylhydrindane, 52
Methyl 3-hydroxyoctadecano-
    ate, 302
3-Methylindanone, 259
Methyl 2-methyl-2-phenyl-
    butanoate, 294
2-Methylnorbornanol, 293
3-Methyl-2-pentanone,
    295

1-Methyl-2-phenylcyclopro-
    pane, 215
Methyl 2-phenylcyclopropane-
    carboxylate, 214
1-(Methylphenyl)ethanol, 293
N-Methyl-1-phenylethylamine,
    294
Methyl 3-phenyl-3-hydroxypen-
    tanoate, 297
3-Methyl-3-phenyl-2-pentan-
    one, 295
Methyl 2-phenylpropionoate,
    295
2-Methylpiperidine, 317
11α-Methylprogesterone, 270
β-Methylstyrene oxide, 295
2-Methyl-1,4,7-triazacyclo-
    nonane, chelate complex
    of, 122
Methyl vinyl ketone, 234
Michael reaction, 234, 235,
    236
Microbiological reductions,
    267
MINDO calculations, see
    Quantum mechanical
    calculations
Molecular mechanics, see
    Force field calculations
Molybdenum catalyst, 251
Mosher's method of enantio-
    meric purity measure-
    ment, 203
MTPA, 202, 204
Mucor rammanianus, 267

NADH, 211
α-Naphthyl methyl ketone
    oxime, 209
Neomenthyldiphenylphosphine,
    239, 249
Neopentyl diazoacetate, 251
2-Nitrobutane, 296, 316
Non-bonded interactions, 17
Norbornadiene, 248
Norbornane, 26,
    56, 57

Norbornene, 248
Nuclear magnetic resonance,
    see Carbon-13; and under
    compound

1,3,6,8,10,13,16,19-Octaaza-
    bicyclo[6,6,6]eicosane-
    cobalt(III), 151
1,7-Octadiene, 268
Optical activation, 259
Optical yield, 201
Orbital, frontier, control,
    197
Oxalato-5,7-dimethyl-1,4,8,11-
    tetraazaundecanecobalt-
    (III), 136
1,3-Oxathiolanes, 28, 69, 78
Oxazolines, chiral, 199, 201,
    220, 221, 224, 226
Oxidation, asymmetric, 251
    of sulfides, 252
Oxolane, 27, 28, 59
Oxolanes, polysubstituted,
    61, 62
Oxymercuration, of 2-ally-
    phenol, 213
    of styrene, 213
Ozonides, 28, 29

Parkinson's disease, 239
Penicillium decumbers, 267
1,4,7,10,13-Pentaazatridec-
    ane, chelate complexes
    of, 141
Pentalane, 20, 22, 51
Perhydroazulene, 20
Perpendicular attack, 187, 188
Pestalotin, 208
Phase angle, of cyclopentane,
    4, 7, 10
    of glycosides, 75
Phase transfer catalysis,
    asymmetric, 236, 238
Phenacyl chloride, 245
Phenylalanine, 234, 261, 266
3-Phenylamino-2-norbornanol,
    208

2-Phenyl-1-butanol, 292
2-Phenyl-2-butanol, 291, 304,
    306, 307, 314, 318
α-Phenylcinnamyl alcohol, 251
Phenylcyclopentane-1-carbox-
    boxylic acid, 42
1-Phenyl-2,2-dimethyl-1-pro-
    panol, 292
1-Phenylethanol, 245, 292,
    317, 318
α-Phenylethylamine, 289, 290,
    291, 293, 294, 305, 317
α-Phenylglycine, 218, 232
Phenylglyoxylic acid, 265, 266
Phenylmethylbenzylphosphine,
    247
1-Phenyl-1-methylcyclohexane,
    193
Phenyloxirane, 236
1-Phenyl-1-propanol, 292
1-Phenylpropene, 215
2-Phenylpropionaldehyde, 296
Phenyltrifluoromethyl ketene,
    181
Phosphinates, 71
Phosphines, chiral, 228, 229,
    238
Phosphinites, 229, 242
Phosphites, cyclic, 71
Phosphonates, 71
Photoactivation, 258
Photoalkylation, 261
Photocyclization, 254, 257
2 + 2 Photocycloaddition, 262
Photodimerization, of
    thymine, 261
Photoinduced asymmetric
    synthesis, 254
Photoisomerization, 255
Photopinacolization, 264, 265
Photopolymerization, 263
Photoreduction, 259, 261
Photosynthesis, diastereo-
    selective, 259
  enantioselective, 254
  of helicene, 261
2,3-Pinanediol, 205

3-Pinanylborane, 203
Platinum catalysts, 248
Point groups, *see* C_2; C_s
Polar effects, 193
Polymer, optically active,
    263
Polymerization, asymmetric,
    252, 263
Polypeptide, asymmetric
    synthesis of, 218
Praseodymium, as shift
    reagent, 306. *See also
    under specific compounds*
Prelog's Rule, 181, 191, 199,
    210, 212
Pr(facam)$_3$, 307
Product development control,
    184
Prolinatotrimethylenetetra-
    minecobalt(III), 133
Proline, 66, 67, 78, 226, 234
Proline dimer, 28
Prolyl-α-amino acids, 220
Pronethalol, 268
2-Propanol, 259
Propylenediamine-N,N'-di-
    methylethylenediamine-
    platinum(II), 115
Prostaglandins, 75, 78, 190,
    203, 217, 232, 243, 266,
    267, 270
Pseudoaxial (a'), 3
Pseudoequatorial (e'), 3
Pseudo-four-membered rings,
    27
Pseudolibration, 17, 36
Pseudomonas olevoraus, 269
Pseudorotation, of cyclo-
    pentane, 4, 7, 8, 9, 10,
    11, 29
  of five-membered rings, 14
  of monosubstituted five-
    membered rings, 7
  restricted, 17, 27
Puckering of cyclopentane, 4,
    5, 10, 13
Pulegone, 217

Pyrazoline, 77, 78
Pyrrolidine, 28, 66, 78

Quadridentate ligands, complexes involving, 125
Quantum mechanical calculations, on cyclopentane, 13, 14, 77
on heterocycles, 27, 78
Quinine, 208, 235, 236, 243, 244, 265
quaternary ammonium salt of, 238
Quinquedentate ligands, complexes involving, 141

Reduction, by chiral Grignard reagents, 210
of ketones, 191, 205, 206, 208, 210, 237
microbiological, 266, 268
of pyruvate, 211
Reformatsky reaction, 212
Rhodium, catalyst, 243, 247
Ruch-Ugi model for asymmetric induction, 179, 181, 182
R-value method, with cyclopentanes, 32, 36, 48

Salem model for asymmetric induction, 182
Sarcosinatobis(ethylenediamine)cobalt(lll), 135, 136, 157
β(N-Sarcosinato)propionate, chelate complex of, 121
Schiff's base, chiral, 228, 231, 243
Selenolane, 70
Sensitizers, chiral, 258, 259
Sexidentate ligands, complexes involving, 142
Shift reagents, see Lanthanide shift reagents
Sigmatropic rearrangement, 191, 192
Silacyclopentane, 72

Spin-spin decoupling, 312
Spiro(4,4)nonane-1,6-dione, 46, 47
Stereoselectivity, of carbonyl addition, 189. See also Asymmetric induction; Diastereoselective synthesis; and Enantioselective synthesis
Steric approach control, 184
Steric congestion, 187, 188
Steric effects, 189, 190. See also Van der Waals forces
Steroids, asymmetric synthesis of, 215, 229, 232, 261
ring D, conformation in, 23, 24, 25, 52, 53, 54, 55
Strain energy, 5, 7
Strecker synthesis, 217
Strychnine, 265
Styrene, 213, 248, 250
Styrene oxide, 295, 317
Sugars, asymmetric synthesis from, 270
conformations of, 73
Sulfides, oxidation of, 252
Sulfinates, 229
Sulfoxides, 252

Tartaric acid, 231, 232, 271
Telomerization of isoprene, 249
Terdentate ligands, 116
Terpenes, asymmetric synthesis from, 270
Tetrahydrofuran, 59
$N,N,N'N'$-Tetrakis-(2'-aminoethyl)-1,2-diaminoethanecobalt(lll), 147, 161
$N,N,N'N'$-Tetrakis-(2'-aminoethyl)-1,2-diaminopropane, 148
$N,N,N'N'$-Tetrakis-(2'-aminoethyl)-1,2-diaminopropanecobalt(lll), 149, 161

$N,N,N'N'$-Tetramethyl-1,2-
    cyclohexanediamine, 215
3,3,5,5-Tetramethylcyclopen-
    tane-1,2-dione, 48
3,3,4,4-Tetramethylcyclopen-
    tanol, 42, 43
1,3,8,10-Tetramethyl-4,7-
    diazadecane-1,10-dia-
    mine, chelate complex
    of, 138
7-Thiabicyclo[2,2,1]heptane,
    70
1,3-Thiazolidines, 69
Thiiranes, 223, 252
Thiolane, 28
Thymine, 261
p-Tolylisopropylsulfoxide,
    259
p-Tolymethylsulfoxide, 259,
    266
Torsional angles, in cyclo-
    pentane, see Cyclopen-
    tane
Torsional effect, in carbonyl
    reduction, 188, 191
Torsional strain, 5, 17
Transamination, 218
Transition metal complexes,
    absolute configuration
    of, 151
  as catalysts, 238
  circular dichroism of, 151
Tribenzo(b,f,j)-(1,5,9)triaza-
    cyclododecahexaene, 121
1,1,2-Trichlorocyclopentane,
    40
Triethylenetetramine, 125
N-Trifluoroacetyl-α-naphthyl-
    ethylamine, 259
1,1-Trifluoro-2-propanol, 292
Trihalogenocyclopentanes,
    40
3,3,5-Trimethylcyclohexanol,
    44, 45
Trimethylenediaminetetra-
    acetatocobaltate(lll),
    150, 161

Trimethylenediaminetetra-
    acetic acid, 150
Trimethyl sulfonium iodide,
    236
1,2,3-Trioxolane, 28, 29, 78
1,2,4-Trioxolane, 28, 29, 65,
    78
Tris(acetylacetonato)cobalt-
    (lll), 113
Tris(alaninato)cobalt(lll),
    113, 157
1,1,1-Tris(aminoethyl)ethane,
    chelate complex of,
    124
Tris(1,2-benezenediolato)arsen-
    ate(V), 112, 165
Tris(bidentate) complexes,
    with five-membered chelate
    rings, 98
  with seven-membered chelate
    rings, 110
Tris(2,2'-bipyridyl)nickel(ll),
    112
Tris(3-t-butylhydroxymethyl-
    ene-d-camphorato)euro-
    pium(lll), 289
Tris(1,4-diaminobutane)cobalt-
    (lll), 111
Tris(1,2-diaminocyclohexane)-
    cobalt(lll), 103, 106,
    153
Tris(1,2-diaminocyclohexane)-
    rhodium(lll), 112
Tris(1,2-diaminocyclopentane)-
    cobalt(lll), 106, 153
Tris(2,4-diaminopentane)co-
    balt(lll), 109, 110, 154
Tris(d,d-dicampholylmethanato)-
    europium(lll), 290
Tris(ethylenediamine)chromium-
    (lll), 101
Tris(ethylenediamine)cobalt-
    (lll), 99, 153
Tris(3-heptafluoro-
    butyryl-d-camphorato)-
    europium(lll),
    289

Tris(malonato)chromium(111),
    110, 158
Tris(oxalato)chromium(111),
    112, 158
Tris(oxalato)cobalt(111),
    112, 158
Tris(phenanthroline)iron(11),
    112
Tris(phenanthroline)nickel(11),
    112
Tris(propylenediamine)cobalt-
    (111), 102, 103, 153
Tris(2,2,6,6-tetramethyl-3,5-
    heptanedionate)euro-
    pium(111), 291
Tris(thiooxalato)cobalt(111),
    112, 158
Tris(3-trifluoroacetyl-*d*-
    camphorato)europium-
    (111), 289

Tris(trimethylenediamine)-
    cobalt(111), 107,
    154

α,β-Unsaturated acids,
    asymmetric reduction
    of, 243, 268
α,β-Unsaturated ketones,
    asymmetric reduction
    of, 215

Van der Waals forces, 189.
    See also Steric effects
Vinyl allyl ether, 191
3-Vinyl-1-cyclooctene,
    248

Yeast, 266
Ytterbium, as shift reagent,
    306

CUMULATIVE INDEX, VOLUMES 1-10

	VOL.	PAGE
Absolute Configuration of Planar and Axially Dissymmetric Molecules *(Krow)*	5	31
Absolute Stereochemistry of Chelate Complexes *(Saito)*	10	95
Acetylenes, Stereochemistry of Electrophilic Additions *(Fahey)*	3	237
Analogy Model, Stereochemical *(Ugi and Ruch)*	4	99
Asymmetric Synthesis, New Approaches in *(Kagan and Fiaud)*	10	175
Atomic Inversion, Pyramidal *(Lambert)*	6	19
Axially and Planar Dissymmetric Molecules, Absolute Configuration of *(Krow)*	5	31
Barton, D.H.R., and Hassel, O. - Fundamental Contributions to Conformational Analysis *(Barton, Hassel)*	6	1
Carbene Additions to Olefins, Stereochemistry of *(Closs)*	3	193
Carbenes, Structure of *(Closs)*	3	193
sp^2-sp^3 Carbon-Carbon Single Bonds, Rotational Isomerism about *(Karabatsos and Fenoglio)*	5	167
Carbonium Ions, Simple, the Electronic Structure and Stereochemistry of *(Buss, Shleyer and Allen)*	7	253
Chelate Complexes, Absolute Stereochemistry of *(Saito)*	10	95
Chirality Due to the Presence of Hydrogen Isotopes at Noncyclic Positions *(Arigoni and Eliel)*	4	127
Chiral Lanthanide Shift Reagents *(Sullivan)*	10	287
Classical Stereochemistry, The Foundations of *(Mason)*	9	1
Conformational Analysis, Applications of the Lanthanide-induced Shift Technique in *(Hofer)*	9	111
Conformational Analysis-The Fundamental Contributions of D.H.R. Barton and O. Hassel *(Barton, Hassel)*	6	1

	VOL.	PAGE
Conformational Analysis of Six-membered Rings (Kellie and Riddell)	8	225
Conformational Analysis and Steric Effects in Metal Chelates (Buckingham and Sargeson)	6	219
Conformational Analysis and Torsion Angles (Bucourt)	8	159
Conformational Changes, Determination of Associated Energy by Ultrasonic Absorption and Vibrational Spectroscopy (Wyn-Jones and Pethrick)	5	205
Conformational Changes by Rotation about sp^2-sp^3 Carbon-Carbon Single Bonds (Karabatsos and Fenoglio)	5	167
Conformational Energies, Table of (Hirsch)	1	199
Conformational Interconversion Mechanisms, Multi-step (Dale)	9	199
Conformations of 5-Membered Rings (Fuchs)	10	1
Conjugated Cyclohexenones, Kinetic 1,2 Addition of Anions to, Steric Course of (Toromanoff)	2	157
Crystal Structures of Steroids (Duax, Weeks and Rohrer)	9	271
Cyclobutane and Heterocyclic Analogs, Stereochemistry of (Moriarty)	8	271
Cyclohexyl Radicals, and Vinylic, The Stereochemistry of (Simamura)	4	1
Double Bonds, Fast Isomerization about (Kalinowski and Kessler)	7	295
Electronic Structure and Stereochemistry of Simple Carbonium Ions (Buss, Schleyer and Allen)	7	253
Electrophilic Additions to Olefins and Acetylenes, Stereochemistry of (Fahey)	3	237
Enzymatic Reactions, Stereochemistry of, by Use of Hydrogen Isotopes (Arigoni and Eliel)	4	127
1,2-Epoxides, Stereochemical Aspects of the Synthesis of (Berti)	7	93
EPR, in Stereochmistry of Nitroxides (Janzen)	6	177
Five-Membered Rings, Conformations of (Fuchs)	10	1
Foundations of Classical Stereochemistry (Mason)	9	1
Geometry and Conformational Properties of Some Five- and Six-Membered Heterocyclic Compounds Containing Oxygen or Sulfur (Romers, Altona, Buys and Havinga)	4	39

	VOL.	PAGE

Hassel, O. and Barton, D.H.R. - Fundamental
   Contributions to Conformational Analysis
    (Hassel, Barton)   6   1

Helix Models, of Optical Activity *(Brewster)*   2   1

Heterocyclic Compounds, Five- and Six-Membered,
   Containing Oxygen or Sulfur, Geometry and
   Conformational Properties of *(Romers,*
    Altona, Buys and Havinga)   4   39

Heterocyclic Four-membered Rings, Stereochemistry
   of *(Moriarty)*   8   271

Heterotopism *(Mislow and Raban)*   1   1

Hydrogen Isotopes at Noncyclic Positions, Chirality Due to the Presence of *(Argoni and Eliel)*   4   127

Intramolecular Rate Processes *(Binsch)*   3   97

Inversion, Atomic, Pyramidal *(Lambert)*   6   19

Isomerization, fast, about Double Bonds *(Kalinowski*
   and Kessler)   7   295

Lanthanide-induced Shift Technique - Applications
   in Conformational Analysis *(Hofer)*   9   111

Lanthanide Shift Reagents, Chiral *(Sullivan)*   10   287

Mass Spectrometry and the Stereochemistry of
   Organic Molecules *(Green)*   9   35

Metal Chelates, Conformational Analysis and
   Steric Effects in *(Buckingham and Sargeson)*   6   219

Metallocenes, Stereochemistry of *(Schlögl)*   1   39

Multi-step Conformational Interconversion Mechanisms *(Dale)*   9   199

Nitroxides, Stereochemistry of *(Janzen)*   6   177

Non-Chair Conformations of Six-Membered Rings
   (Kellie and Riddell)   8   225

Nuclear Magnetic Resonance, ^{13}C, Stereochemical
   Aspects of *(Wilson and Stothers)*   8   1

Nuclear Magnetic Resonance, for Study of Intra-
   Molecular Rate Processes *(Binsch)*   3   97

Nuclear Overhauser Effect, some Chemical Applications of *(Bell and Saunders)*   7   1

Olefins, Stereochemistry of Carbene Additions to
   (Closs)   3   193

Olefins, Stereochemistry of Electrophilic Additions to *(Fahey)*   3   237

Optical Activity, Helix Models of *(Brewster)*   2   1

	VOL.	PAGE
Optical Circular Dichroism, Recent Applications in Organic Chemistry *(Crabbé)*	1	93
Optical Purity, Modern Methods for the Determination of *(Raban and Mislow)*	2	199
Optical Rotatory Dispersion, Recent Applications in Organic Chemistry *(Crabbé)*	1	93
Overhauser Effect, Nuclear, some Chemical Applications of *(Bell and Saunders)*	7	1
Phosphorus Chemistry, Stereochemical Aspects of *(Gallagher and Jenkins)*	3	1
Piperidines, Quaternization Stereochemistry of *(McKenna)*	5	275
Planar and Axially Dissymmetric Molecules, Absolute Configuration of *(Krow)*	5	31
Polymer Stereochemistry, Concepts of *(Goodman)*	2	73
Polypeptide Stereochemistry *(Goodman, Verdini, Choi and Masuda)*	5	69
Pyramidal Atomic Inversion *(Lambert)*	6	19
Quaternization of Piperidines, Stereochemistry of *(McKenna)*	5	75
Radicals, Cyclohexyl and Vinylic, The Stereochemistry of *(Simamura)*	4	1
Resolving-Agents and Resolutions in Organic Chemistry *(Wilen)*	6	107
Rotational Isomerism about sp^2-sp^2 Carbon-Carbon Single Bonds *(Karabatsos and Fenoglio)*	5	167
Stereochemical Aspects of ^{13}C Nmr Spectroscopy *(Wilson and Stothers)*	8	1
Stereochemistry, Classical, The Foundations of *(Mason)*	9	1
Stereochemistry, Dynamic, A Mathematical Theory of *(Ugi and Ruch)*	4	99
Stereochemistry of Chelate Complexes *(Saito)*	10	95
Stereochemistry of Cyclobutane and Heterocyclic Analogs *(Moriarty)*	8	271
Stereochemistry of Nitroxides *(Janzen)*	6	177
Stereochemistry of Organic Molecules, and Mass Spectrometry *(Green)*	9	35
Stereoisomeric Relationships, of Groups in Molecules *(Mislow and Raban)*	1	1
Steroids, Crystal Structures of *(Duax, Weeks and Rohrer)*	9	271

	VOL.	PAGE
Structures, Crystal, of Steroids *(Duax, Weeks and Rohrer)*	9	271
Torsion Angle Concept in Conformational Analysis *(Bucourt)*	8	159
Ultrasonic Absorption and Vibrational Spectroscopy, Use of, to Determine the Energies Associated with Conformational Changes *(Wyn-Jones and Pethrick)*	5	205
Vibrational Spectroscopy and Ultrasonic Absorption, Use of, to Determine the Energies Associated with Conformational Changes *(Wyn-Jones and Pethrick)*	5	205
Vinylic Radicals, and Cyclohexyl, The Stereochemistry of *(Simamura)*	4	1
Wittig Reaction, Stereochemistry of *(Schlosser)*	5	1